S0-CDW-983

Comparative Ultrastructural Pathology of Selected Tumors in Man and Animals

Editor
Hildegard M. Schuller
Department of Pathobiology
College of Veterinary Medicine
University of Tennessee

CRC Press, Inc.
Boca Raton, Florida

Library of Congress Cataloging-in-Publication Date

Comparative ultrastructural pathology of selected tumors in man and animals/[edited by] Hildegard M. Schuller. p. cm.
Includes bibliographies and index.
ISBN 0-8493-5662-8
1. Tumors--Histopathology. I. Schuller, Hildegard M.
[DNLM: 1. Neoplasms--pathology. I.2. Neoplasms--ultrastructure.
QZ 200 U47]
RC296.U57 1989
616.99'2071--dc19
DNLM/DLC
For Library of Congress 88-36603
CIP

Direct all inquiries to CRC Press, Inc., 2000 Corporate Blvd., N.W., Boca Raton, Florida, 33431.

International Standard Book Number 0-8493-5662-8

Library of Congress Card Number 88-36603
Printed in the United States

PREFACE

Cancer is among the most common chronic diseases in man. While certain types of leukemias can be cured today, solid tumors in general have a poor prognosis unless detected at an early stage. Such early detection, however, is only possible when the diagnosing pathologist is familiar with the morphology of precancerous lesions. While most of our knowledge on the biology and response to therapy of the fully developed disease is derived from studies in human cancer patients, a systematic investigation of the various precancerous stages of the disease is only feasible in animal experiments. Interpretation of such experimental data requires a detailed knowledge of the interspecies differences in anatomy, cellular and subcellular morphology, and response to carcinogenic stimuli.

Complex organs, like the lungs or the pancreas, are comprised of many cell types, some of which can only be identified with certainty by electron microscopy. Likewise, the tumors arising in such organs may be comprised of many different cell types which in turn may differ in differentiation. Since the cellular composition and degree of differentiation of a tumor exerts a great deal of influence on its biological behavior and response to therapy, it seems desirable to arrive at a fully detailed analysis of the precancerous and cancerous lesion to facilitate the selection of proper therapy.

This book focuses on comparative aspects of the biology, pathology, and ultrastructure of selected solid tumors which are among the most common malignancies in man. The information provided on tumors of the lungs, pancreas, liver, kidneys, and disperse endocrine system should facilitate the classification of cancerous and precancerous lesions in these organs and stimulate a critical analysis of published data in these areas of research.

THE EDITOR

Hildegard M. Schuller, D.V.M., Ph.D., Certified experimental oncologist, is Chief, Experimental Oncology Laboratory, and Professor of Pathobiology, Department of Pathobiology, College of Veterinary Medicine, University of Tennessee, Knoxville, Tennessee.

Dr. Schuller obtained her D.V.M. degree in 1971 from the Justus Liebig University, Federal Republic of Germany, her Ph.D. degree in pathology in 1972 from the College of Veterinary Medicine, Hannover, Federal Republic of Germany, and her board certification in experimental oncology in 1977 from the College of Medicine, University of Hannover, Federal Republic of Germany.

Dr. Schuller is a member of the American Association for Cancer Research, United States Canadian Division of the International Academy of Pathology, Society of Toxicology, and American College of Toxicology.

Dr. Schuller has published 122 research articles in international and national journals. She is the author of 6 books and has contributed chapters to 20 books. She has presented two to four lectures annually at international and national meetings since 1972.

Dr. Schuller's research has consistently focused on respiratory tract carcinogenesis. Her work has involved experimentation in laboratory rodents, studies in human cancer patients, as well as *in vitro* experimentations on human cancer cell lines. Her major research interests involve the elucidation of mechanisms of lung carcinogenesis integrated with studies on tumor prevention and target oriented cancer therapy.

CONTRIBUTORS

William H. Butler
Assistant Director and Chief Pathologist
Department of Pathology
British Industrial Biological Research Association
Carshalton, Surrey, England

Gordon C. Hard
Professor, MRC Toxicology Unit
Medical Research Council Laboratories
Carshalton, Surrey, England

Rudolph Hauser
Department of Toxicology
Sandoz, AG
Basel, Switzerland

Takafumi Ichida
Lecturer
Third Department of International Medicine
Niigata University School of Medicine
Niigata City, Japan

Hideki Mori
Professor
Department of Pathology
Gifu University School of Medicine
Gifu, Japan

Parviz M. Pour
Professor
The Eppley Institute and Department of Pathology and Microbiology
University of Nebraska Medical Center
Omaha, Nebraska

Hildegard M. Schuller
Professor
Department of Pathobiology
College of Veterinary Medicine
University of Tennessee
Knoxville, Tennessee

Takuji Tanaka
Assistant Professor
Department of Pathology
Gifu University School of Medicine
Gifu City, Japan

Timothy J. Triche
Pathologist-in-Chief
Childrens Hospital Los Angeles
Professor, Pediatrics and Pathology
University of Southern California
Los Angeles, California

Gary M. Williams
Director of Medical Sciences
Naylor Dana Institute
American Health Foundation
Valhalla, New York

TABLE OF CONTENTS

Chapter 1

COMPARATIVE ULTRASTRUCTURAL PATHOLOGY OF LUNG TUMORS

H. M. Schuller

TABLE OF CONTENTS

I. INTRODUCTION

Lung cancer is one of the leading and most dreadful malignancies in man. Despite worldwide efforts, the disease remains generally incurable, and successful therapy with more than 3 years survival is the exception rather than the rule.

As is true for all other cancers, diagnosis at an early stage of the disease is vital to ensure successful response to therapy. However, in order to be able to recognize the early stages of the disease, the pathologist needs to be familiar with the morphology of such early "precancerous" lesions. Most of our knowledge of such lesions is derived from animal experiments. Such data generated in one or the other "animal model" are then generally extrapolated to some lesions accompanying malignant growth in the human lung. The real question here is: to what extent — if at all — do such experimentally induced precancerous and cancerous lesions in animals resemble the respective neoplastic processes in man? One major difficulty linked with this problem is the interspecies differences in the cellular and subcellular constitution of the lungs. It is important to remember that the distribution of cell types in the respiratory airways of our laboratory animals differs quite considerably from that in man. The interested reader is referred to a recent publication on the details.[1] Briefly, the small lungs of our laboratory animals resemble in many aspects the peripheral parts of the human lung, while central parts of the human lung are generally finding their counterparts in the extrapulmonary bronchi and trachea of the laboratory rodents.[1] Such differences have to be kept in mind particularly when dealing with pathogenesis type experiments aimed at elucidating the mode of development of human lung cancers.

This chapter will focus on comparative aspects of the most common lung cancer types in man, but will also discuss some rare lung cancer types as well as data on pathogenesis. Ultrastructural studies can add significantly to an understanding of cellular and subcellular events. They are even unreplaceable in the case of a complex organ like the lungs which is comprised of about 40 different cell types, most of which can only be identified with certainty by electron microscopy.

II. ADENOCARCINOMA

Adenocarcinomas of the lung used to be an uncommon malignancy in man[2] while it is the predominating lung tumor type (spontaneous and induced) in experimental and domesticated animals.[3-8] However, in recent years, a pronounced shift in the incidence of this tumor type in man has been observed. Adenocarcinoma is today the leading type of lung cancer while squamous cell carcinoma has decreased in incidence.[9] It appears quite possible that this shift in the incidence of histological types of lung cancer is attributable to changes in the major identifiable cause of lung cancer: cigarettes.[9] Soon after the identification of polycyclic aromatic hydrocarbons as potent inducers of respiratory tract tumors in experimental animals, the cigarette industry has made significant efforts to minimize the contamination of cigarette smoke with this class of chemical carcinogens. The proportion of tar — which is rich in polycyclic aromatic hydrocarbons (PAH) — per cigarette has been substantially reduced. Moreover, filters prevent the inhalation of most of the particulate matter of smoke which again reduces the exposure to this class of chemical carcinogens. However, the levels of nicotine consumed by the individual smoker are generally kept on a constant level (needed by the nicotine addict to reach individual satisfaction) regardless of the contents of nicotine per cigarette. Accordingly, the "safer" cigarettes may protect against exposure to PAH but through compensatory measures (inhale deeper, smoke more cigarettes) the smoker retains a certain level of exposure to nicotine which is the major source of carcinogenic nitrosamines in cigarettes.[10] Interestingly, nitrosamines induce adenomas and adenocarcinomas of the lung in the majority of laboratory animals[8] while PAH consistently induce

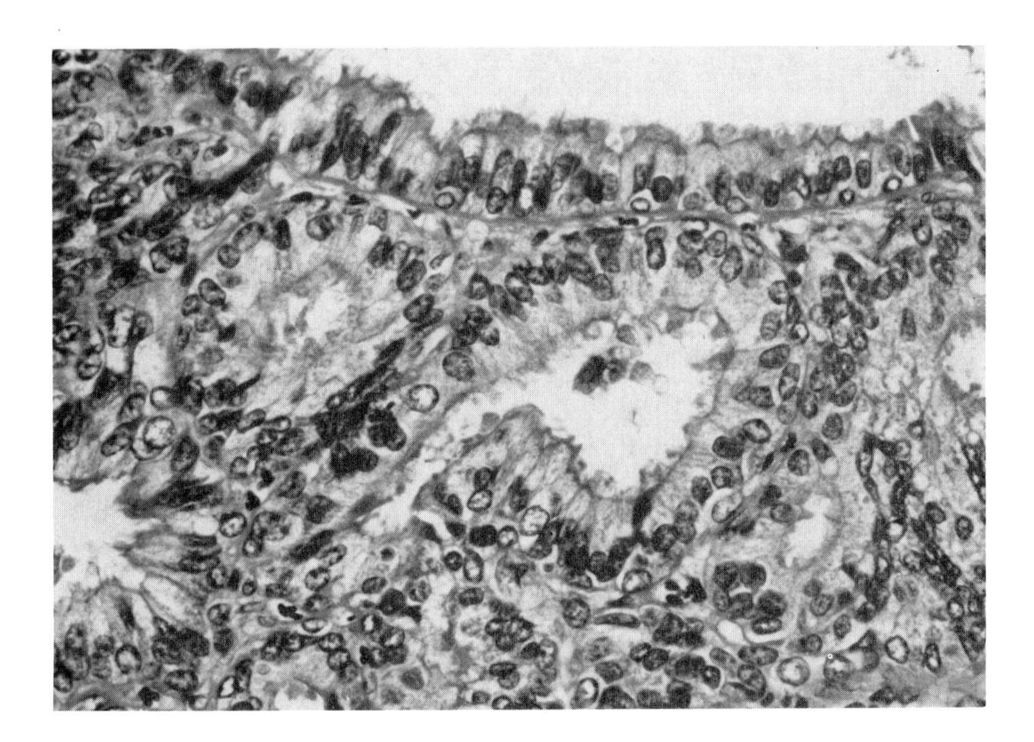

FIGURE 1. Histopathology of a bronchogenic adenocarcinoma induced in a European hamster with *N*-nitroso-heptamethyleneimine. Note gland-like growth pattern. (Hematoxylin/eosin; magnification × 480.)

squamous cell carcinomas.[11] With respect to this, the recent dramatic rise in the incidence of pulmonary adenocarcinoma in man may in fact be a reflection of a predominance of nitrosamines as carcinogenic stimuli in the "safer" cigarettes.

A. BRONCHOGENIC ADENOCARCINOMA

Bronchogenic adenocarcinomas arise at the level of the bronchi. Accordingly, they are comprised of cell types which are normally found at this level. Such cells include mucous cells, ciliated cells, basal cells, and, occasionally, a few neuroendocrine cells. Because of interspecies differences in the types of cells constituting the respiratory epithelia at the various airway levels[1] bronchogenic adenocarcinomas are rare in rats and hamsters and they are not found at all in mice. In man, such tumors may develop in areas of central airways down to the level of subsegmental bronchi. Moreover, adenocarcinomas involving bronchial submucous glands are not uncommon in man while they are rare in the laboratory rodents (which have only few submucous glands at the most central bronchi).

The main diagnostic criterion for the identification of adenocarcinomas is their gland-like growth pattern (Figure 1). In well differentiated tumors, this pattern is easily visible by light microscopy. In poorly differentiated tumors, it may become difficult to detect acinar lumina by light microscopy. In such cases, electron microscopy can help to clarify the diagnosis. Even narrow acinar lumina are easily found with this methodology. Other diagnostic features at the ultrastructural level include junctional complexes and microvilli at cell surfaces (Figure 2). The production of mucous can very often not be detected by light microscopy, including histochemistry. The reason for this is either a paucity of substrate (sparce production of mucous by the tumor cells) or the production of immature or chemically altered substrate. As in many other tumors, the cancer cells in an adenocarcinoma may perform certain functions of their counterparts only incompletely or sporadically.

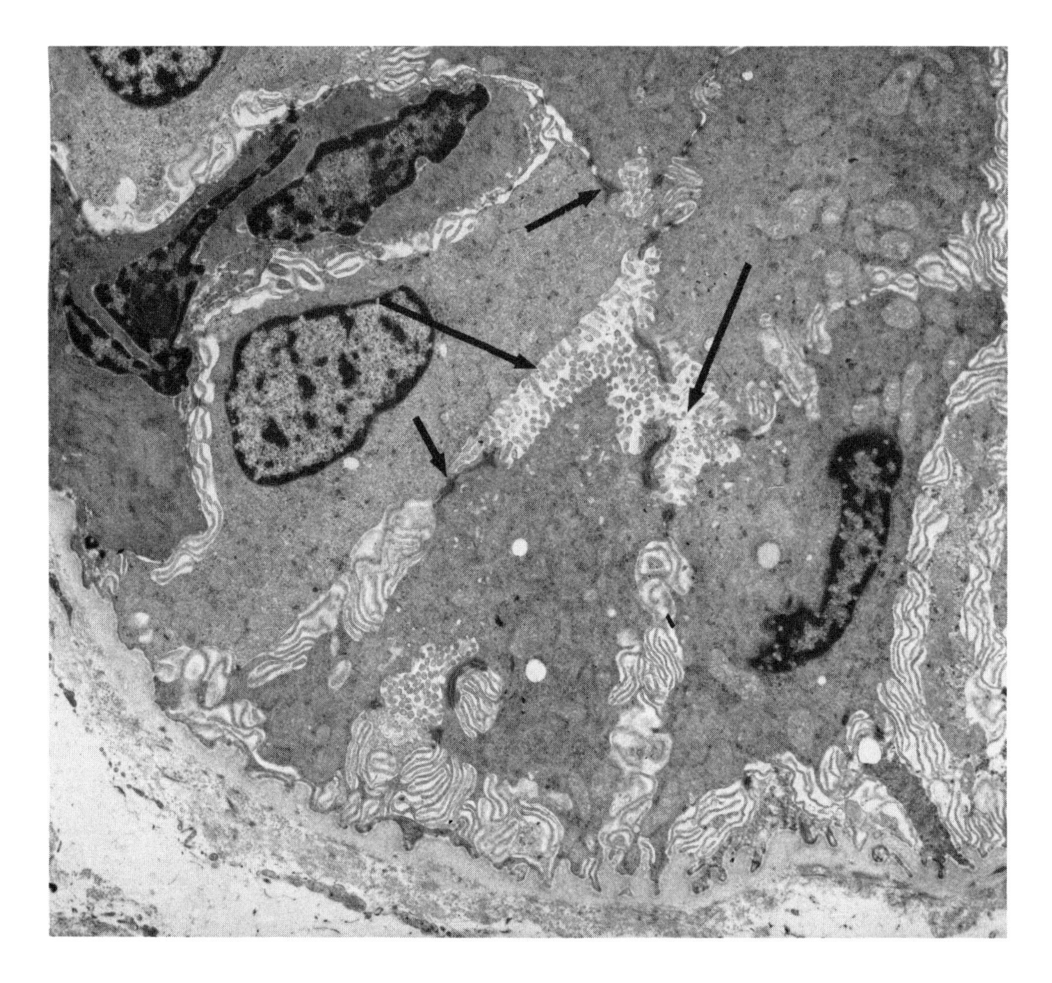

FIGURE 2. Electron micrograph of a bronchogenic adenocarcinoma in the human lung. The tumor cells are arranged around a narrow acinar lumen which is lined by numerous microvilli (long arrows); junctional complexes (short arrows) connect the apical portions of the tumor cells. (Uranyl acetate, lead citrate; magnification × 5200.)

B. BRONCHIOLO-ALVEOLAR CARCINOMA

Bronchiolo-alveolar carcinomas arise from bronchiolar or alveolar epithelia. A bronchiolar epithelium comprised of nonciliated Clara cells and occasional ciliated cells is found in the peripheral bronchioles in man while it coats more centrally located parts of the airways in the laboratory rodents as well.[1] In mice, for example, this type of epithelium coats all intrapulmonary airways while an epithelium compatible with the human bronchial epithelium is only found in the trachea (mucous cells, basal cells, ciliated cells). Accordingly, all pulmonary adenocarcinomas in mice have to be addressed as bronchiolo-alveolar carcinomas because they are comprised of cell types which in man are restricted to the bronchiolo-alveolar region. Mouse lung adenomas and adenocarcinomas are hence a good model system for the study of bronchiolo-alveolar tumors while they are not suited for research on bronchogenic adenocarcinoma.

Bronchiolo-alveolar carcinomas are the most common lung tumors found in laboratory animals and domesticated animals (both spontaneous and induced). In man, bronchiolo-alveolar carcinoma used to be an uncommon lung tumor type.[2] However, the recently reported dramatic increase in the incidence of adenocarcinomas includes bronchiolo-alveolar carcinomas in particular.[9]

Although the pathogenesis of bronchiolo-alveolar carcinomas is still a matter of debate

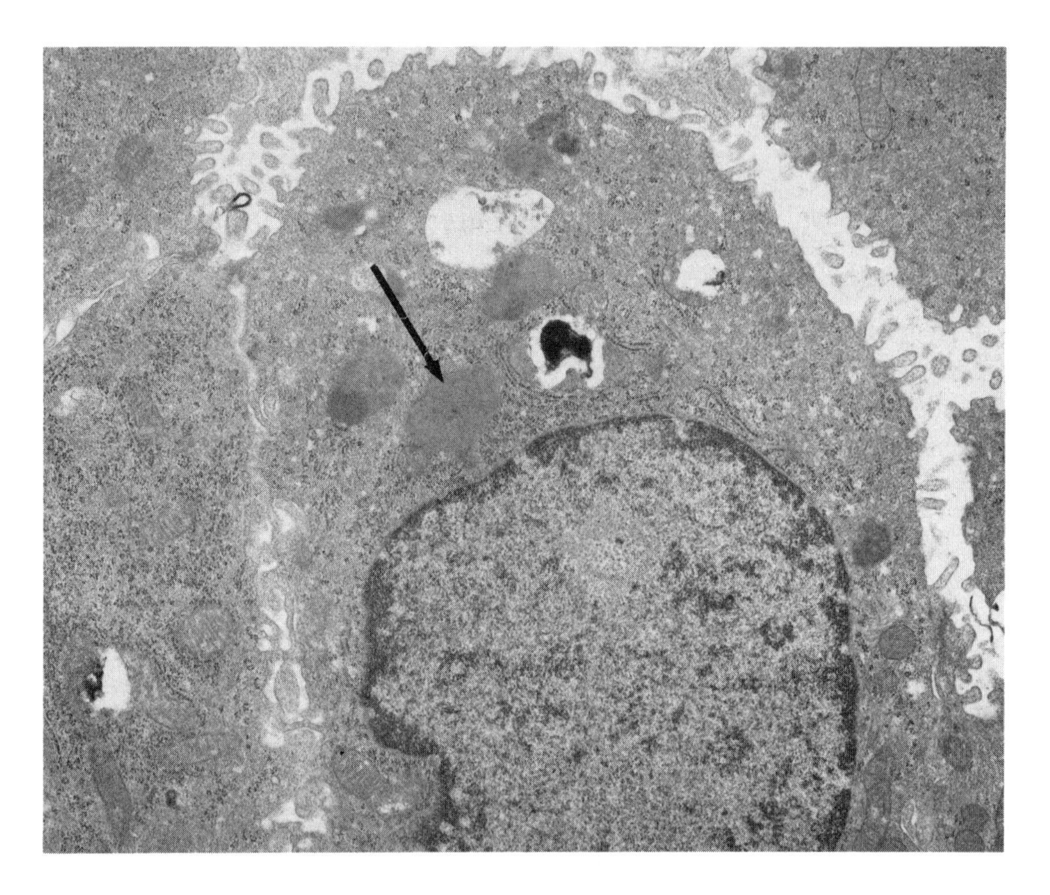

FIGURE 3. Electron micrograph illustrating typical features of Clara cell differentiation in a pulmonary adenoma induced in a Syrian golden hamster with *N*-nitrosodiethylamine. The tall columnar tumor cells demonstrate rough and smooth endoplasmic reticulum, occasional electron-dense secretion granules (arrowed), and microvilli. (Uranyl acetate, lead citrate; magnification × 14,000.)

(see below) nonciliated Clara cells (Figure 3) and alveolar type II cells (Figure 4) are generally the major constituents of this tumor type. Well differentiated tumors of this type do not pose any diagnostic difficulties. However, tumors of poorer differentiation may require electron microscopy for a verification of the diagnosis and determination of cell type. This is especially true for cases which are detected at a late stage when the tumor occupies large portions of the lung. Immunocytochemical methods, which are liked by many investigators today, are not reliable in this aspect because the substrate that yields a positive immunoreaction may not be present in sufficient quantity. For example, alveolar type II cells are identified by the positive immunoreaction with an antibody to the phospholipids which they produce.[12] However, in poorly differentiated tumor cells of this type the production of phospholipids may be rudimentary. By electron microscopy, even remnants of the lamellar bodies which are the storage site of phospholipids in such cells (Figure 5) or immature lamellar bodies (Figure 6) are easy to identify. In conjunction with the presence of other diagnostic features, such as microvilli and junctional complexes, such findings help to establish the diagnosis. Bronchiolo-alveolar carcinomas comprised of Clara cells are generally less common than such tumors comprised of alveolar type II cells. In particular, well differentiated tumors of this composition are rare in man while in the laboratory rodents some adenomas may display this type of morphology with the more advanced adenocarcinomas expressing dual differentiation (Clara cell and type II cell) or type II cell morphology. Clara cells are characterized by their tall columnar shape, dome-shaped apex, presence of smooth and rough endoplasmic

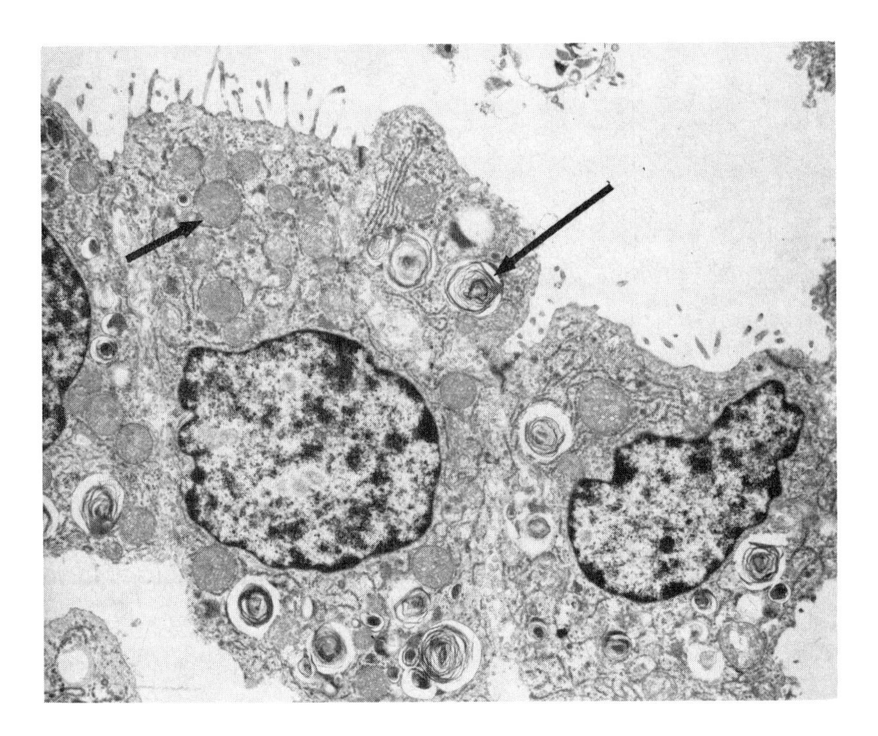

FIGURE 4. Electron micrograph illustrating typical features of alveolar type II cell differentiation in a bronchiolo-alveolar carcinoma in the human lung. Numerous osmiophilic lamellar bodies (long arrows) are the storage site of phospholipids produced by the tumor cells. Note also mitochondria with tubular cristae (small arrow). Mitochondria of this type are not found in normal alveolar type II cells but are common in cells involved in the synthesis of steroid hormones. (Uranyl acetate, lead citrate; magnification × 5200.)

reticulum and presence of electron-dense secretion granules.[1,13] Most tumor cells of Clara-cell derivation exhibit only some of these features (Figure 3). In particular, secretion granules are generally rare. Because of such diagnostic difficulties, the pathogenesis of at least some bronchiolo-alveolar carcinomas from Clara cells has only been firmly established by serial sacrifice type experiments in laboratory animals (see below).

The histological appearance of bronchiolo-alveolar carcinomas can vary considerably. The textbook gland-like or papillary growth pattern is, in reality, only found in well differentiated tumors of this category. The majority of cases demonstrates a more compact arrangement of cells (Figure 7). A large number of chemical carcinogens induce bronchiolo-alveolar carcinomas in laboratory rodents.[14] Among these are in particular such agents like nitrosamines which require metabolic activation in the host organism.[15,16] Such metabolic activation is mediated by cytoplasmic enzyme systems (see below) which are abundant in Clara cells and alveolar type II cells.

C. PATHOGENESIS

The pathogenesis of tumors can only be investigated in animal models through serial sacrifice type experiments which monitor the sequential development of pathological lesions in the organ under study. Numerous statements on the "pathogenesis" of tumors have been reported which are only based on the study of lesions occurring simultaneously with a fully developed tumor. Such interpretations are undesirable and can in fact be very misleading. For example, squamous cell carcinomas of the lung were for decades believed to originate from basal cells which first undergo hyperplasia, then squamous metaplasia, and finally form squamous cell carcinomas.[11] This "dogma" was based on the observation that focal areas of basal cell hyperplasia and squamous metaplasia are frequently found in human

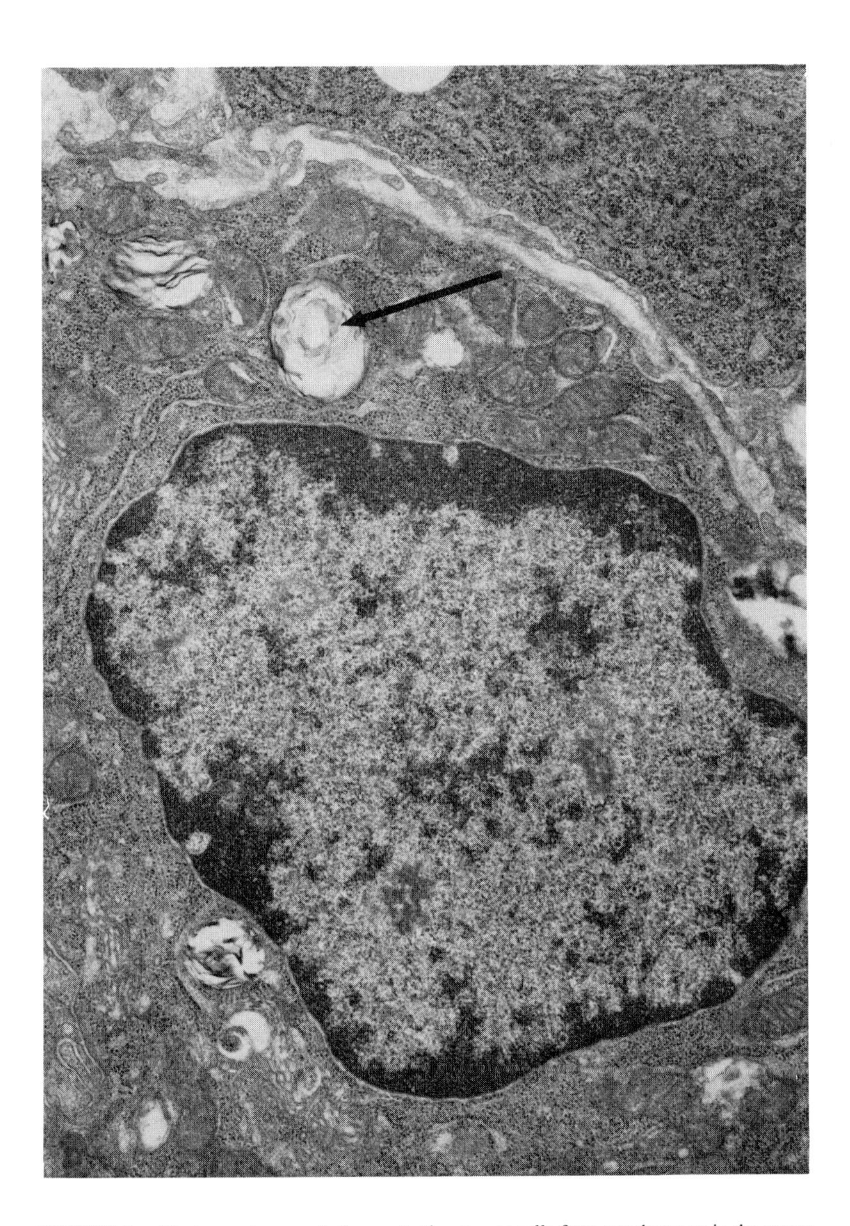

FIGURE 5. Electron micrograph demonstrating tumor cells from another area in the tumor of Figure 4: the tumor cells are less well differentiated than in Figure 4. Most notably, only remnants of osmiophilic lamellar bodies (arrowed) are detectable. (Uranyl acetate, lead citrate; magnification × 12,600.)

patients with squamous cell carcinoma of the lung. For many years, no attempts were made to find experimental proof for this interpretation. On the contrary, animal experiments aimed at furthering our understanding of pulmonary squamous cell carcinoma were conducted in such a way that they did not allow any insight into the pathogenesis of the chemically induced lung tumors, and the data were reported in such a way that they would parallel the findings in humans.[17] Only many years later and against considerable resistance from peer reviewers who were biased by the widely accepted dogma, serial sacrifice type experiments were conducted in laboratory rodents which showed that squamous cell carcinoma of the lung can arise from a variety of cell types including basal cells, mucous cells, and neuroendocrine cells.[18-21]

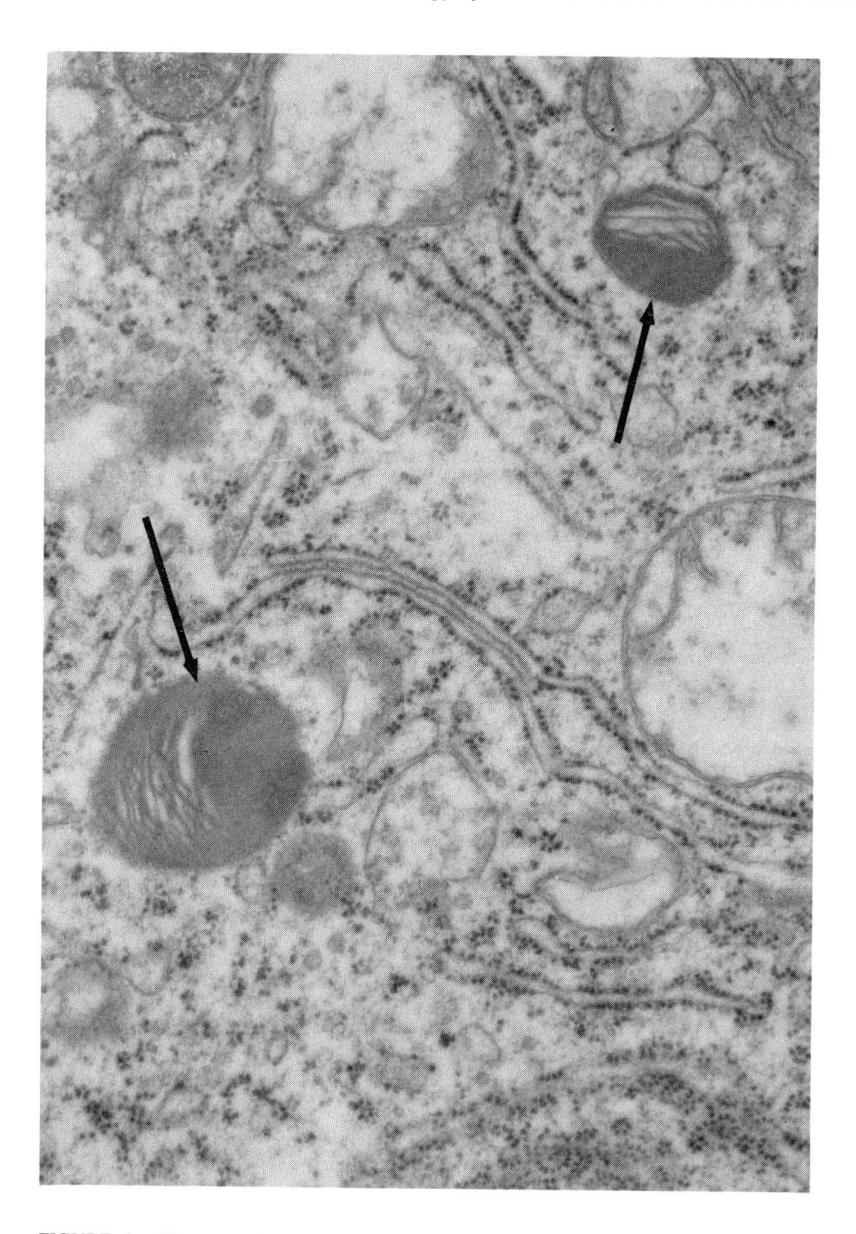

FIGURE 6. Electron micrograph illustrating immature osmiophilic lamellar bodies (arrowed) in a pulmonary adenocarcinoma induced with *N*-nitrosodibutylamine in a European hamster. (Uranyl acetate, lead citrate; magnification × 30,000.)

As mentioned in the previous paragraphs, bronchogenic adenocarcinoma is not as frequent in laboratory rodents as in man. There is only one report in the literature which describes the sequential development of such lung tumors in Syrian golden hamsters.[20] In that experiment, the animals received multiple intratracheal instillations of the carcinogenic polycyclic aromatic hydrocarbon benzo(a)pyrene. The earliest morphologically detectable change in the lungs was a pronounced vacuolization of the cytoplasm in all cell types of the bronchial epithelium (Figure 8). Electron microscopy revealed that these vacuoles were predominantly swollen mitochondria (Figure 9). This nonspecific toxic response was followed by a reactive increase in the number of lysosomes (Figure 10). For several weeks, these remained the only detectable changes. About 10 weeks after the occurrence of the first

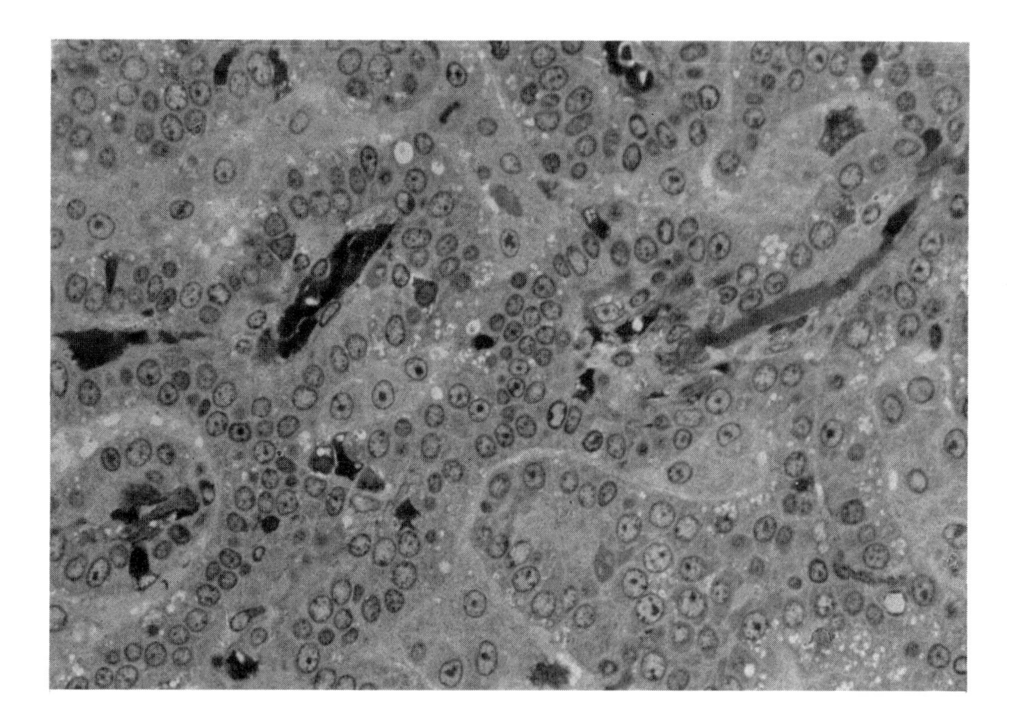

FIGURE 7. Histopathology of a bronchiolo-alveolar carcinoma in a mouse. Although by electron microscopy the tumor cells were diagnosed as alveolar type II cells, their growth pattern is compact and is lacking the gland-like arrangement and/or papillary growth pattern found in tumors of this category which are comprised of Clara cells. (Hematoxylin/eosin; magnification × 160.)

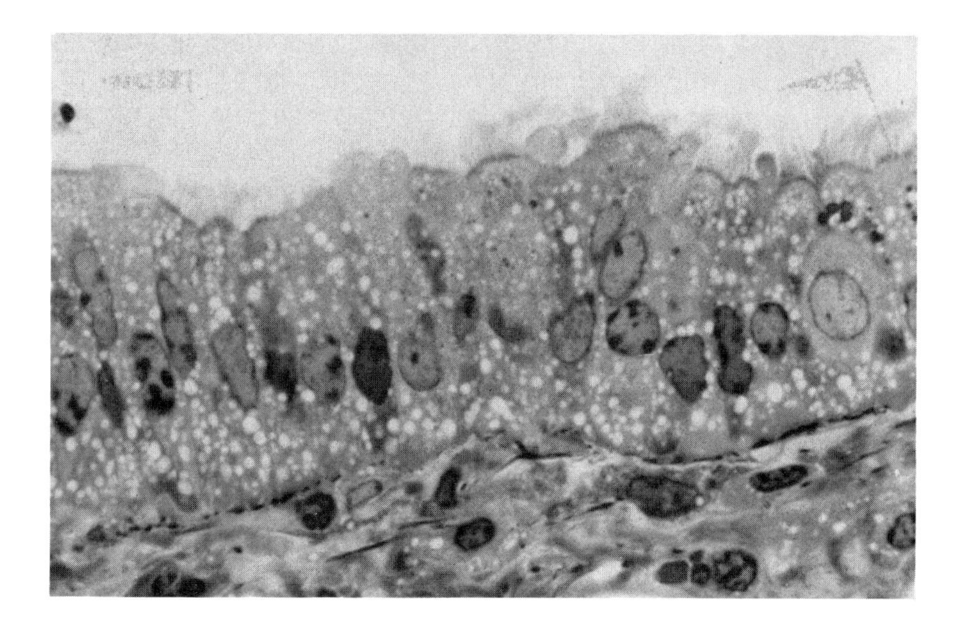

FIGURE 8. Histopathology of bronchial epithelium in a Syrian golden hamster lung after 2 weeks of intratracheal instillations with benzo(a)-pyrene. Note nonspecific vacuolization of all epithelial cell types. (Toluidine blue; magnification × 820.)

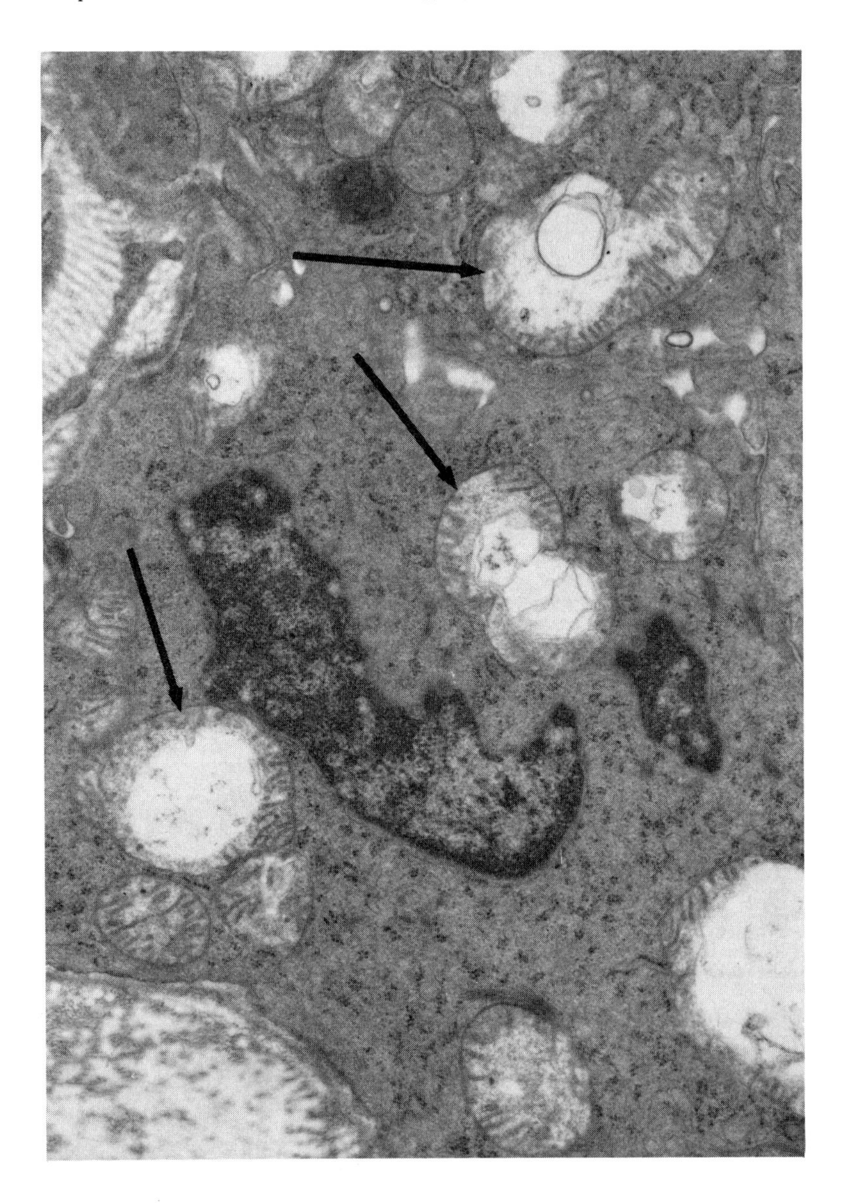

FIGURE 9. Electron micrograph illustrating part of the bronchial epithelium of Figure 8. It is evident that the cytoplasmic vacuoles are mostly mitochondria (arrowed). (Uranyl acetate, lead citrate; magnification × 20,000.)

toxic effects, hyperplastic areas formed in basal regions of the bronchial epithelium. The proliferated cells were poorly differentiated and were classified (because of the lack of unequivocal markers of other cell types) as "basal cells". Narrow cytoplasmic processes of such cells invaded the basement membrane and soon after that, entire strands of bronchial epithelium including ciliated, mucous, and basal cells were seen to grow through such gaps in the basement membrane into the adjacent lung (Figure 11). The resulting small adenomas demonstrated a typical glandular growth pattern with central acini surrounded by respiratory epithelium (Figure 12). Under continued treatment with the chemical carcinogen, they increased in size, and developed features of invasive growth, thus gradually progressing into malignant adenocarcinomas.

The pathogenesis of bronchiolo-alveolar carcinomas has been a matter of debate for

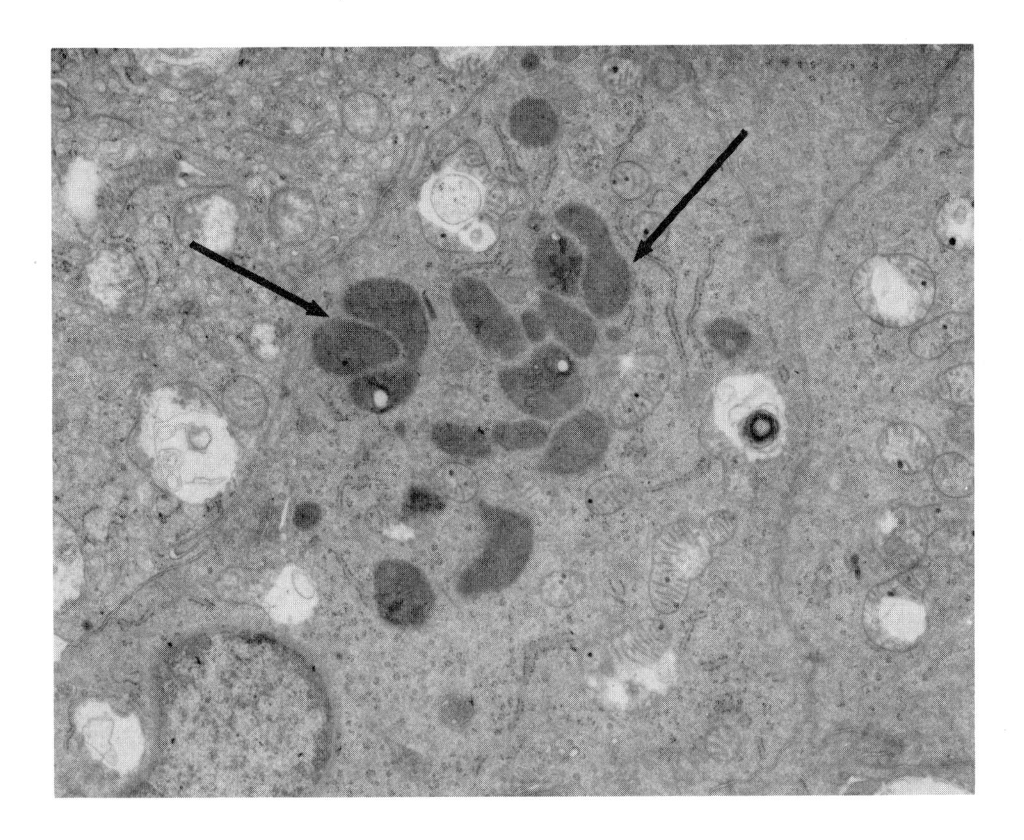

FIGURE 10. Electron micrograph exemplifying the abundance of lysosomes (arrowed) in the bronchial epithelium of a Syrian golden hamster after continued administration of benzo(a)pyrene. (Uranyl acetate, lead citrate; magnification × 14,000.)

many years. While some investigators interpreted their data as indicating an origin of this tumor type from alveolar type II cells,[12,22] others claimed an origin of the same tumor type from bronchiolar Clara cells.[23] Serial sacrifice type experiments have been conducted in Syrian golden hamsters using several different *N*-nitrosamines as tumor-inducing agents.[8] With all nitrosamines, the sequential morphological changes were identical, although the latency periods differed considerably. The most potent carcinogen in this respect was the simple alipatic *N*-nitrosodiethylamine (DEN) which induced bronchiolo-alveolar adenomas in 60% of the animals within 20 weeks of multiple subcutaneous DEN injections.[14,24] Progression of such lesions into malignant adenocarcinomas was observed within 1 year after start of the experiment. The earliest detectable morphological changes were found selectively in Clara cells and neuroendocrine cells.[21] In the Clara cells, a pronounced hypertrophy of smooth endoplasmic reticulum compatible with an induction of cytochrome P450 enzymes was noticeable after only a few DEN injections. Subsequently, the Clara cells within the bronchial and bronchiolar epithelium started to produce cytoplasmic lamellar bodies (Figure 13) compatible with the synthesis of phospholipids, which is normally limited to alveolar type II cells. Clara cells later invaded through the bronchial and bronchiolar basement membranes and formed small gland-like structures with all the typical features of adenomas (Figure 14). The cells constituting such lesions exhibited either ultrastructural features of Clara cells (Figure 15) or of alveolar type II cells (Figure 16). Since alveolar type II cells had not demonstrated any lesions during the early period of DEN treatment, the data were interpreted as suggesting an origin of the induced tumors from Clara cells.[24] In keeping with

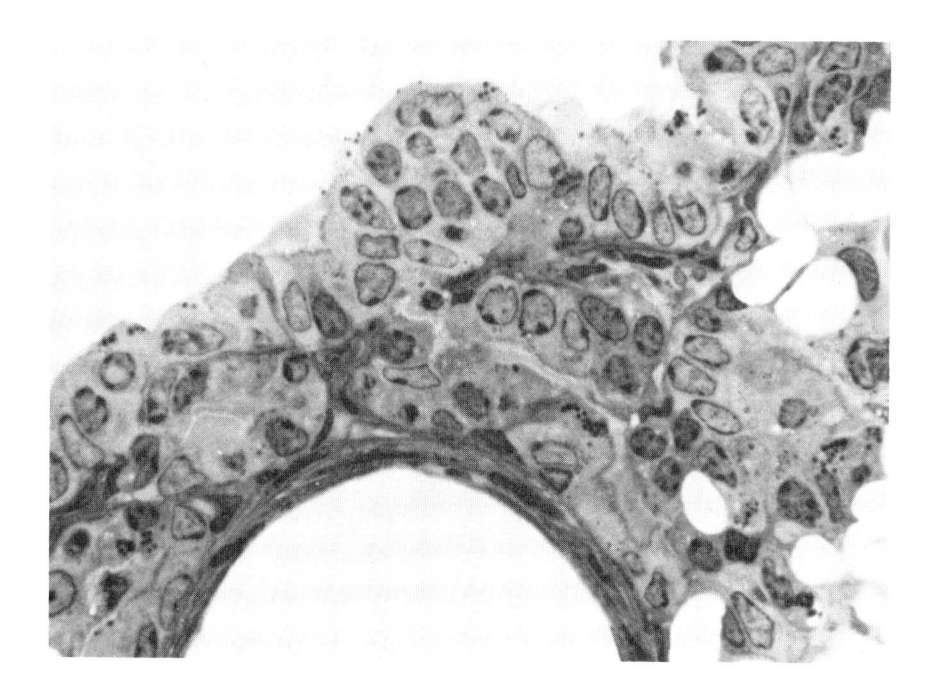

FIGURE 11. Histopathology of a bronchogenic lesion in a Syrian golden hamster after 14 weeks of benzo(a)pyrene: strands of epithelial cells comprised of ciliated, mucous, and basal cells are growing through the basement membrane into the lung parenchyma. (Toluidine blue; magnification × 820.)

this suggestion, autoradiographic studies provided evidence for Clara cells as being the principal site of metabolic activation of the nitrosamine in the hamster lung.[25,26] Moreover, administration of the cytochrome P450 inhibitor piperonylbutoxide before each DEN injection inhibited the metabolic activation and prevented the development of lung tumors.[27] These data are in strong support of the assumption that the nitrosamines are metabolically activated via Cytochrome P450 mediated mechanism in Clara cells under these conditions. Since the Clara cell is at the same time a stem cell responsible for cell renewal in the lung periphery, it of course also possesses the proliferative capacity to respond to a carcinogenic metabolite by proliferation and ultimately uncontrolled cell growth.

The pathogenesis data generated in the hamster nitrosamine model appear to be of considerable relevance to cellular reactions involved in the development of this tumor type in man. A cell line derived from a well differentiated human bronchiolo-alveolar carcinoma and demonstrating features of Clara cells was reported to metabolize DEN via cytochrome P450 enzymes while tumor cells with features of other cell types metabolized the compound via other enzyme systems[28,29] or not at all.[28]

The pathology and pathogenesis of bronchiolo-alveolar tumors in mice have been the subject of extensive research. Investigations on the sequential development of pathological lesions in the lungs of adult mice exposed to multiple administrations of dibenz(a,h)anthracene or urethane identified alveolar type II cells as tumor origin.[22] On the other hand, bronchiolo-alveolar tumors induced in mice via transplacental administration of ethylnitrosourea were reported to have derived from Clara cells in the case of papillary adenomas while alveolar type II cells were reportedly giving rise to alveolar adenomas.[22,30] The most recent study which used the transplacental ethylnitrosourea mouse model applied histopathology, electron microscopy, and immunocytochemistry to a serial sacrifice type experiment.[31] The smallest detectable tumors as well as more advanced lesions showed negative immunoreactivity to a rabbit anti-rat Clara cell antiserum while they gave a positive

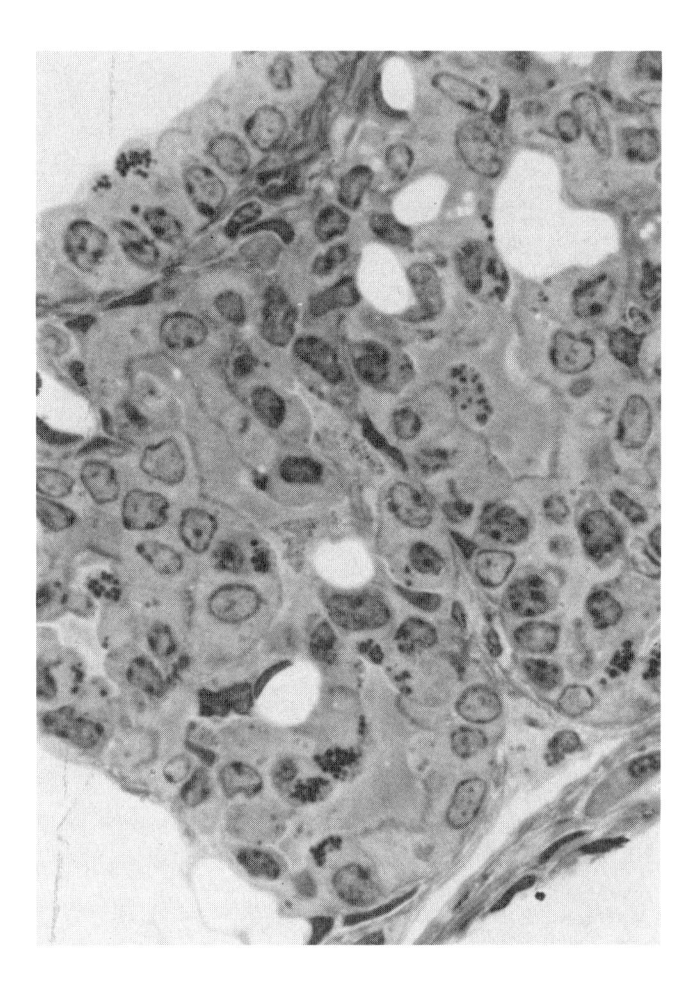

FIGURE 12. Histopathology of a small bronchogenic adenoma in a Syrian golden hamster after 20 weeks of benzo(a)pyrene: the tumor cells are growing in a glandular pattern. (Toluidine blue; magnification × 820.)

reaction with a rabbit antiserum against mouse lung surfactant apoprotein. Moreover, the tumor cells demonstrated ultrastructural features of alveolar type II cells but no Clara cell features. The authors conclude that the tumors are derived from alveolar type II cells. Although this study has devoted an enormous amount of time and effort to the project, it is hampered by the fact that no pretumorous lesions have been observed. As the authors state, already at the earliest investigated sacrifice time, at an age of only 1 week, the mice already had small tumors. Since no earlier time point was investigated, the actual development of such tumors from one or the other cell type could not be investigated. With respect to the pathogenesis findings on nitrosamine-induced bronchiolo-alveolar tumors in Syrian golden hamsters[24] it appears at least theoretically possible that a neoplastic transformation of Clara cells has preceded the development of tumors in the transplacental mouse model. In the hamster, such neoplastic transformation changed the morphology of Clara cells in such a way that they became virtually indistinguishable from alveolar type II cells, including the formation of phospholipid containing lamellar bodies. With respect to this, the debate on the pathogenesis of bronchiolo-alveolar tumors in the transplacental mouse system has not been settled by the cited experiments.[31] Further studies are needed to focus on the pretumorous stage of neoplastic development, meaning the mouse lungs have to be investigated during the prenatal and immediate postnatal age period.

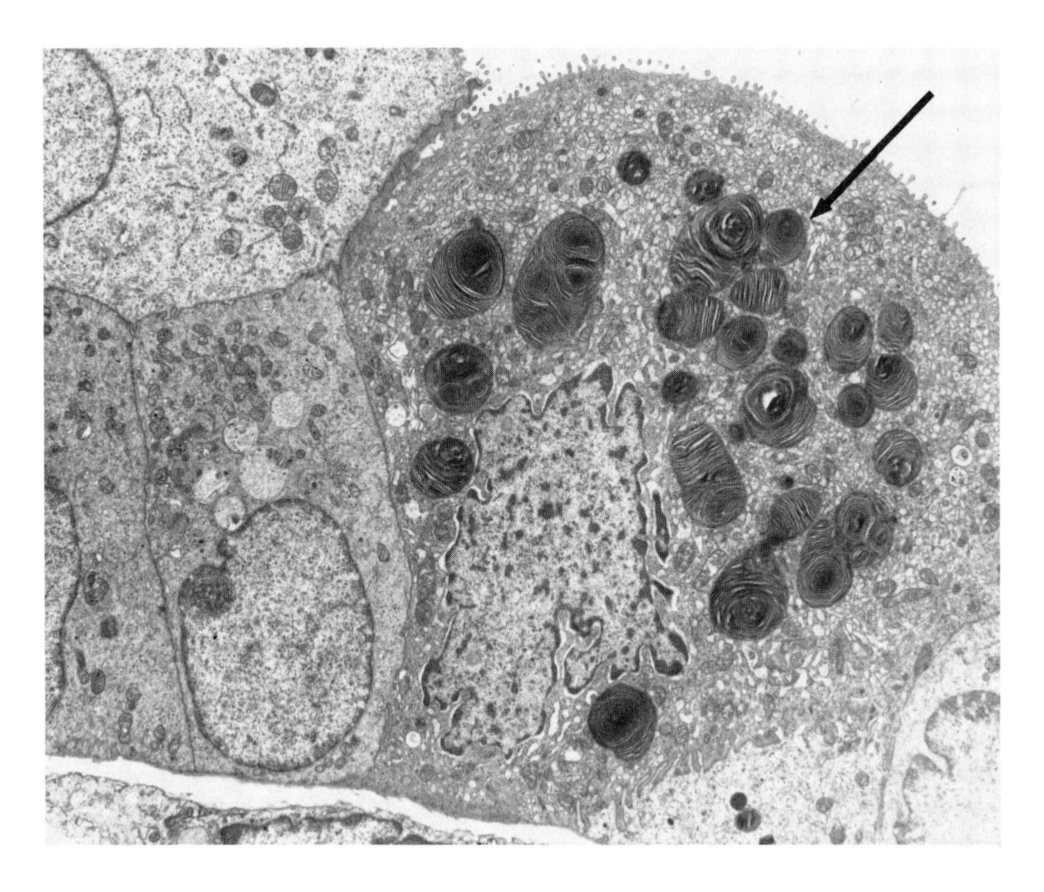

FIGURE 13. Electron micrograph of segmental bronchus in a Syrian golden hamster treated for 3 weeks with *N*-nitrosodiethylamine: a Clara cell (arrowed) contains numerous osmiophilic lamellar bodies, which in the healthy mammalian lung are produced by alveolar type II cells, exclusively. (Uranyl acetate, lead citrate; magnification × 4200.)

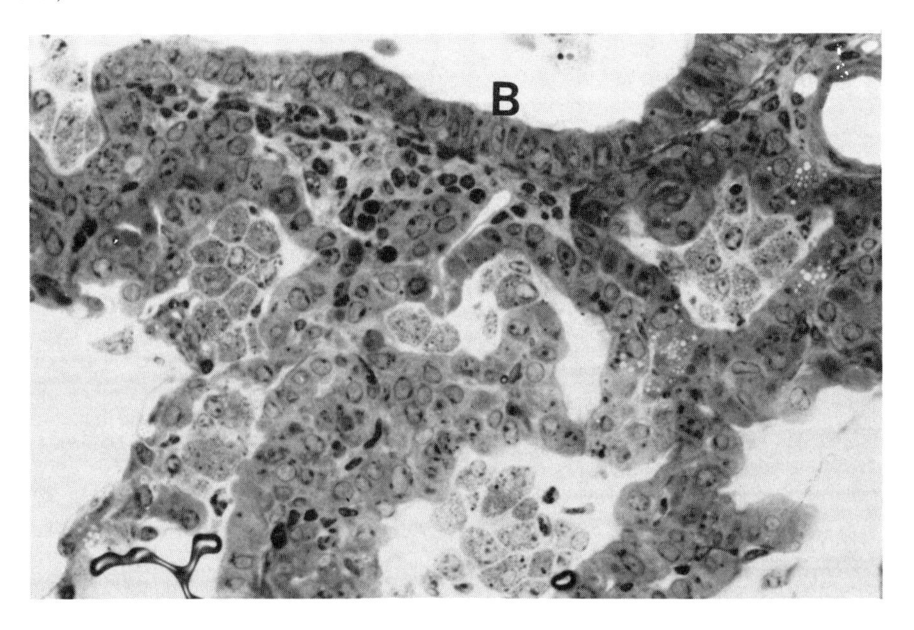

FIGURE 14. Histopathology of a small adenoma in the lungs of a Syrian golden hamster after 20 weeks of treatment with *N*-nitrosodiethylamine: the tumor cells which are associated with a segmental bronchus (B) grow in a gland-like pattern. (Toluidine blue; magnification × 240.)

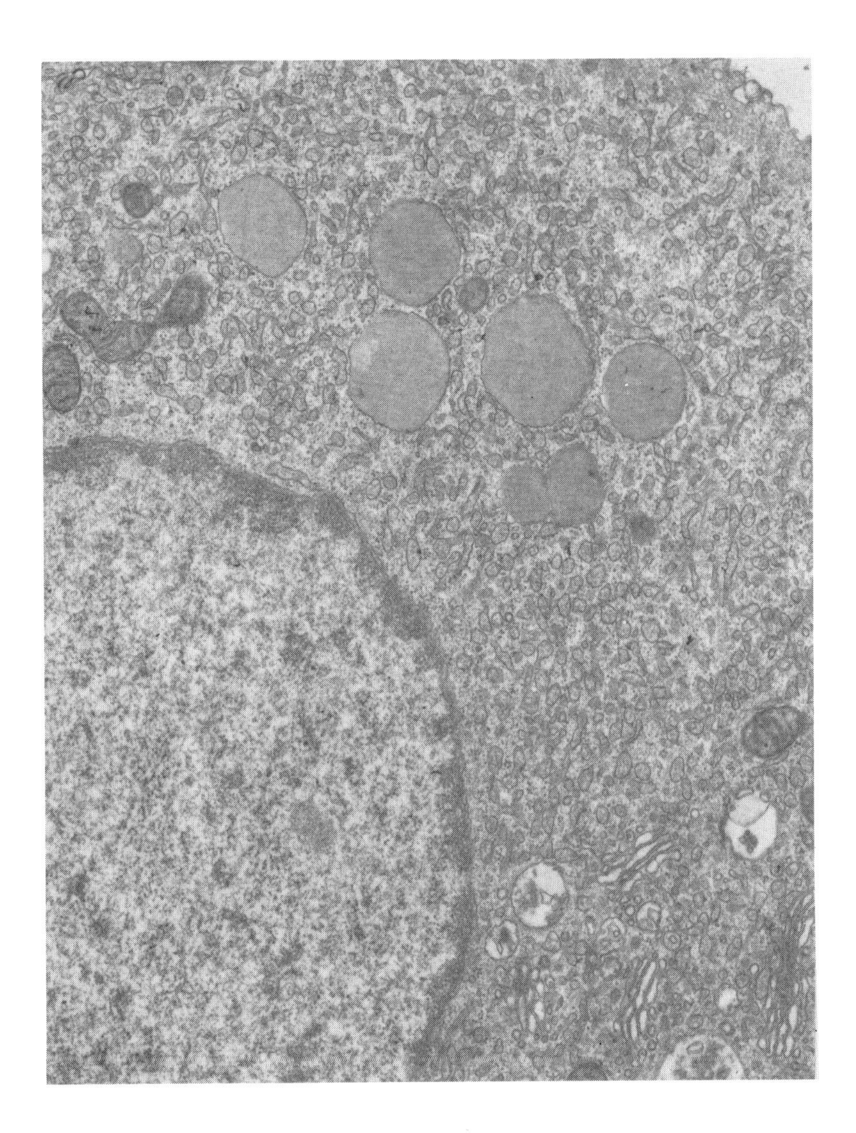

FIGURE 15. Electron micrograph of a tumor cell in a pulmonary adenoma induced in a Syrian golden hamster by *N*-nitrosodiethylamine. The cell demonstrates abundant smooth endoplasmic reticulum and homogeneously electron-dense secretion granules, both features of Clara cells. (Uranyl acetate, lead citrate; magnification × 20,000.)

III. NEUROENDOCRINE CARCINOMA

Neuroendocrine carcinomas of the lung are among the most common types of lung cancer in man and demonstrate a strong epidemiological link with cigarette smoking.[2] This tumor category is virtually not found in nonsmokers, is extremely rare as a spontaneous tumor in all animal species investigated, and has, until recently, not been induced at a significant incidence in animal experiments.[32,33] All efforts of the cigarette industry to develop a safer cigarette have apparently not affected the induction of this cancer type by cigarette smoke because its incidence in smokers has remained virtually unchanged (at about 30%) for several decades.[2,9]

Neuroendocrine tumors of the lung are usually subdivided into carcinoids and small cell cancer. While carcinoids are generally well differentiated and rarely demonstrate metastatic spread into extrapulmonary organs, small cell cancer metastasizes at an early stage of the

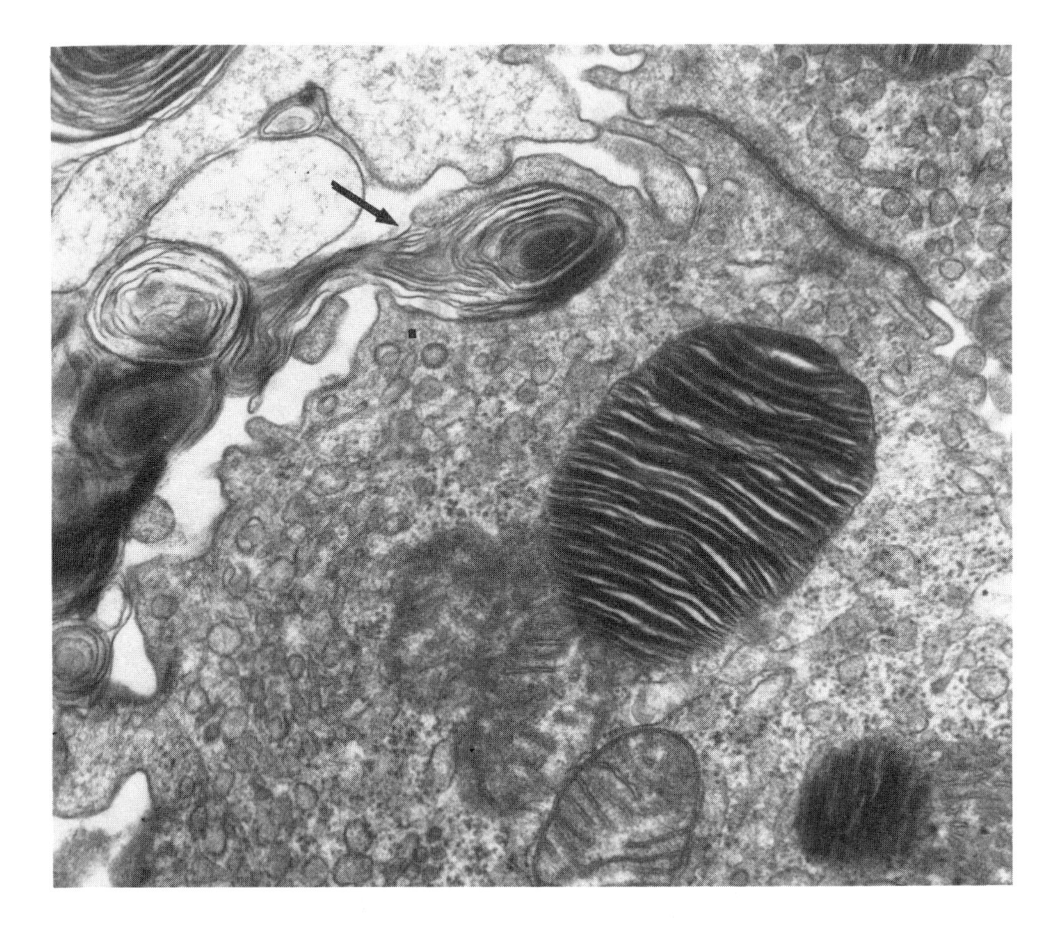

FIGURE 16. Electron micrograph illustrating tumor cells in a pulmonary adenoma induced in a Syrian golden hamster by *N*-nitrosodiethylamine: the cells demonstrate numerous osmiophilic lamellar bodies (the hallmark of alveolar type II cells in the healthy mammalian lung). Note merocrine secretion of the contents of a lamellar body (arrowed). (Uranyl acetate, lead citrate; magnification × 30,000.)

disease, and the tumor cells are poorly differentiated by morphological criteria. Both tumor types express a variety of morphological and biochemical features of neuroendocrine function. Among these are the APUD characteristics (amine precursor uptake and decarboxylase activity),[34] synthesis, storage, and secretion of peptide hormones and humoral substances,[35] expression of neuron specific enolase,[36] and the presence of dense-cored granules, detectable by electron microscopy and believed to store the peptide hormones.[37] The fact that similar neuroendocrine features are also expressed by the pulmonary neuroendocrine cell[38] has led to the assumption that neuroendocrine lung tumors are derived from this cell type.[37]

A. CARCINOIDS

Carcinoids of the lung are extremely rare in animals. In man, they are found at a much lesser frequency than the more malignant small cell cancer.[39] The majority of carcinoids is located centrally in the lung and linked with the larger bronchi. Peripheral carcinoids do occur but only on rare occasions. The bronchial carcinoids grow frequently below the intact epithelial surface. They demonstrate a well differentiated morphology at the light and electron microscopic level. The tumor cells may be spindle shaped, cuboidal, or columnar and may grow in a glandular, rosette-like or palisading pattern. They contain significant amounts of dense-cored secretion granules (Figure 17) along with a well developed smooth and rough endoplasmic reticulum (Figure 17). The coexpression of neuroendocrine and mucous pro-

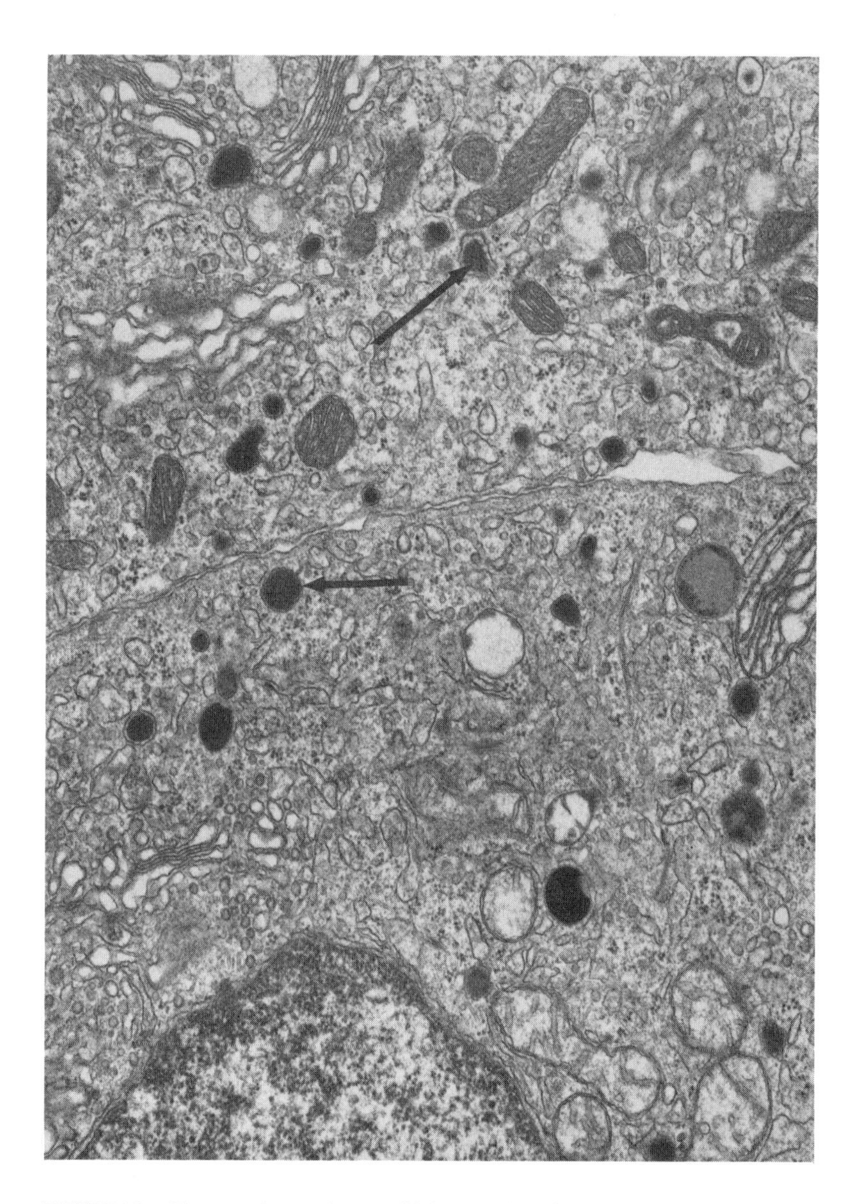

FIGURE 17. Electron micrograph exemplifying a portion of a carcinoid tumor in the human lung: the tumor cells demonstrate well developed endoplasmic reticulum, active Golgi fields, and numerous dense-cored neuroendocrine secretion granules (arrowed). (Uranyl acetate, lead citrate; magnification × 16,600.

ducing function is not uncommon (Figure 18). We have also seen a few cases in which features of neuroendocrine and oncocytoma cells were coexpressed (Figure 19). The margination and clumping of heterochromatin which is typical for normal pulmonary neuroendocrine cells is also evident in most carcinoids.

B. SMALL CELL CARCINOMA

As already mentioned, small cell cancer of the lung is one of the most common lung cancers found in man. The single and unequivocally identified cause of this cancer type appears to be cigarette smoking.[2] With respect to this, it is hard to accept that all international efforts to induce this cancer type in animals have failed.[32,33] Neither exposure to cigarette smoke itself nor treatment with any of the known constituents of cigarette smoke has

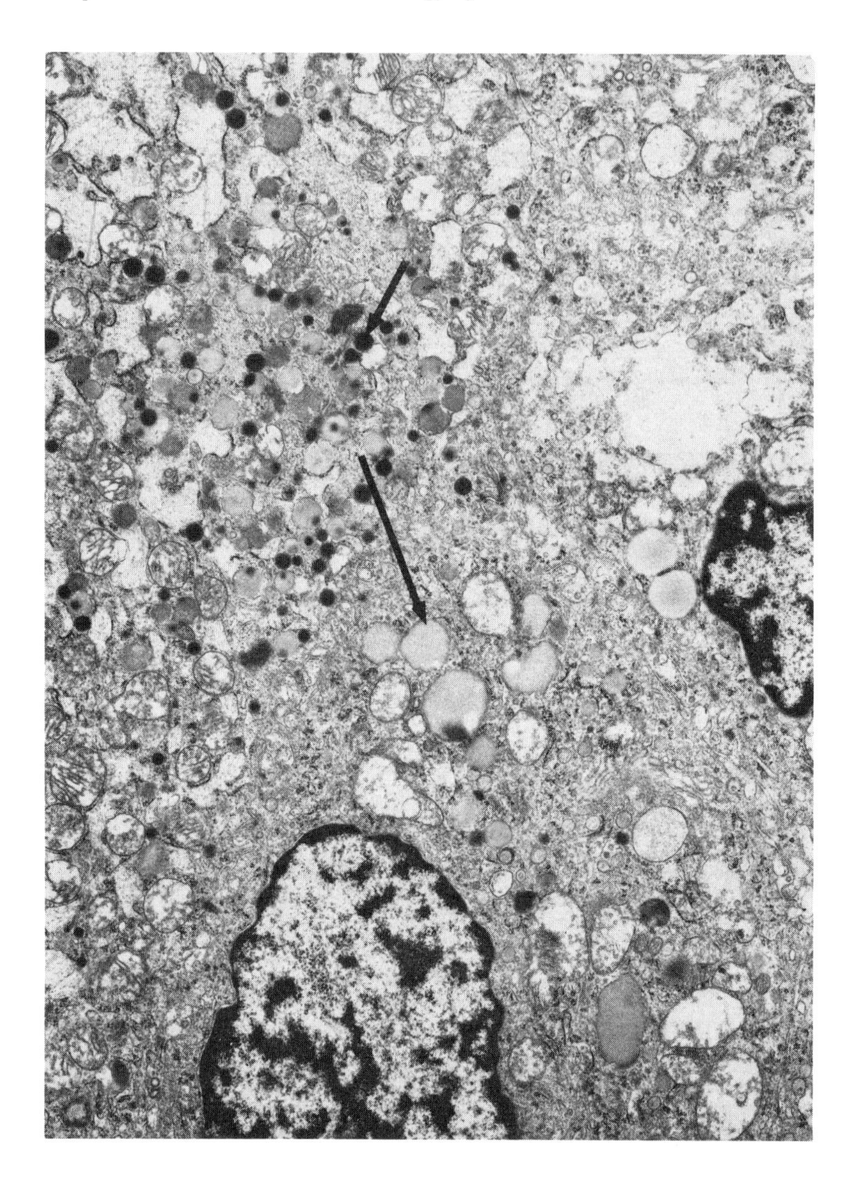

FIGURE 18. Electron micrograph of a human lung carcinoid tumor: the tumor cells exhibit mucous granules (long arrows) and neuroendocrine secretion granules (short arrows). (Uranyl acetate, lead citrate; magnification × 10,000.)

succeeded in inducing this cancer type in any animal species.[32,33] Obviously, all of these experiments have been lacking a critical element which appears to be uniquely prominent in smokers, and which seems to be necessary to produce a tumor type of neuroendocrine differentiation in the lung. Only very recently has an animal model finally been developed which allows the study of this important lung cancer type in detail.[41] This novel model and its implications are discussed in detail in the pathogenesis section of this chapter.

Because of the very recent availability of an animal model for neuroendocrine lung cancer,[41] most of our current knowledge of the pathology, biology, and response to therapy of this cancer type is based on experience with human cancer patients and cell lines derived from such human lung tumors.

The majority of human pulmonary small cell cancers are localized centrally in the lungs where they are linked with major bronchi, although they do occur in peripheral lung areas

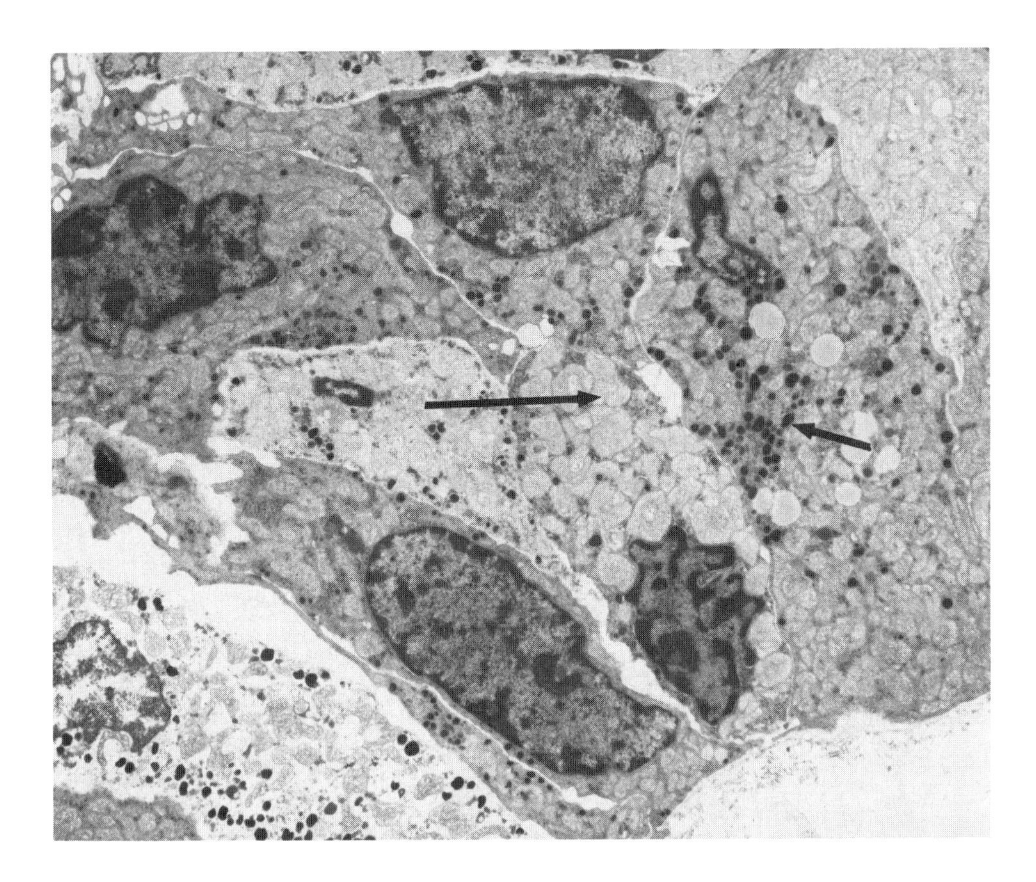

FIGURE 19. Electron micrograph of human lung carcinoid tumor: the tumor cells contain neuroendocrine secretion granules (short arrows) along with abundant mitochondria (long arrow). (Uranyl acetate, lead citrate; magnification × 5400.)

as well.[39] This cancer type metastasizes rapidly via blood vessels and lymphatics even at an early stage when the primary tumor measures just a few millimeters in diameter (and thus may escape clinical detection by conventional methods). Unlike most other lung cancer types, most small cell cancers initially respond well to radio- and chemotherapy. However, the vast majority of these cancers will recur after initial remission and then remain nonresponsive to any further therapy.[39]

Small cell cancer of the lung used to be subdivided into a lymphocyte-type (''oat cell carcinoma'') and an intermediate type. However, this subclassification was based merely on cell and nuclear sizes and thus was subject to artefacts. Since neither biological differences nor differences in their clinical course of response to therapy justifies this subclassification,[39] the term ''neuroendocrine carcinoma of the lung'' is finding increasingly more advocates.

Small cell cancer of the lung may demonstrate a variety of growth patterns. All of such tumors are generally highly cellular, with their cells demonstrating relatively little pale-staining (hematoxylin/eosin stain) cytoplasm. The nuclei may be hyperchromatic, or demonstrate the margination of heterochromatin typical of neuroendocrine cells. All small cell cancers express a variety of neuroendocrine features (see introductory part of neuroendocrine carcinoma). Their most reliable diagnostic features are the presence of dense-cored secretion granules and the expression of L-dopa decarboxylase activity. The dense-cored granules occur in small clusters (Figure 20). They may differ quite considerably from their counterparts in normal pulmonary neuroendocrine cells in that they are very variable in size, form, and shape (Figure 20). The most commonly expressed peptide hormone in small cell cancers

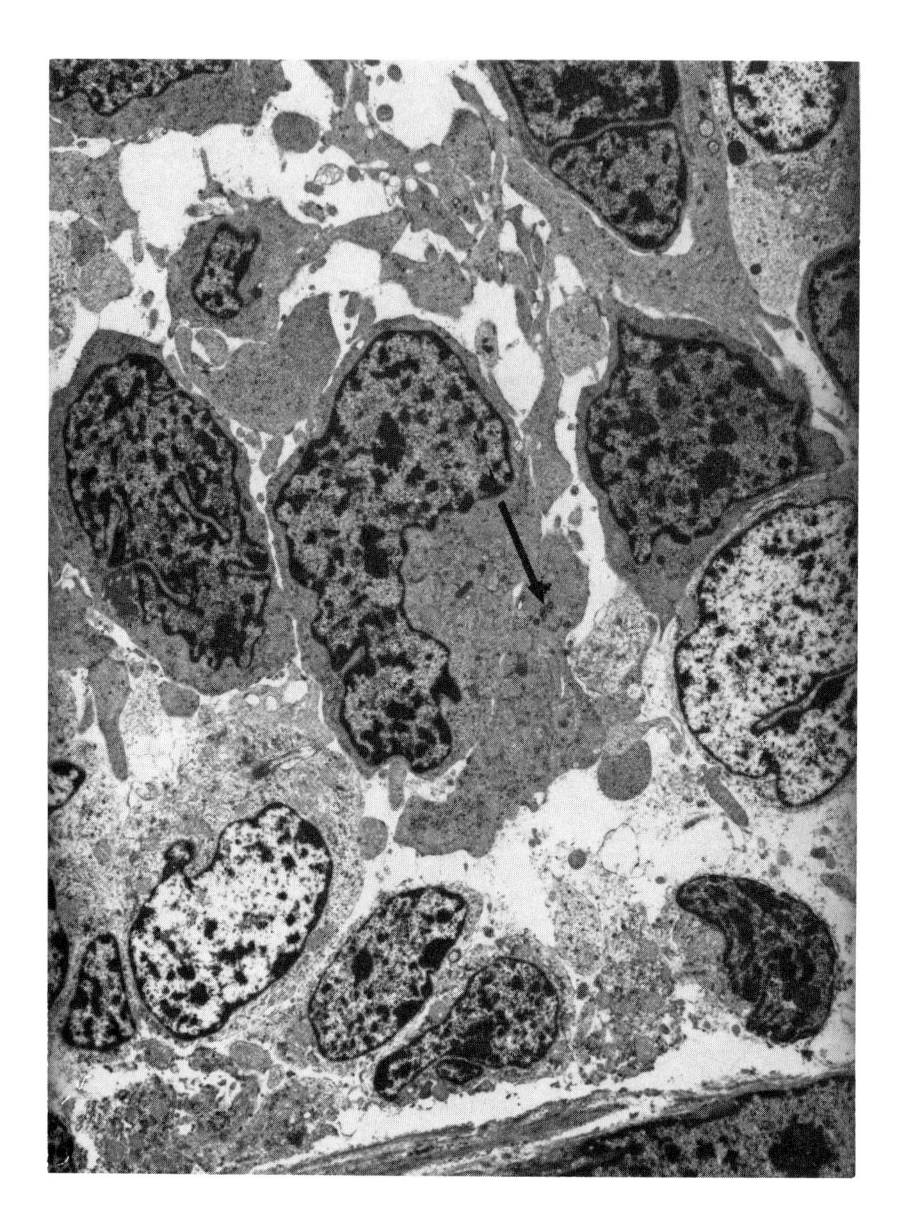

FIGURE 20. Electron micrograph illustrating the poor cytoplasmic differentiation of small cell carcinoma of the human lung. Note the variable size and shape of the tumor cells, a general paucity of cytoplasm, lack of endoplasmic reticulum, and occasional clusters of dense-cored neuroendocrine granules (arrowed). (Uranyl acetate, lead citrate; magnification × 5400.)

and cell lines derived from them is mammalian bombesin.[39] The cytoplasm of tumor cells is generally poor in organelles with sparse mitochondria and virtually no endoplasmic reticulum (Figure 20). In keeping with this, the tumor cells constituting pulmonary small cell carcinoma are virtually devoid of enzyme systems linked with mitochondria (e.g., monoamine oxidase) or endoplasmic reticulum (e.g., cytochrome P450).

C. PATHOGENESIS

Because of their morphological and biological resemblance with pulmonary neuroendocrine cells, it has long been postulated that neuroendocrine tumors of the lung are derived from this cell type.[37] Pulmonary neuroendocrine cells (PNE cells) are sparse in healthy adult mammals but occur in great abundance during the immediate perinatal age period.[38] Because the levels of peptide hormones and humoral substances such as mammalian bombesin and serotonin are significantly higher in PNE cells and serum during this age period,[42] it has been postulated that PNE cells play a major role in helping the lungs adjust to the drastic changes in pulmonary oxygen levels from the relatively hypoxic intrauterine to the normoxic extrauterine life.[42] Upon critical review, it becomes apparent that these changes in pulmonary oxygen levels are in fact quite dramatic with the partial oxygen pressure in the pulmonary arteries rising to a level three times as high as during the intrauterine period. Moreover, experiments conducted in rabbits and rats have provided evidence that the reactive numerical increase of PNE cells upon exposure to deviations from preexisting ''normal'' pulmonary oxygen levels is mediated via receptors.[43] With respect to all of these data, it was assumed that the one factor which is important to render PNE cells into a proliferative state, and which is uniquely present in a smoker's lung, but yet has never been incorporated in any of the numerous unsuccessful animal experiments aimed at inducing neuroendocrine lung cancer, is a significant deviation from preexisting normal pulmonary oxygen levels.[41] As is well established, smokers gradually develop a host of chronic diseases including chronic obstructive lung disease, emphysema, and cardiovascular disease.[44] In conjunction with the acute effects of such noxious agents as carbon monoxide contained in tobacco smoke, these conditions result in a pronounced impairment of pulmonary ventilation which ultimately leads to a reduced oxygen supply in the affected tissue segments. Accordingly, a continuous and pronounced deviation from normal pulmonary oxygen levels should be combined with simultaneous multiple exposures to chemical carcinogens contained in cigarette smoke if neuroendocrine tumors of the lungs were to be induced in experimental animals. Although tobacco smoke contains a host of different chemical carcinogens, the choice of the ''right'' chemical was easy in this case. Pathogenesis studies in Syrian golden hamsters had already shown in 1976 that several nitrosamines which are lung carcinogens in this species cause a pronounced and selective hyperplasia of pulmonary neuroendocrine cells[21,45,46] (Figure 21) during the first few weeks of the experiments. However, upon continuation of the nitrosamine administrations, such hyperplastic PNE cells gradually lost their neuroendocrine granules and underwent squamous metaplasia[26,46] (Figure 22). Even when the nitrosamines were administered throughout the lifespan of the animals, no neuroendocrine tumors developed.[8,32,33] With respect to the well-established ability of PNE cells for the selective uptake of simple amines and their precursors (e.g., serotonin and 5-hydroxytryptophan)[34] it was already speculated at that time that structural similarities between such endogenous amines and the nitrosamines were one factor involved in the selective stimulation of PNE cells by nitrosamines. For many years, no further information on the mechanisms involved could be generated due to the lack of a suitable model system. However, recent experiments using a unique well-differentiated cell line (Figure 23) derived from a human lung carcinoid has supplied additional information needed to design and understand the animal experiments that finally lead to the successful induction of pulmonary neuroendocrine cancer. These *in vitro* experiments have provided evidence that PNE cells are able to metabolically activate nitrosamines, and that such metabolism is mediated via monoamine oxidase and prostaglandin synthetase enzymes,[29] both of which are involved in the metabolism and degradation of serotinin in normal PNE cells.[35,47] Moreover, subsequent experiments using the same *in vitro* system are suggesting that the nitrosamines are selectively picked up by PNE cells via the same receptors which are involved in the uptake of physiological substrates.[59] Based on all this information, the simple aliphatic nitrosamine *N*-nitrosodiethylamine (DEN) became the chemical carcinogen of choice for the animal experiments aimed at inducing neuroendocrine lung tumors. In order to create a truly dramatic deviation from preexisting normal pulmonary oxygen levels simultaneously with the exposure to DEN, it was decided to use

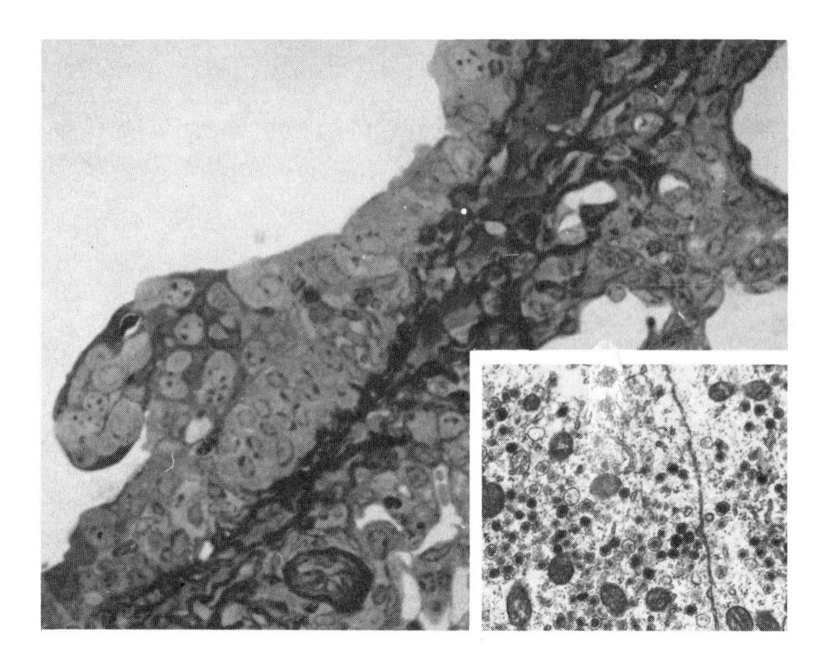

FIGURE 21. Histopathology of a focal hyperplasia of neuroendocrine cells in a Syrian golden hamster segmental bronchus after 4 weeks of treatment with *N*-nitrosodiethylamine. (Toluidine blue; magnification × 820.) Insert: Electron micrograph illustrating cytoplasmic detail of the hyperplastic neuroendocrine cells: Note abundant dense-cored neuroendocrine secretion granules. (Uranyl acetate, lead citrate; magnification × 12,600.)

a drastic hyperoxia (70% oxygen) rather than a hypoxia, in which case a decrease to 10% would have been the maximum tolerable level. In support of the theory,[41] such simultaneous exposure of hamsters to hyperoxia and DEN in fact resulted in the induction of neuroendocrine lung tumors within 8 weeks of treatment at an incidence of 100%.[41] The tumors occurred at all levels of the airways, including central bronchi, peripheral bronchioles, or alveoles. They were highly cellular with only little stroma interspersed (Figure 24), and exhibited a variety of growth patterns including rosette-like formations, palisading of cells, or sporadically, formation of small acinar structures. All of the tumors demonstrated a strong positive immunoreactivity to mammalian bombesin as assessed by immunocytochemistry (Figure 25), as well as positive immunoreactivity to calcitonin and neuron specific enolase. All of the tumors, which ranged in size between 2 mm and 3.5 cm grew invasively into surrounding lung parenchyma. Even the smallest lesions were never confined (Figure 26) as is usually seen with precancerous lesions of other types. Moreover, in one case, infiltration of regional blood vessels with tumor cells was observed. Taking into consideration that the animals had to be sacrificed to avoid unnecessary suffering after only 8 weeks of the experiment, (which really is not much time at all for a tumor to metastasize) these facts have to be viewed as distinctive indicators of a pronounced metastatic potential of the induced tumors.

In summary, the data presented show that two joint factors are necessary to induce neuroendocrine lung tumors: (1) chronic exposure to unphysiological pulmonary oxygen levels and (2) simultaneous exposure to a chemical carcinogen like the nitrosamines that can be taken up and metabolically activated by the PNE cells present in abundance in response to the altered oxygen levels.

It has yet to be established if the neuroendocrine lung tumors do in fact arise from preexisting PNE cells, or if other cell types are possibly able to change differentiation under the altered pulmonary oxygen conditions. Further studies are underway to clarify this point.

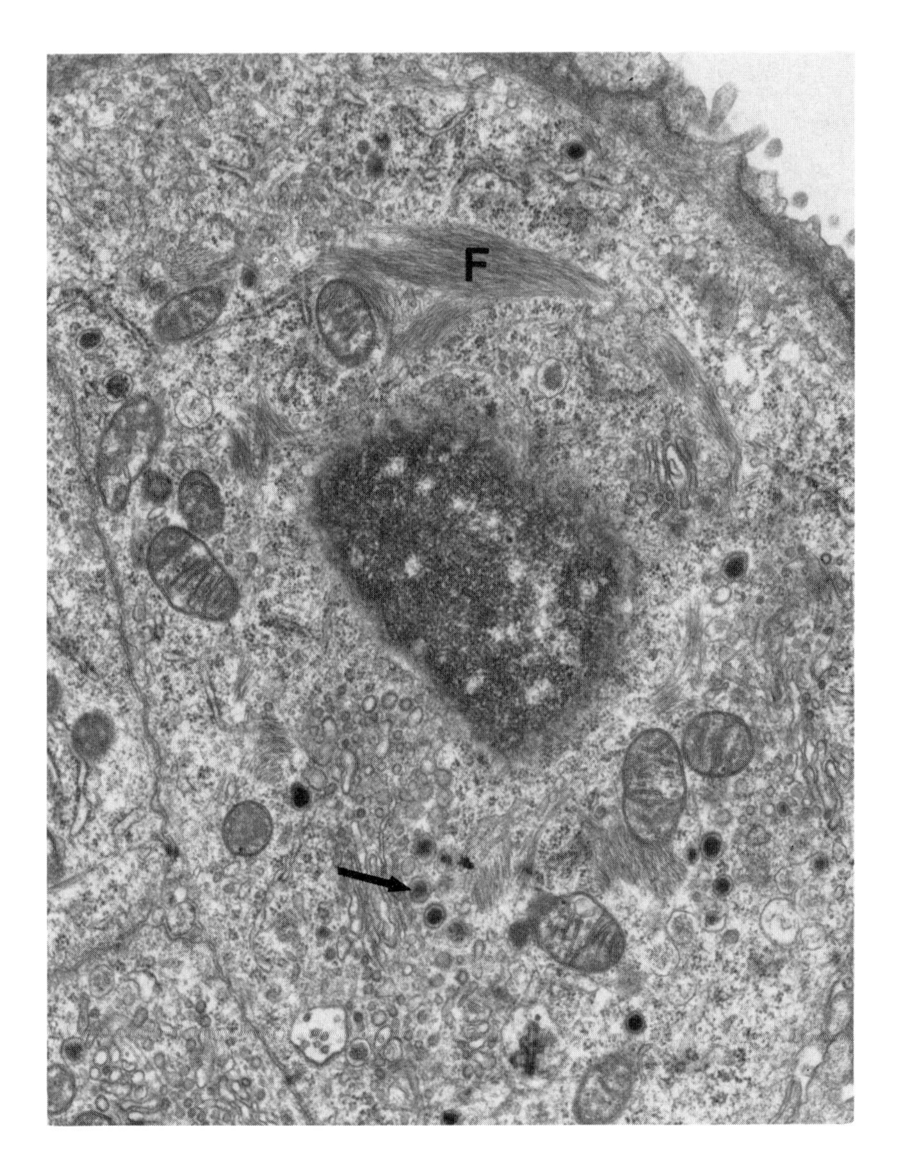

FIGURE 22. Electron micrograph exemplifying the development of squamous metaplasia of hyperplastic neuroendocrine cells in the hamster lung after prolonged nitrosamine treatment: the number of neuroendocrine secretion granules (arrowed) has decreased while bundles of cytoplasmic tonofilaments (F) indicate the beginning of immature keratin synthesis and hence the onset of squamous metaplasia. (Uranyl acetate, lead citrate; magnification × 18,000.)

In any event, the data presented in this section not only advance our current knowledge of the mechanisms involved in the induction of this deadly lung cancer type, but they also suggest that everybody exposed to higher or lower than normal pulmonary oxygen levels by habit, profession, or disease is at higher risk than the average population to develop pulmonary neuroendocrine cancer if exposed simultaneously to chemical carcinogens (most of which are inseparably linked with modern life style and behavior). In keeping with this interpretation, chronic obstructive lung disease has recently been identified as a strong risk factor for the development of lung cancer.[48] However, futher epidemiological studies using the data and interpretations presented in this chapter as a guideline are clearly necessary.

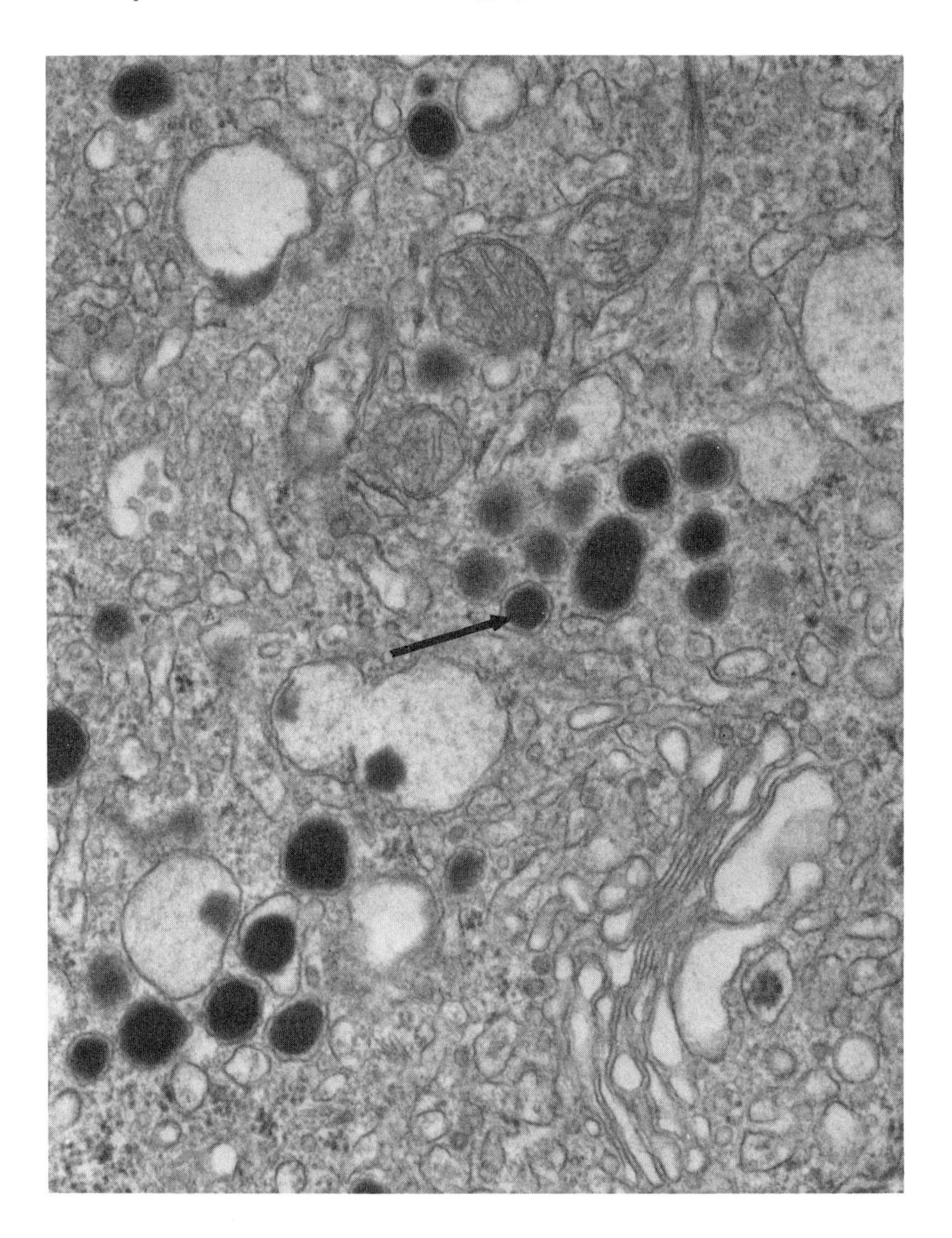

FIGURE 23. Electron micrograph demonstrating the well-differentiated ultrastructure of cell line NCI-H 727 (stock material kindly supplied by Dr. A. F. Gazdar, NCI-Navy, Clinical Oncology Branch, Bethesda, MD). This cell line is derived from a human lung carcinoid tumor. Note well-developed endoplasmic reticulum, Golgi fields, and abundant dense-cored neuroendocrine secretion granules (arrowed). (Uranyl acetate, lead citrate; magnification × 12,600.)

IV. SQUAMOUS CELL CARCINOMA

For decades, squamous cell carcinoma was the leading type of lung cancer in man.[2,9,10] It is, therefore, not surprising that most of the early work in experimental lung cancer research was focused on this tumor type.[11] In fact, the pressure through political channels (e.g., funding agencies) and the public media on the scientific community was so powerful that support for a project on lung cancer research as well as recognition of publishable data in this field seemed to depend on whether or not the model system in question was compatible

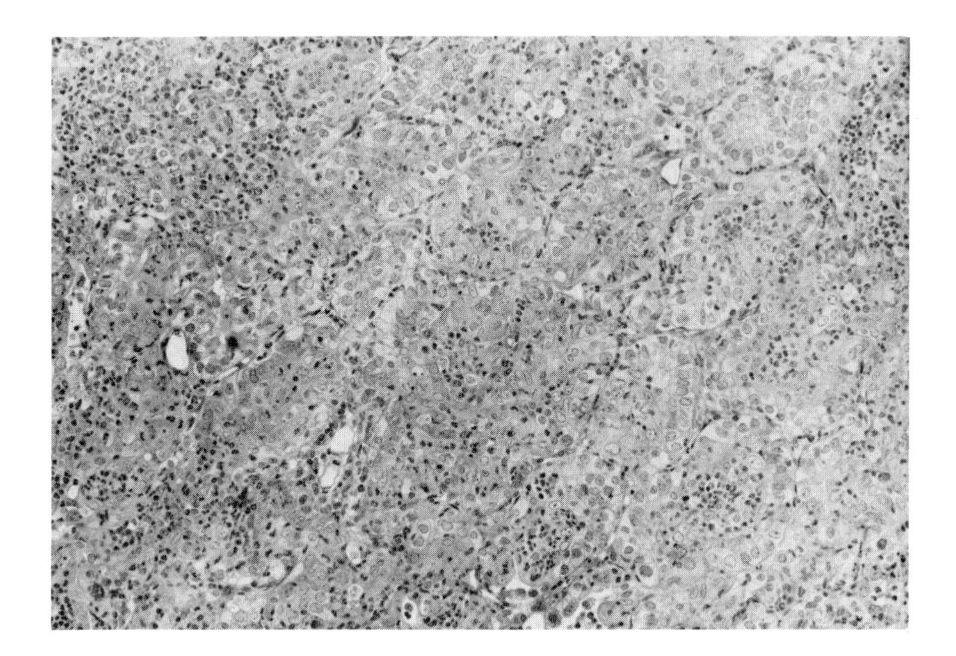

FIGURE 24. Histopathology of neuroendocrine lung tumor induced in a Syrian golden hamster by simultaneous exposure to hyperoxia and *N*-nitrosodiethylamine: note high cellularity and variable growth patterns. (Hematoxylin/eosin; magnification × 250.)

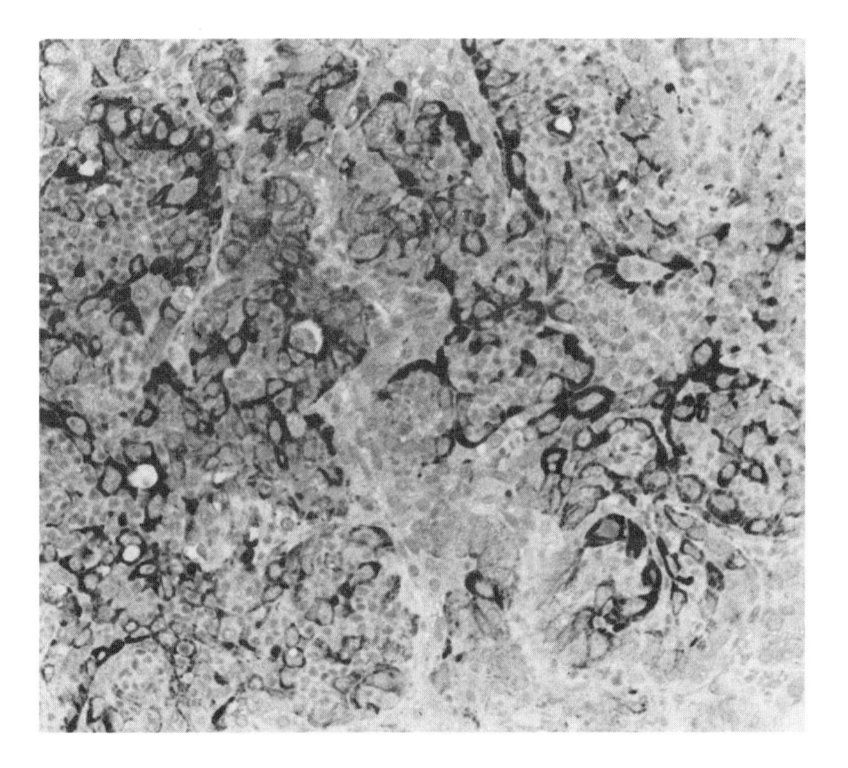

FIGURE 25. Neuroendocrine lung tumor induced in a Syrian golden hamster by simultaneous exposure to hyperoxia and *N*-nitrosodiethylamine. Immunocytochemical stain for mammalian bombesin (1:1000; 2 h). Large areas of the tumor tissue demonstrate a strong positive reaction. (Vecta-Stain ABC Kit, immunoperoxidase, Mayer's hematoxylin; magnification × 250.)

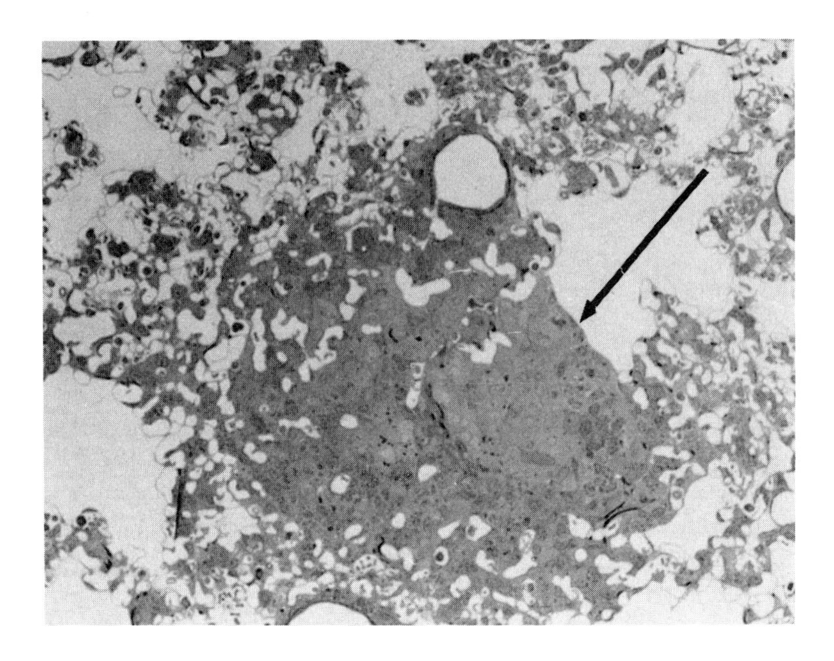

FIGURE 26. Small hyperplastic focus of neuroendocrine cells (arrowed) in the lungs of a Syrian golden hamster exposed for 2 weeks to hyperoxia and *N*-nitrosodiethylamine: although the lesion is only microscopically visible it grows invasively into the lung parenchyma where it appears to spread along preexisting structures like alveolar walls. (Toluidine blue; magnification × 160.)

with squamous cell carcinoma in the human lung. Regrettably, the flood of research projects and articles published during this time, and which all claimed to be using a "model system for human squamous cell carcinoma of the lung" would in many cases not withstand the scrutiny of modern diagnostics and knowledge.

Pulmonary squamous cell carcinoma in man demonstrates a prominent epidemiological link with cigarette smoking,[2] although, unlike neuroendocrine carcinoma, it may also occur in nonsmokers, especially in some population groups exposed by occupation to chemical carcinogens such as the PAH.[2] The contents of PAH in cigarette smoke used to be fairly high before low tar and filter cigarettes became available. However, the reduction in tar (which is the principal source of PAH in tobacco smoke) in combination with the use of filters has significantly reduced the carcinogenic burden by this class of chemical carcinogens in smokers. In response to this, the percentage of human lung tumors diagnosed as squamous cell carcinomas has begun to steadily decline.[9,10] Unfortunately, we are paying a high price for this "success" in that the incidence of pulmonary adenocarcinoma has increased more than proportionately (see Section II).[9,10]

In man, squamous cell carcinoma of the lung is a centrally arising malignancy associated with the larger airways. In contrast to this, lung tumors of squamous differentiation in the rat are generally peripherally located without association to central bronchi.[3,14,49,50] With respect to the anatomical differences outlined in the introductory section, any squamous cell carcinoma developing in the mouse lung would only be compatible with such tumor type arising in the lung periphery of the human lung (a situation which virtually does not exist). In Syrian golden hamsters, squamous cell carcinomas compatible with human lung carcinomas of this type can be induced by a variety of carcinogenic agents in the trachea, extrapulmonary stem bronchi, and the uppermost part of the lobar bronchi.[8,11] The situation is similar in European hamsters, in which tumors of this type are inducible at the level of

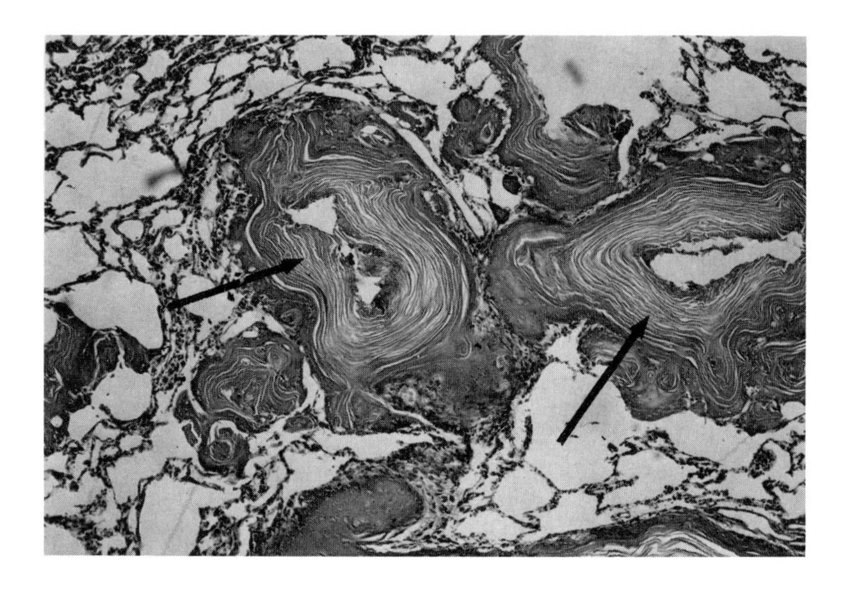

FIGURE 27. Histopathology of pulmonary squamous cell carcinoma in F344 rat after treatment with *N*-nitrosohepatamethyleneimine: some of the tumor cells produce mature keratin (arrowed). (Hematoxylin/eosin; magnification × 80.)

extrapulmonary stem bronchi and lobar bronchi.[8] However, it is important to note that this hamster species does not develop tumors in the trachea.[8] In man and hamsters, the majority of pulmonary squamous cell carcinomas are single neoplasms (one per organ), while in rats and mice, they occur usually as multiple lesions. In household pets (e.g., dog and cat) pulmonary squamous cell carcinomas are rare. Although the lung cancer incidence in these two species is very low anyway, most of the few lung tumors found are of the bronchiolo-alveolar variety.[51]

The morphology of squamous cell carcinomas of the lung is similar in all species investigated, although the pathogenesis may vary considerably (see below). Well-differentiated squamous cell carcinomas produce mature keratin (Figure 27) while poorly differentiated tumors of this category demonstrate either formation of immature keratin (Figures 28 and 29) or no keratin at all. By electron microscopy, mature keratin forms very electron-dense deposits which may form distinctive layers at the surface of tumor cells (Figure 30) or patchy intracytoplasmic deposits of similar density. Tumor cells which contain such mature keratin would also yield a positive staining reaction with immunostains after exposure to antisera to keratin. However, tumors which produce either immature or no keratin will generally yield a negative reaction with this procedure. Immature keratin is detectable by transmission electron microscopy as bundles of cytoplasmic tonofilaments. Pending on the degree of differentiation such filament bundles can be abundant (Figure 29) or sparse (Figure 28). There is always a pronounced concentration of tonofilament bundles in the perinuclear cytoplasmic area (Figure 29). The perinuclear area is also the site where the first of such filaments become noticeable during the early stages of squamous differentiation. Accordingly, this is the area to be thoroughly searched for tonofilaments in poorly differentiated tumors. Tonofilament bundles are also closely associated with desmosomes (Figures 31 and 32), the latter being another important diagnostic feature for the identification of squamous differentiation. The desmosomes with their inserting tonofilament bundles form areas of strong intercellular attachment while other parts of apposed cell membranes are unconnected, thus forming gaps between the desmosomes (Figure 32). This very peculiar morphological appearance was recognized many years ago as one of the typical features of squamous cell carcinomas detectable by light microscopy and was termed ''intercellular bridges''.

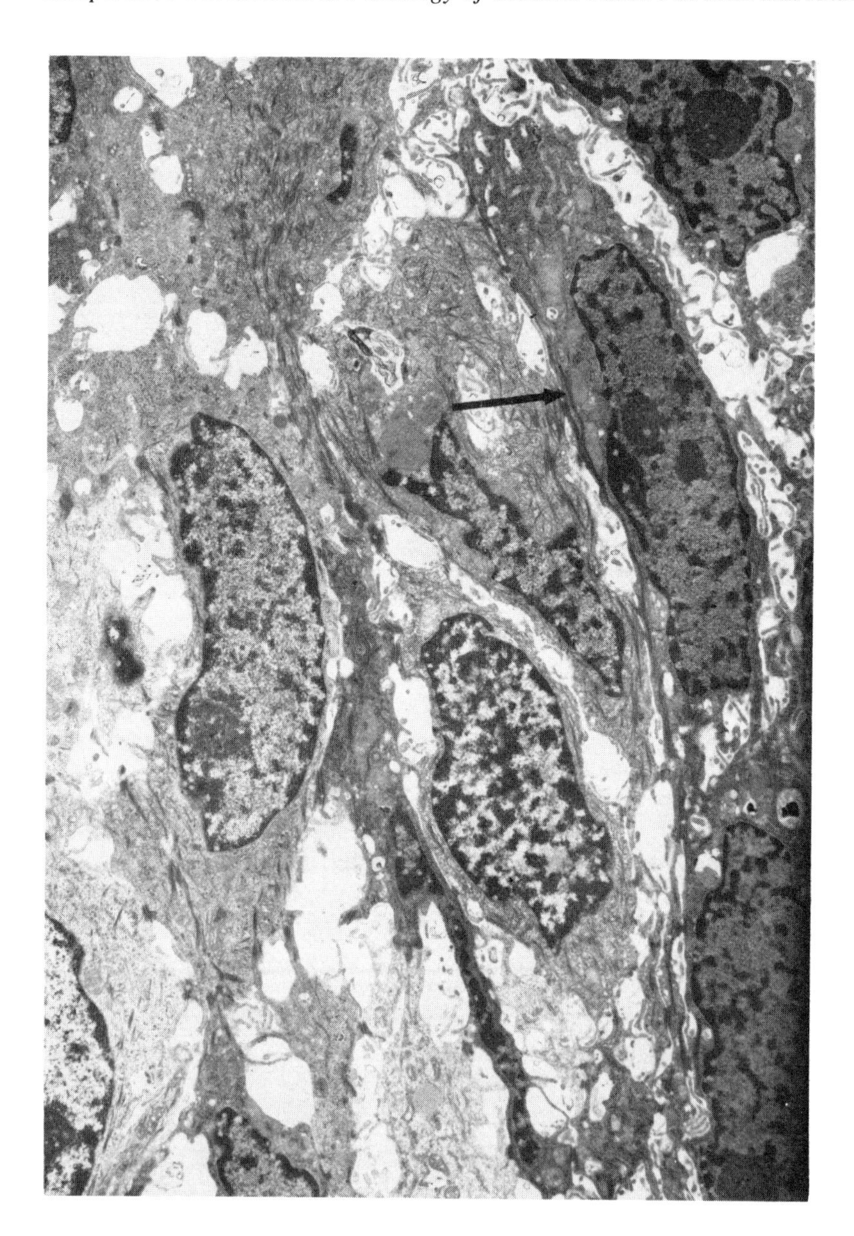

FIGURE 28. Electron micrograph of a poorly differentiated squamous cell carcinoma in the human lung: the tumor cells contain sparse, delicate bundles of cytoplasmic tonofilament bundles (arrowed), the morphological expression of immature keratin synthesis. (Uranyl acetate, lead citrate; magnification × 5200.)

In many cases, squamous cell carcinomas of the lung coexpress several types of cellular differentiation. It is, for example, very common to see mucous-producing cells in a squamous cell carcinoma, or even tumor cells which coexpress mucous production and keratin synthesis in the same cell (Figure 33). Moreover, the coexpression of neuroendocrine and squamous features in the same tumor (Figure 34) or even in the same tumor cell (Figure 22) is not uncommon. Such mixed differentiation patterns are to be understood as the result of different pathogenetic pathways (for details see Section IV. A).

A large number of chemicals, many of which are contained in cigarette smoke, have been shown to induce squamous cell carcinomas of the lung and/or upper respiratory tract

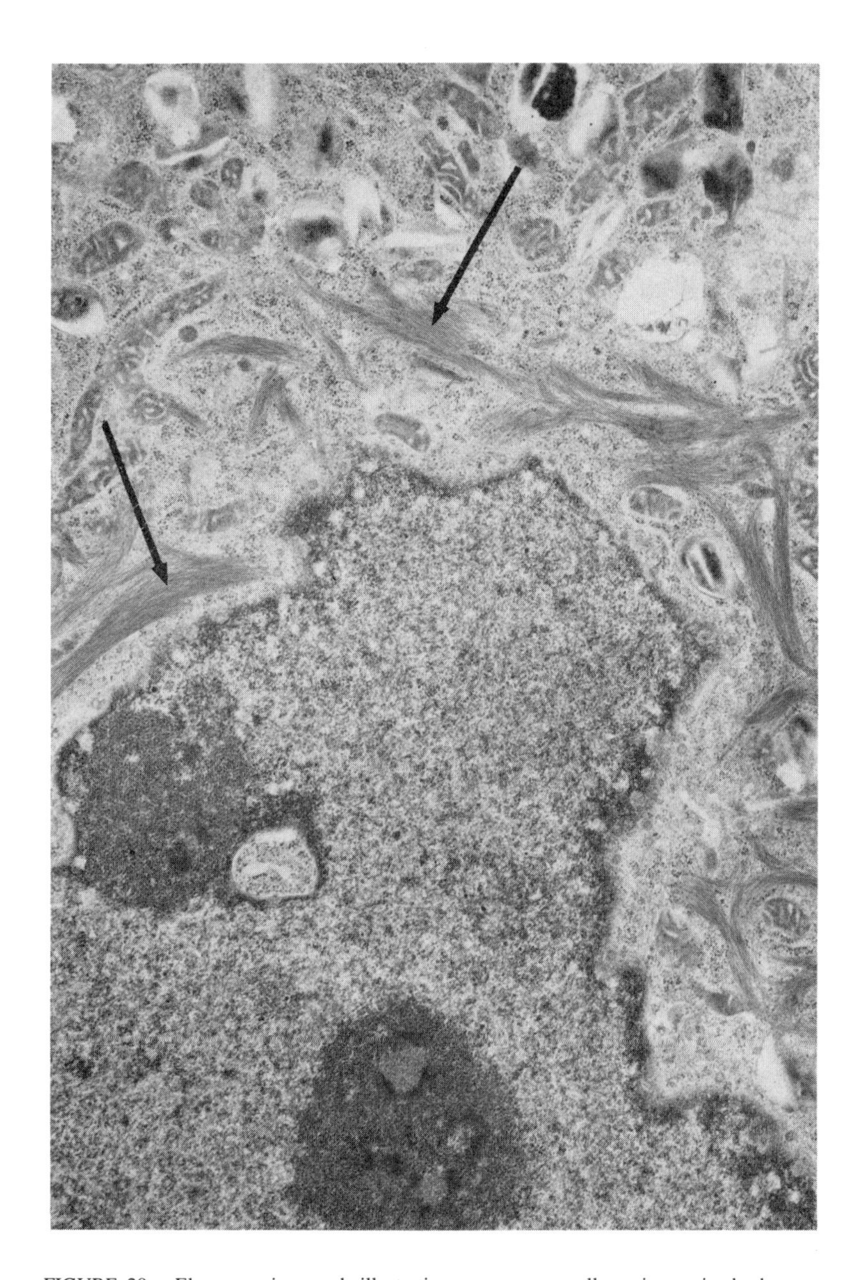

FIGURE 29. Electron micrograph illustrating a squamous cell carcinoma in the human lung: thick bundles of cytoplasmic tonofilament bundles (arrowed) are abundant, indicating a more active keratin synthesis and hence better differentiation than the tumor in Figure 28. (Uranyl acetate, lead citrate; magnification × 18,000.)

in rodents.[11] Among these are agents to which a large proportion of the general population is exposed either by profession, habit, or life style. Such agents include PAH,[11] *N*-nitrosamines,[8] and formaldehyde.[52]

A. PATHOGENESIS

Although a considerable number of publications include statements on the pathogenesis or histogenesis of squamous cell carcinomas in the human and animal lung, very few of these investigations have truly studied the sequential development of this tumor type in a

FIGURE 30. Electron micrograph demonstrating mature keratin (short arrows) and keratohyalin granules (long arrows) in a squamous cell carcinoma induced in the Syrian golden hamster trachea by multiple intratracheal instillations of benzo(a)pyrene. (Uranyl acetate, lead citrate; magnification × 20,000.)

serial sacrifice type experiment. In the human lung, squamous cell carcinomas are generally associated with bronchi lined by an epithelium comprised of basal cells, mucous cells, ciliated cells, and occasional neuroendocrine cells. This type of respiratory epithelium is found in all central bronchi as well as in segmental and subsegmental bronchi in this species.[1,53] In none of the commonly used laboratory rodents does this type of epithelium reach as far down into the lung periphery as in man.[1] The extreme in this aspect is the mouse, which demonstrates an epithelium of this composition only in the trachea while all

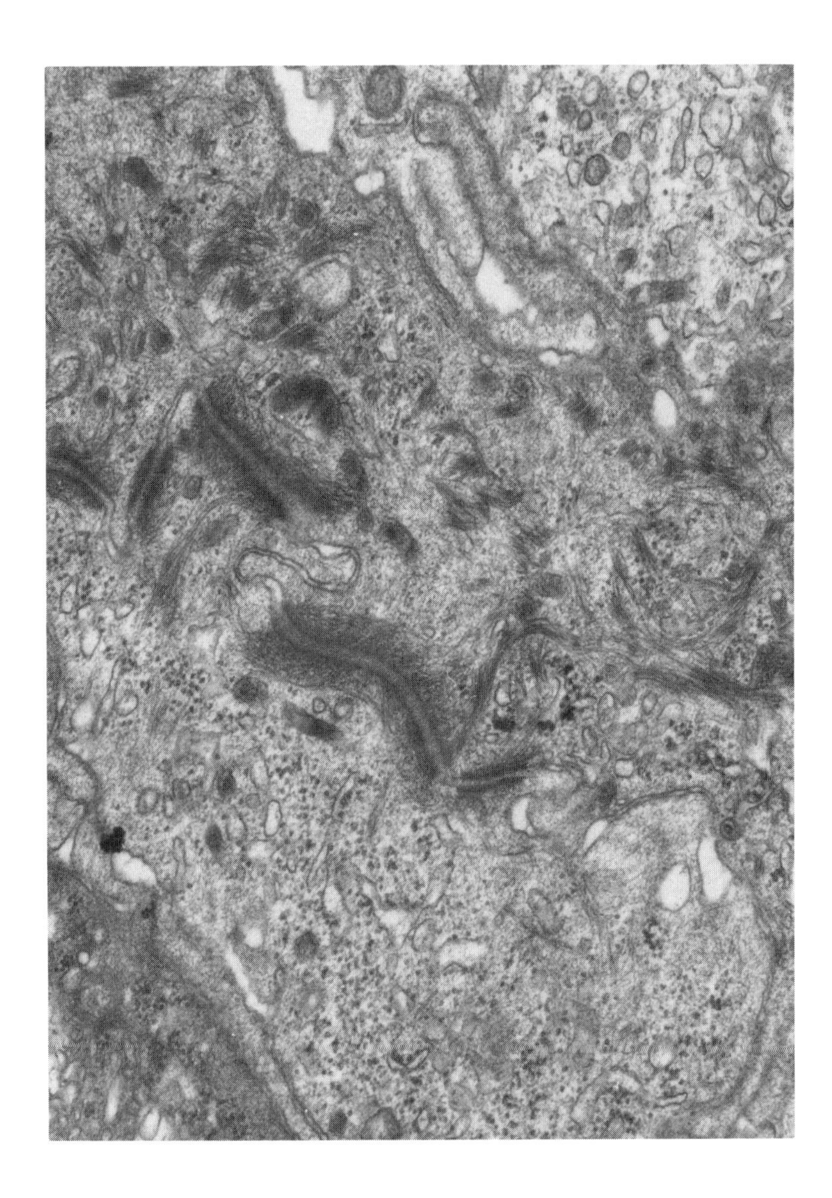

FIGURE 31. Electron micrograph of a focal area of poor squamous differentiation in a Syrian golden hamster lung tumor induced by *N*-nitrosodiethylamine. Tonofilament bundles are associated with desmosomes. (Uranyl acetate, lead citrate; magnification × 38,000.)

intrapulmonary airways are lined by an epithelium comprised of Clara cells and ciliated cells and is thus compatible with the epithelium found in the peripheral bronchioles of the human lung.[1,53] With respect to this, the mouse is not a suitable model to study the biology and pathogenesis of squamous cell carcinoma as related to this malignancy in man. As far as other laboratory rodents are concerned, it is important to remember the above described interspecies differences. Moreover, there are some species, like the rat, who consistently develop squamous cell carcinomas in the bronchiolo-alveolar region at which site tumors of this category are virtually never found in man. Accordingly, the rat would be useless as a model to study the biology and pathogenesis of squamous cell carcinomas as related to this tumor type in the human lung.

Squamous cell carcinomas in the human lung are frequently accompanied by multiple foci of basal cell hyperplasia and squamous cell metaplasia. Since it is well known that

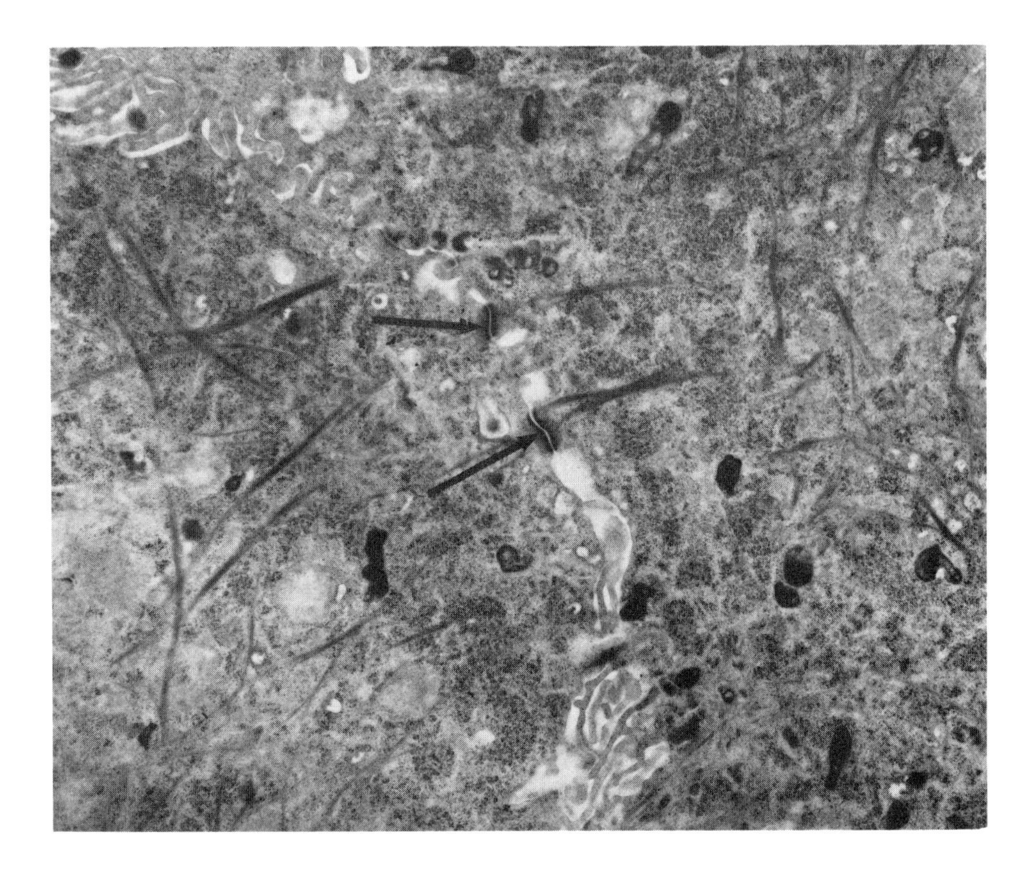

FIGURE 32. Electron micrograph of a poorly differentiated squamous cell carcinoma in the human lung: sparse bundles of cytoplasmic tonofilaments are mostly associated with desmosomes (arrowed). Note intercellular gaps at areas not connected by desmosomes. (Uranyl acetate, lead citrate; magnification × 10,000.)

basal cells serve as stem cells for cell renewal in those portions of the airways which are coated by an epithelium containing basal cells[1] (see above); this observation has lead to the "dogma" that all pulmonary squamous cell carcinomas are derived from basal cells which first undergo hyperplasia, then squamous metaplasia, and finally progress into squamous cell carcinoma. Animal experiments have shown that this pathogenetic pathway does, in fact, exist.[18] However, it has also been demonstrated that squamous cell carcinoma of the lung can arise from cell types other than basal cells.[21]

A pathogenesis of squamous cell carcinoma from basal cells has been demonstrated in the trachea of Syrian golden hamsters which received multiple intratracheal instillations of the PAH benzo(a)pyrene.[18] On the other hand, experiments in the same animal species using *N*-nitrosamines as cancer-inducing agents have provided evidence that pulmonary neuroendocrine cells can gradually lose their neuroendocrine properties and transform into squamous cells (Figure 22). Observations in human lung cancer patients have lead to the suggestion that mucous cells of the respiratory epithelium may yet be another possible source of tumors with squamous differentiation.[54] This theory was substantiated by experiments in hamsters which demonstrated the development of squamous metaplasia in mucous cells of the respiratory epithelium of the nasal cavity during the early stages of nitrosamine-induced carcinogenesis.[55] A very unique pathway of pathogenesis of squamous cell carcinoma has been reported in rats.[49,50] In this species, multiple exposure to *N*-nitroso-heptamethyleneimine resulted in early changes suggestive of an enzyme induction in Clara cells while newly formed basal cells in the bronchiolar epithelium (at which site basal cells are *not* found in

FIGURE 33. Electron micrograph of a tumor cell which coexpresses features of mucous production (long arrows) and squamous differentiation (bundles of cytoplasmic tonofilaments, short arrows). (Uranyl acetate, lead citrate; magnification × 30,000.)

healthy rats) then gave rise to hyperplasia (Figure 35), squamous metaplasia, and finally progressed into squamous cell carcinoma (Figure 27).

In keeping with the different pathogenetic pathways that can be involved in the formation of squamous cell carcinomas, many such tumors will demonstrate focal areas of differentiation of one or the other cell types mentioned above. In both man[54,56] and laboratory rodents,[21,55] cells with mucous production, neuroendocrine features, or basal cell features have been described to coexist in squamous cell carcinomas.

V. LARGE CELL CARCINOMA

Large cell carcinoma is among the common lung tumor types in man. Contrary to its misleading name, this tumor is not generally comprised of large cells. A lung tumor is

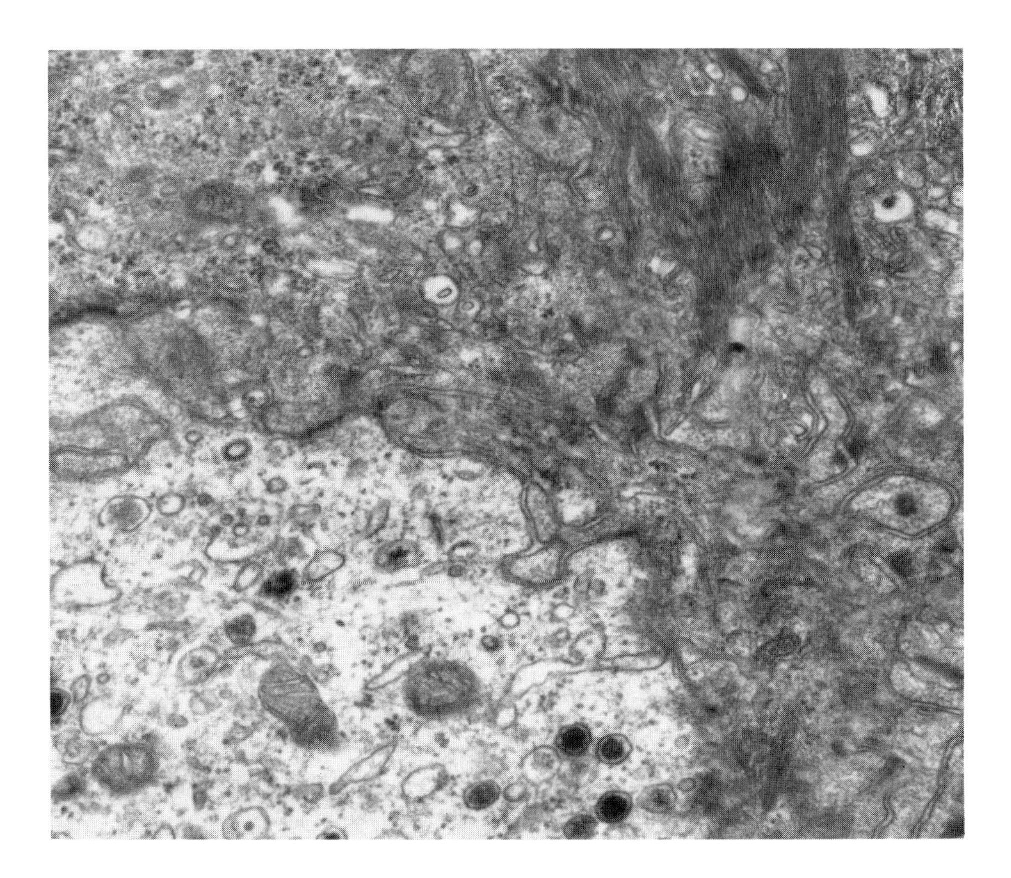

FIGURE 34. Electron micrograph of a lung tumor induced by *N*-nitrosodiethylamine in a Syrian golden hamster: the tumor cell in the upper portion of the picture demonstrates squamous differentiation while the cell in the lower left corner shows neuroendocrine differentiation. (Uranyl acetate, lead citrate; magnification × 30,000.)

usually diagnosed as a large cell carcinoma when it is so poorly differentiated that it cannot be placed with certainty under any of the known histological tumor types. Accordingly, large cell carcinoma may well represent a dedifferentiated developmental stage of many other lung tumor types. In keeping with this interpretation, electron microscopy may reveal a variety of rudimentary features of various cell types such as occasional neuroendocrine granules (Figure 36), cytoplasmic tonofilaments, abundant mitochondria (Figure 37), or morphological features of fetal epithelial cells (Figure 38).

Large cell carcinoma is rarely ever diagnosed in animals. One possible reason for this is that most animals under study are often sacrificed upon the occurrence of clinical symptoms to avoid unnecessary suffering. Moreover, the life-span of laboratory rodents may be too short for a tumor to progress to such an extremely dedifferentiated stage.

VI. ONCOCYTOMA

Although oncocytoma of the lung is rarely ever diagnosed in man or animals, this tumor type may, in fact, be more common than generally assumed. Upon electron microscopic reexamination of human lung tumors of all of the major histological tumor types, three of ten large cell carcinomas were diagnosed as oncocytomas in this author's laboratory while no features of this tumor type were revealed in any neoplasm belonging to the other histological categories.

Oncocytomas are characterized by an abundance of mitochondria (Figure 37). Although

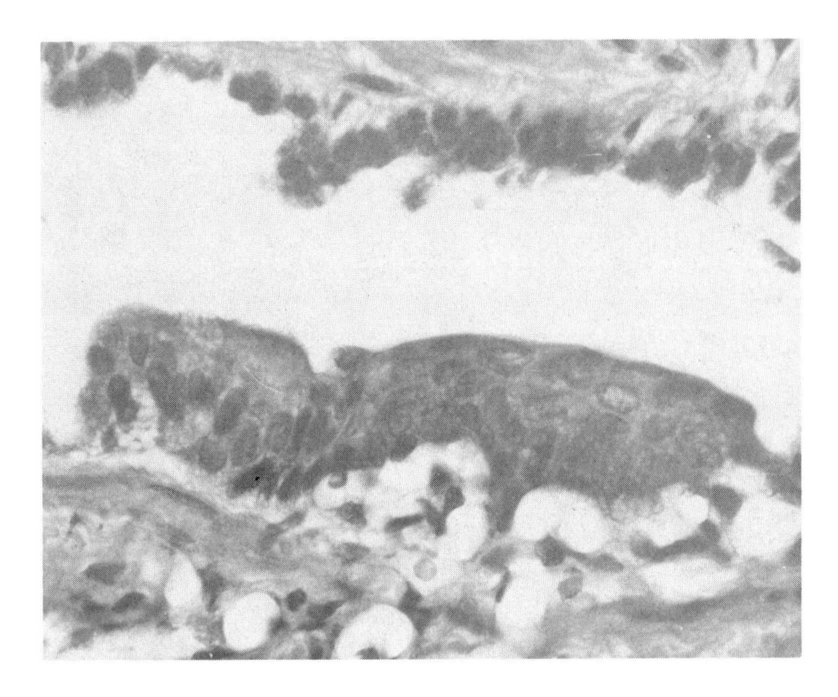

FIGURE 35. Histopathology of basal cell hyperplasia in a rat (F344 rat) bronchiole induced by *N*-nitrosoheptamethylenimine. (Toluidine blue; magnification × 820.)

tumors of this type are common in the salivary glands, they may develop in other organs as well, including the liver[57,58] and lung.

A. PATHOGENESIS

The origin of oncocytes (individual epithelial cells which demonstrate excessive numbers of mitochondria) in the lungs is not known and no model system is available to study the pathogenesis of pulmonary oncocytoma in a serial sacrifice type experiment. However, studies conducted in this author's laboratory have shown that chronic administration of *N*-nitrosamines causes a transformation of some Clara cells into oncocytes (Figure 39). Although the significance of this observation is not known, this finding suggests Clara cells as a potential origin for tumors with oncocytic differentiation in the lung.

VII. CONCLUSIONS

The study of lung cancer is complicated by the pronounced complexity of the lung. Moreover, most of the epithelial lung cell types cannot be diagnosed with certainty by routine light microscopy (paraffin embedded tissue sections stained with hematoxylin and eosin). Although more advanced techniques are available (immunocytochemistry, electron microscopy) cost-efficiency considerations generally prohibit their application on a routine basis. Moreover, with respect to the life-threatening nature of the disease, physician and patient alike want a fast diagnosis so that therapy can start as early as possible. However, the disappointing failure of lung cancer therapy in general should finally teach us a lesson.

Experimental lung cancer research has, by now, provided ample evidence that the reaction of cells to chemicals and drugs is largely regulated by functional pathways which are dependent on cell type and degree of differentiation.[8,14,28,29,41] A cure for lung cancer can, therefore, only be developed when each individual cancer case is thoroughly investigated by the most advanced techniques so that different treatment regimens can be developed for

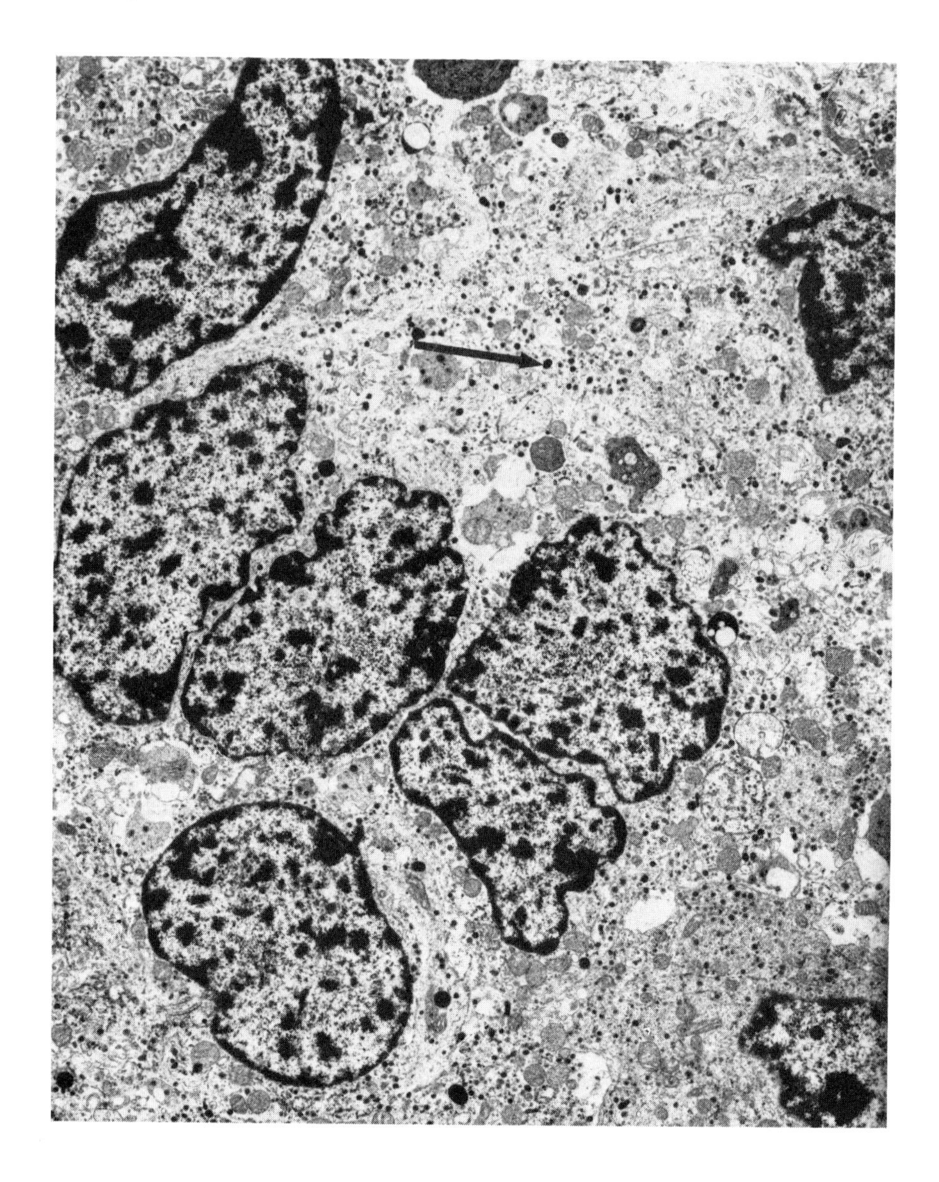

FIGURE 36. Electron micrograph of human lung tumor diagnosed as large cell carcinoma by histopathology: the tumor contains focal areas of cells with neuroendocrine secretion granules (arrowed). (Uranyl acetate, lead citrate; magnification × 5200.)

well defined (by cell type and degree of differentiation) subclasses of the existing lung tumor cateogries. It will not save any lung cancer victim when millions of dollars are spent on sometimes highly artificial, drug screening systems while the treatment of lung cancer patients is based on a simplified tumor classification into "small cell and non-small cell cancer". The therapy of leukemias has only become a success after scientists finally gave up on their efforts to develop "a cure for leukemia in general" and started to thoroughly subclassify each case based on cellular criteria so that "custom tailored" forms of therapy could be developed for each subclass of the disease. As long as a similar concept is not adopted, all efforts to arrive at a successful therapy of lung cancer are doomed for failure.

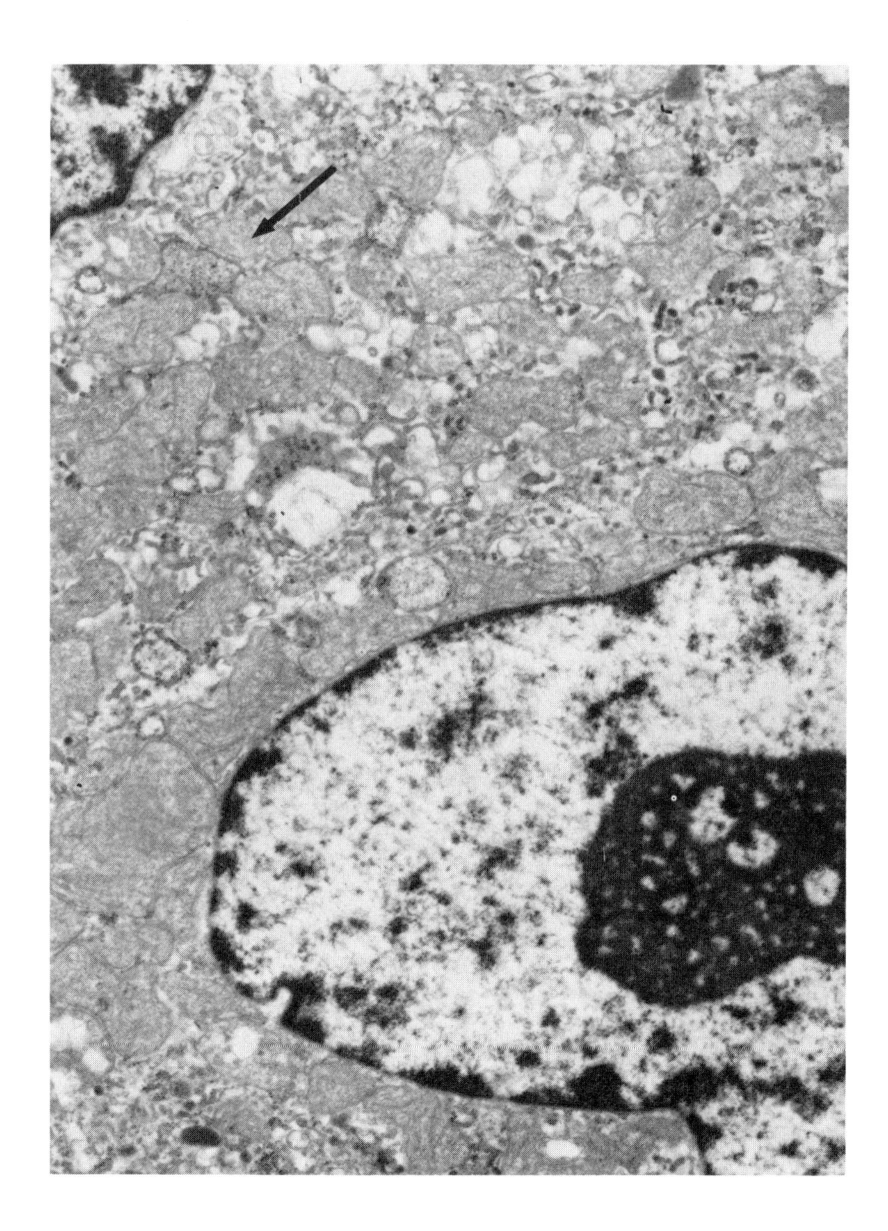

FIGURE 37. Electron micrograph of a human lung tumor diagnosed as large cell carcinoma by histopathology: the tumor cells contain excessive numbers of mitochondria (arrowed), the morphological hallmark of oncocytes. Accordingly, the tumor had to be reclassified as oncocytoma. (Uranyl acetate, lead citrate; magnification × 16,600.)

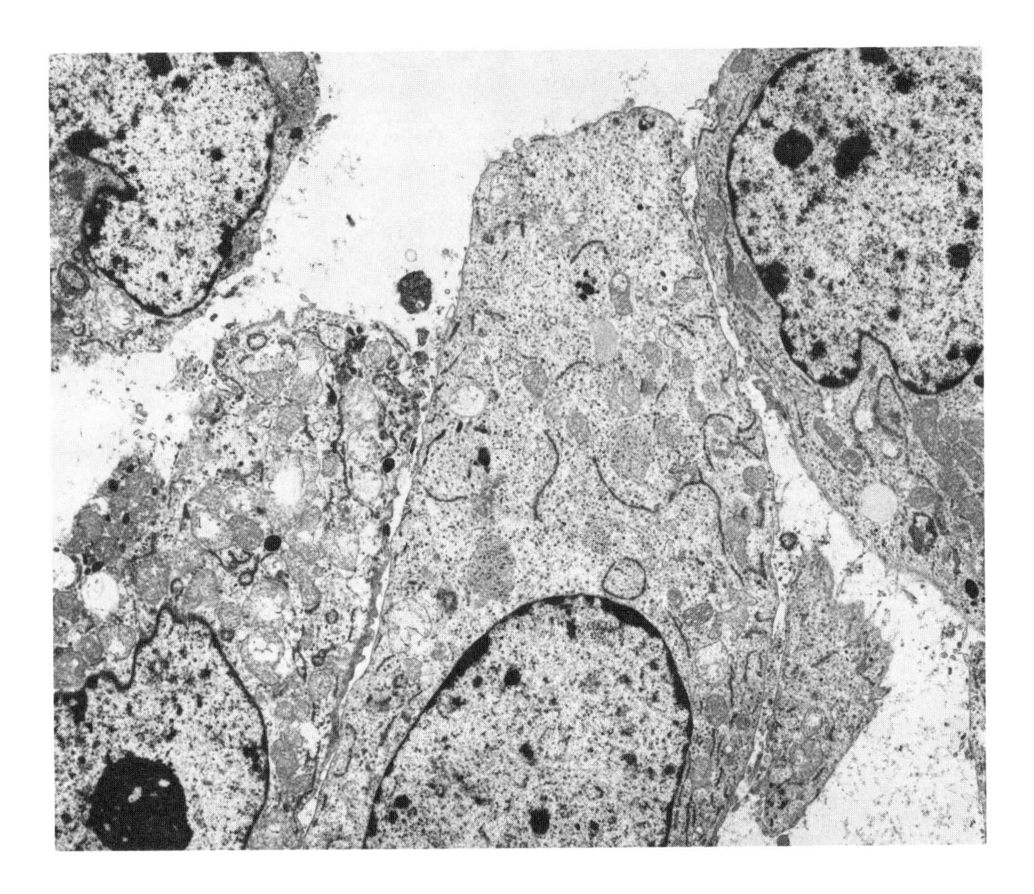

FIGURE 38. Electron micrograph of a human lung tumor diagnosed as large cell carcinoma by histopathology. Some of the tumor cells demonstrate glycogen throughout the cytoplasm along with a general paucity of other cytoplasmic organelles, both features of fetal respiratory epithelia. (Uranyl acetate, lead citrate; magnification × 5200.)

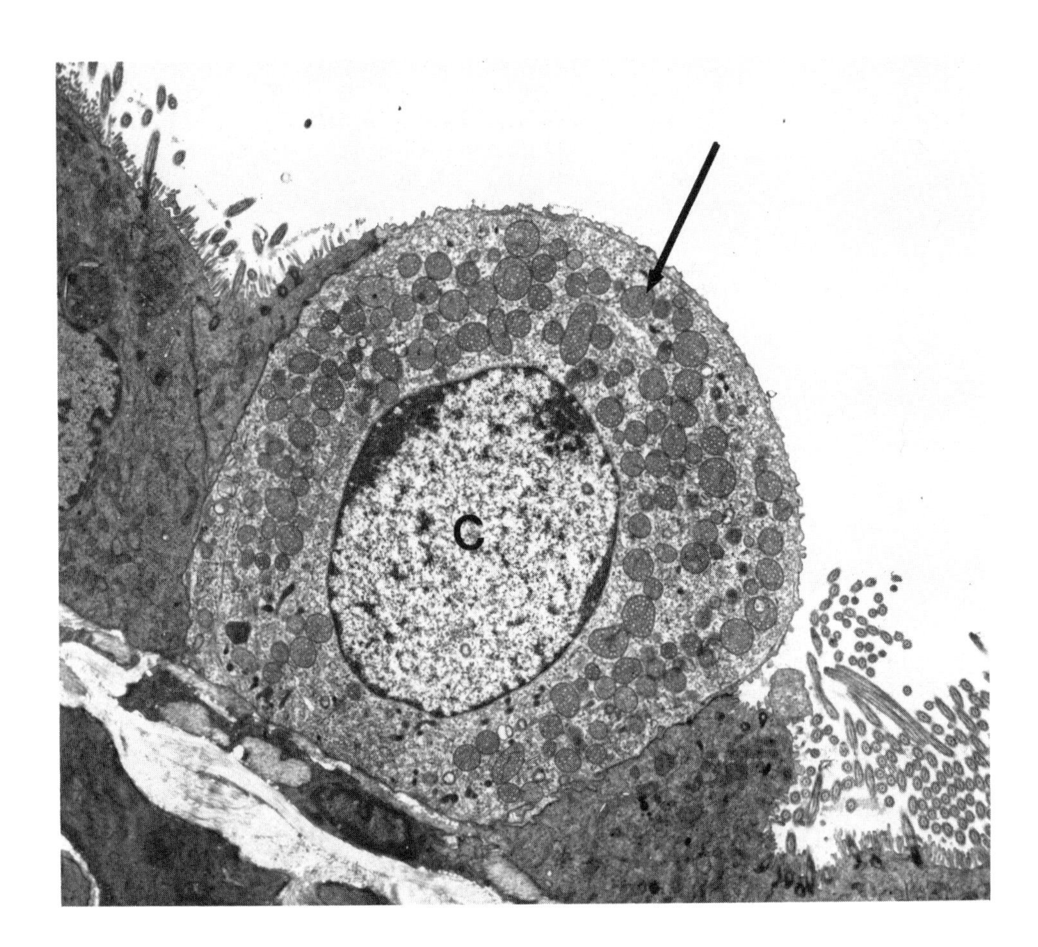

FIGURE 39. Electron micrograph of a Syrian golden hamster bronchus after 8 weeks of treatment with *N*-nitrosodiethylamine: a Clara cell (C) contains excessive numbers of mitrochondria (arrowed), the morphological marker of oncocytes. (Uranyl acetate, lead citrate; magnification × 12,600.)

REFERENCES

1. **Reznik-Schüller, H. M.,** The respiratory tract in rodents, in *Comparative Respiratory Tract Carcinogenesis,* Vol. I, Reznik-Schüller, H. M., Ed., CRC Press, Boca Raton, FL, 1983, 79.
2. **Weiss, W.,** Epidemiology of lung cancer, in *Comparative Respiratory Tract Carcinogenesis,* Vol. I, Reznik-Schüller, H. M., Ed., CRC Press, Boca Raton, FL, 1983, 1.
3. **Reznik, G.,** Spontaneous primary and secondary lung tumors in the rat, in *Comparative Respiratory Tract Carcinogenesis,* Vol. I, Reznik-Schüller, H. M., Ed., CRC Press, Boca Raton, FL, 1983, 95.
4. **Reznik, G.,** Spontaneous respiratory tract tumors in the mouse, in *Comparative Respiratory Tract Carcinogenesis,* Reznik-Schüller, H. M., Ed., Vol. I, CRC Press, Boca Raton, FL, 1983, 117.
5. **Pour, P.,** Spontaneous respiratory tract tumors in Syrian hamsters, in *Comparative Respiratory Tract Carcinogenesis,* Vol. I, Reznik-Schüller, H. M., Ed., CRC Press, Boca Raton, FL, 1983, 131.
6. **Hoch-Ligeti, C., Argus, M. F., and Strandberg, J. D.,** Primary pulmonary tumors and adenomatosis in the lung of the guinea pig, in *Comparative Respiratory Tract Carcinogenesis,* Vol. I, Reznik-Schüller, H. M., Ed., CRC Press, Boca Raton, FL, 1983, 171.
7. **Stünzi, H. and Hauser, B.,** Lung cancer in domestic animals, in *Comparative Respiratory Tract Carcinogenesis,* Vol. I, Reznik-Schüller, H. M., Ed., CRC Press, Boca Raton, FL, 1983, 171.
8. **Reznik-Schüller, H. M.,** Cancer induced in the respiratory tract of rodents by N-nitroso compounds, in *Comparative Respiratory Tract Carcinogenesis,* Vol. II, Reznik-Schüller, H. M., Ed., CRC Press, Boca Raton, FL, 1983, 109.

9. **Wynder, E. L., Goodman, M. T., and Hoffman, D.,** Lung cancer etiology: challenges of the future, in *Carcinogenesis, A Comprehensive Survey,* Vol. 8, Mass, M. J., Kaufman, D. G., Siegfried, J. M., Steele, V. E., and Nesnow, S., Raven Press, New York, 1985, 39.
10. **Hoffman, D., Melikian, A., Adams, J. D., Brunnemann, K. D., and Haley, N. J.,** New aspects of tobacco carcinogenesis, in *Carcinogenesis, A Comprehensive Survey,* Vol. 8, Mass, M. J., Kaufman, D. G., Siegfried, J. M., Steele, V. E., and Nesnow, S., Raven Press, New York, 1985, 239.
11. **Stinson, S. F. and Saffiotti, U.,** Experimental respiratory tract carcinogenesis with polycyclic aromatic hydrocarbons, in *Experimental Respiratory Tract Carcinogenesis,* Vol. II, Reznik-Schüller, H. M., Ed., CRC Press, Boca Raton, FL, 1983, 75.
12. **Katyal, S. L. and Singh, G.,** An immunologic study of the apoproteins of rat lung surfactant, *Lab. Invest.,* 40, 562, 1979.
13. **Plopper, C. G., Mariassy, A. T., and Hill, L. H.,** Ultrastructure of the nonciliated bronchiolar epithelial (Clara) cell of mammalian lung, *Exp. Lung Res.,* 1, 139, 1980.
14. **Reznik-Schüller, H. M.,** Experimental carcinogenesis of peripheral lung tumors, in *Current Problems in Tumor Pathology,* Vol. 3, McDowell, E. M., Ed., Churchill-Livingstone, Edinburgh, 1987, 348.
15. **Magee, P. N.,** Activation and inactivation of chemical carcinogens and mutagens in the mammal, *Essays Biochem.,* 10, 105, 1974.
16. **Lijinsky, W.,** Significance of *in vivo* formation of N-nitroso compounds, *Oncology,* 37, 223, 1980.
17. **Dontenwill, W. and Mohr, U.,** Carcinome des Respiratonstraktes nach Behandlung von Goldhamstern mit Diathylnitrosamin, *Z. Krebsforsch.,* 64, 305, 1961.
18. **Reznik-Schüller, H. and Mohr, U.** Investigations on the carcinogenic burden by air pollution in man. IX. *Zbl. Bakteriol. Hyg. I. Abt. Orig. B.,* 159, 493, 1974.
19. **Reznik-Schüller, H. and Mohr, U.,** Investigations on the carcinogenic burden by air pollution in man. X. *Zbl. Bakteriol. Hyg. I. Abt. Orig. B.,* 159, 503, 1974.
20. **Reznik-Schüller, H. and Mohr, U.,** Investigations on the carcinogenic burden by air pollution in man. XII. *Zbl. Bakteriol. Hyg. I. Abt. Orig. B.,* 160, 108, 1975.
21. **Reznik-Schüller, H.,** Ultrastructural alterations of APUD-type cells during nitrosamine-induced lung carcinogenesis, *J. Pathol.,* 121, 79, 1977.
22. **Stewart, H. L., Dunn, T. B., Snell, K. C., and Deringer, M. K.,** Tumours of the respiratory tract, *IARC Sci. Publ.,* 23, 251, 1979.
23. **Kauffman, S. L.,** Histogenesis of the papillary Clara cell adenoma, *Am. J. Pathol.,* 103, 174, 1981.
24. **Reznik-Schüller, H.,** Ultrastructural alterations of nonciliated cells after nitrosamine treatment and their significance for pulmonary carcinogenesis, *Am. J. Pathol.,* 85, 549, 1976.
25. **Reznik-Schüller, H. and Lijinsky, W.,** *In vivo* autoradiography and nitrosoheptamethyleneimine carcinogenesis in the hamster, *Cancer Res.,* 39, 72, 1979.
26. **Reznik-Schüller, H. M. and Hague, B. F., Jr.,** Autoradiographic study of bound radioactivity in the respiratory tract of Syrian hamsters given N-[^{3}H]-nitrosodiethylamine, *Cancer Res.,* 41, 2147, 1981.
27. **Schüller, H. M. and McMahon, J. B.,** Inhibition of N-nitrosodiethylamine induced respiratory tract carcinogenesis by piperonylbutoxide in hamsters, *Cancer Res.,* 45, 2807, 1985.
28. **Falzon, M., McMahon, J. B., Gazdar, A. F., and Schüller, H. M.,** Preferential metabolism of N-nitrosodiethylamine by two cell lines derived from human pulmonary adenocarcinomas, *Carcinogenesis,* 7, 17, 1986.
29. **Hegedus, T. J., Falzon, M., Margaretten, N., Gazdar, A. F., and Schüller, H. M.,** Inhibition of N-diethylnitrosamine metabolism in human lung cancer cell lines with features of well differentiated pulmonary endocrine cells, *Biochem. Pharmacol.,* 36, 3399, 1987.
30. **Kauffman, S. L. and Sato, T.,** Bronchiolar adenoma, lung, mouse, in *Monographs on Pathology of Laboratory Animals, Respiratory System,* Jones, T. C., Mohr, U., and Hunt, R. D., Eds., Springer-Verlag, Berlin, 1985, 107.
31. **Rehm, S., Ward, J. M., Ten Have-Opbroek, A. A. W., Anderson, L. M., Singh, G., Katyal, S. L., and Rice, J. M.,** Mouse papillary lung tumors transplacentally induced by N-nitrosoethylurea: evidence of alveolar type II cell origin by comparative light microscopic, ultrastructural and immunohistochemical studies, *Cancer Res.,* 48, 1481, 1988.
32. **Reznik-Schüller, H. M.,** An overview of experimental carcinogenesis and endocrine tumors of the lung, in *The Endocrine Lung in Health and Disease,* Becker, K. L. and Gazdar, A. F., Eds., W. B. Saunders, Philadelphia, 1984, 338.
33. **Reznik-Schüller, H. M.,** Carcinogens, the pulmonary endocrine cell, and lung cancer, in *The Endocrine Lung in Health and Disease,* Becker, K. L. and Gazdar, A. F., Eds., W. B. Saunders, Philadelphia, 1984, 345.
34. **Pearse, A. G. E.,** The cytochemistry and ultrastructure of polypeptide hormone-producing cells of the APUD-series and the embryologic, physiologic, and pathologic implications of the concept, *J. Histochem. Cytochem.,* 17, 303, 1969.

35. **Becker, K. L.,** Peptide hormones and their possible function in the normal and abnormal lung, *Rec. Results Cancer Res.*, 99, 17, 1985.
36. **Tapia, F. J., Polak, J. M., Barbosa, A. J. A., Bloom, S. R., Marangos, P. J., and Dermody, C.,** Neuron specific enolase is produced by neuroendocrine tumors, *Lancet*, 11, 808, 1981.
37. **Bensch, K. G., Corrin, B., Pariente, R., and Spence, H.,** Oat cell carcinoma of the lung: its origin and relationship to bronchial carcinoid, *Cancer*, 22, 1162, 1968.
38. **Hage, E.,** Histochemistry and fine structure of endocrine cells in foetal lungs of the rabbit, mouse, and guinea pig, *Cell Tiss. Res.*, 149, 513, 1974.
39. **Becker, K. L. and Gazdar, A. F.,** What can the biology of small cell cancer of the lung teach us about the endocrine lung?, *Biochem. Pharmacol.*, 34, 155, 1985.
40. **Weiss, W.,** The epidemiology of lung cancer, in *Comparative Respiratory Tract Carcinogenesis*, Vol. I, Reznik-Schüller, H. M., Ed., CRC Press, Boca Raton, FL, 1983, 1.
41. **Schüller, H. M., Becker, K. L., and Witschi, H.-P.,** An animal model for neuroendocrine lung cancer, *Carcinogenesis*, 7, 293, 1988.
42. **Cutz, E., Gillan, J. E., and Track, N. S.,** Pulmonary endocrine cells in the developing human lung and during neonatal adaptation, in *The Endocrine Lung in Health and Disease*, Becker, K. L. and Gazdar, A. F., Eds., W. B. Saunders, Philadelphia, 1984, 210.
43. **Lauweryns, J. M. and Cookelaere, M.,** Hypoxia sensitive neuroepithelial bodies. Intrapulmonary secretory neuroreceptors modulated by the CNS, *Z. Zellforsch.*, 145, 521, 1973.
44. **Janoff, A., Pryor, W. A., and Bengali, Z. H.,** NHLBI Workshop Summary: Effects of tobacco smoke components on cellular and biochemical processes in the lung, *Am. Rev. Resp. Dis.*, 136, 1058, 1987.
45. **Reznik-Schüller, H. M.,** Proliferation of endocrine (APUD) type cells during early, DEN-induced lung carcinogenesis in hamsters, *Cancer Lett.*, 1, 255, 1976.
46. **Reznik-Schüller, H. M.,** Sequential morphological alterations in the bronchial epithelium of Syrian golden hamsters during N-nitrosomorpholine induced pulmonary tumourigenesis, *Am. J. Pathol.*, 89, 59, 1977.
47. **Douglas, W. W.,** Histamine and 5-hydroxytryptamine (serotonin) and their antagonists, in *The Pharmacological Basis of Therapeutics*, Goodman Gilman, A., Goodman, L. S., Rall, T. W., and Murad, F., Eds., Macmillan, New York, 1980, 605.
48. **Stockman, M. S., Anthonisen, N. R., Wright, E. C., and Donithan, M. G.,** Airway obstruction and the risk for lung cancer, *Ann. Intern. Med.*, 106, 512, 1987.
49. **Reznik-Schüller, H. M. and Gregg, M.,** Ultrastructure of nitrosoheptamethyleneimine induced lung tumors in Fischer rats, *Anticancer Res.*, 6, 381, 1983.
50. **Reznik-Schüller, H. M. and Gregg, M.,** Pathogenesis of lung tumors induced by N-nitrosoheptamethyleneimine in F-344 rats, *Virch. Arch. Abt. A. Pathol. Anat. Histol.*, 393, 333, 1981.
51. **Stünzi, H. and Hauser, B.,** Lung cancer in domesticated animals, in *Comparative Respiratory Tract Carcinogenesis*, Vol. I., Reznik-Schüller, H. M., Eds., CRC Press, Boca Raton, FL, 1983, 195.
52. **Swenberg, J. A., Kerns, W. D., Mitchell, R. E., Grall, E. J., and Pavcov, K. L.,** Induction of squamous cell carcinoma of the rat nasal cavity by inhalation of formaldehyde vapor, *Cancer Res.*, 401, 3398, 1980.
53. **Bartels, H.,** The human lung, in *Comparative Respiratory Tract Carcinogenesis*, Vol. I, Reznik-Schüller, H. M., Ed., CRC Press, Boca Raton, FL, 1983, 19.
54. **McDowell, E. M.,** Bronchogenic carcinomas, in *Current Problems in Tumor Pathology: Lung Carcinomas*, McDowell, E. M., Ed., Churchill-Livingstone, Edinburgh, 1987, 255.
55. **Reznik-Schüller, H. M.,** Nitrosamine-induced nasal cavity carcinogenesis in rodents, in *Nasal Tumors in Animals and Man*, CRC Press, Boca Raton, FL, 1983, 47.
56. **Warren, W. H., Memoli, V. A. and Gould, V. E.,** Immunohistochemical and ultrastructural analysis of bronchopulmonary neuroendocrine neoplasms. II. Well differentiated neuroendocrine carcinomas, *Ultrastruct. Pathol.*, 7, 185, 1984.
57. **Reznik-Schüller, H. M. and Reuber, M. D.,** Ultrastructure of liver, tumors induced in F-344 rats by methapyrilene, *J. Environ. Pathol. Toxicol.*, 8, 501, 1981.
58. **Reznik-Schüller, H. M. and Linjinsky, W.,** Morphology of early changes in liver carcinogenesis induced by methapyrilene, *Arch. Toxicol.*, 4, 79, 1981.
59. **Schüller, H. M. and Hegedus, T. J.,** Selective uptake/metabolism of nitrosamines by neuroendocrine (NC) cells of the lung, *Proc. Adv. Biol. Chem. of N-Nitroso and Related Compounds*, Omaha, NE, May 1988.

Chapter 2

HISTOGENESIS OF PSEUDODUCTULAR FORMATION DURING PANCREATIC CARCINOGENESIS IN HAMSTERS: ELECTRON MICROSCOPIC AND IMMUNOHISTOCHEMICAL STUDIES

Parviz M. Pour and Rudolph Hauser

TABLE OF CONTENTS

I. INTRODUCTION

Different animal models for pancreatic cancer have been developed in the last several years.[1-5] Although these models have contributed to our understanding of some aspects of human disease, the histogenesis of pancreatic cancer has become a controversial issue. It is generally accepted that in the rat model induced tumors are of acinar cell origin. However, in the hamster model, in which the induced tumors are of a ductal/ductular type and resemble in many aspects the equivalent human tumor,[6] different views have surfaced regarding the origin of induced pancreatic lesions. Several studies in the hamster model have suggested ductal/ductular cells as the primary source of tumors.[7-11] Others argue with this view and believe that acinar cells are the tumor progenitor cells. The latter researchers base their view on the dedifferentiation theory and suggest that during carcinogenesis, acinar cells undergo retrodifferentiation and assume a duct-cell phenotype.[12-17] A third group of investigators consider both of the above possibilities.[18,19]

The present study describes electron microscopic and immunohistochemical findings during pancreatic carcinogenesis in the hamster model and concentrates specifically on the development of pseudoductular or tubular structures, which are the primary events in tumor formation and thus the key to understanding tumor histogenesis.

II. MATERIALS AND METHODS

A. ANIMALS

Syrian golden hamsters from the Eppley colony were used. They were housed in plastic cages in groups of five by sex on granular cellulose bedding. They were kept under standard laboratory conditions, given a pelleted diet (Allied Mills, Chicago, IL), and water *ad libitum*. They were checked twice daily. All hamsters were 8 to 10 weeks old at the start of the carcinogenesis study.

B. CARCINOGEN

N-nitrosobis(2-oxopropyl)amine (BOP) was synthesized as described,[20] dissolved in physiological saline, and injected subcutaneously.

C. CARCINOGENICITY STUDY

Hamsters were treated with BOP either once at 20 mg/kg body weight (b.w.) or weekly for 6 weeks at 10 mg/kg b.w. From each group, three hamsters each were euthanized at 12, 20, 28, 32, and 40 weeks after the beginning of BOP treatment. After complete autopsies, samples were taken from several areas of the pancreas and examined by electron microscopy and immunohistochemistry.

D. ELECTRON MICROSCOPY

Samples were prefixed immediately in 2.5% glutaraldehyde in 0.13 *M* balanced phosphate buffered 1.3% recycled osmium tetroxide at 4°C, and dehydrated through a gradient series of ethanol and propylene oxide prior to embedding in BEEM capsules using Epon 812. During dehydration, the samples were stained with 2% uranyl acetate at the 80% ethanol step. Ultrathin sections (60 to 80 nm) were stained on the grid with 2% uranyl acetate and 0.4% lead citrate and were examined by a Siemens Elmiskop 101 with an instrumental magnification of 2400 to 10,000.

E. IMMUNOHISTOCHEMISTRY

Samples were taken from different areas of the pancreas and fixed in Bouin's solution for 6 h, and then were dehydrated and embedded in paraplast according to conventional methods. Sections were treated with monoclonal antibodies against blood group antigens A, B, and H, as described.[21,22]

III. RESULTS

Detailed histological, immunohistochemical, and electron microscopic characteristics of the hamster pancreas were described earlier.[10,21-26] In untreated hamsters, acinar cells showed a well-developed granular endoplasmic reticulum (ER), electron-dense zymogen granules of various sizes in the apical portion of the cell, usually oval-shaped mitochondria, and few, if any, autophagic vacuoles (Figure 1). The centroacinar and terminal ductular cells are characterized by their sparsity of organelles, scattered small mitochondria, some free ribosomes, irregularly shaped nuclei, and presence of tight junctions between themselves and acinar cells (Figure 1).

Degeneration of a single or group of acini with concomitant proliferation and hypertrophy of terminal ductular and centroacinar cells was the earliest alteration seen after carcinogen treatment. Acinar cells showed swelling of the mitochondria, distension of the ER, alteration in number and density of zymogen granules, and formation of autophagic vacuoles (Figure 2). The centroacinar cells were increased in number (Figure 2), and showed vacuolization of the cytoplasm, swelling of the mitochondria, and in some cases a few granules of an endocrine type, suggesting differentiation toward endocrine cells (Figures 2 and 3).

At a later point in the carcinogenic process, centroacinar cells displayed irregular, long, and tiny cytoplasmic processes (cytp) extending to the surface of adjacent acinar cells or between these cells (Figures 4 to 7). The cytp, which exhibited a number of microvilli, covered part of or most surface areas of the acinar cells, in which granules were found to be localized beneath the remaining free luminal cell surface (Figure 6). Such acinar cells displayed signs of degeneration, the degree of which correlated with the extent of the cell surface occupied by the cytp of centroacinar cells (Figures 4 to 7). In many cases only a portion of cytp of an acinar cell was seen in a given section, obviously due to the irregular shape and course of the cytp. The presence of desmosomes in the cytp covering the acinar cells (Figures 5 to 7) indicated that several altered centroacinar cells take part in this process. Desmosomes were also found between the cytp and the underlying surface of the acinar cells (Figures 6 and 7).

At this stage, although monoclonal anti-A and anti-B did not react with any normal pancreatic cells,[22] they did bind to the cytp, to centroacinar cells, and to the surface of hyperplastic ductal cells (Figures 8 and 9). Detailed immunohistochemical findings with polyclonal and monoclonal antibodies generated against blood group antigens A, B, H, and Lewis antigens during pancreatic carcinogenesis are reported,[22] as are their specific binding patterns to ductal/ductular cell tumors.[21]

The acinar cells, the surfaces of which were entirely covered with cytp of the centroacinar cells, show an advanced degree of degeneration, with dilated ER and alterations in density and quantity of zymogens, autophagic vacuoles and pyknotic nuclei. Such altered acinar cells could occasionally be observed within the lumen of the gland (Figure 10).

In advanced stages of the carcinogenic process, most or all of these acinar cells were replaced by the elongated centroacinar cells (Figures 4 and 11), thus culminating in the formation of pseudoductules or "tubular structures".

Although some pseudoductular structures, lined with a single layer of elongated cells (Figure 4), retained their benign cytologic appearance during the entire carcinogenesis process, others showed an increase in cell size with some degree of pleomorphism. The cells took an "upright" position, i.e., the cell axis became directed toward the lumen, instead of parallel to the lumen (Figure 11), as was the case in normal ducts and ductules. Moreover, they showed irregular and rather plump microvilli and nuclear pleomorphism (Figure 11).

With increasing size, numbers, pleomorphism, and atypia of cells, patterns consistent with carcinoma *in situ* developed.

In no circumstances did we observe flattening of acinar cells and their gradual transformation to ductular type cells. However, concomitant proliferation of endocrine cells was consistently seen during all stages of ductular proliferation, as described previously in detail.[11,25]

Changes in ductal cells were comparable to those in pseudoductules, and some ductal cells were found to be unaltered during the carcinogenesis process, whereas others displayed increasing cell hypertrophy, hyperplasia, pleomorphism, and atypia paralleling that in some pseudoductules.

The speed and degree of malignant change were found to be related to the frequency (dose) of carcinogen application and to the time lapse. In animals receiving repeated carcinogen doses, toxic degenerative changes accompanied hyperplastic and neoplastic ductal/ductular alterations and neoplastic lesions, whereas single doses of the carcinogen resulted in the formation of more benign pseudoductular structures.

IV. DISCUSSION

It has been shown by us and others[10,11,23,24] that the formation of pseudoductular structures is associated with pancreatic carcinogenesis in the hamster model, although these structures are not necessarily precursors of a ductular type of cancer. Although the hyperplastic and metaplastic processes of ductal epithelium represent a straightforward event, those of ductules have not been well understood. Opinions on pseudoductular or tubular structures are still divided, in terms of their origin from dedifferentiated acinar cells or directly from ductular (centroacinar) cells.

Our electron microscopic and immunohistochemical findings strongly support the view that centroacinar cells are the foundation for pseudoductules in the hamster pancreatic cancer model. Our data clearly demonstrate that the primary event during carcinogenesis is proliferation of centroacinar cells, which, contrary to the changes occurring in ductal cells, undergo a unique modification in structure and behavior. The nature of these rather dynamic changes can be understood by serial examination of tissues during the course of carcinogenesis and is characterized by hypertrophy and hyperplasia of centroacinar cells, with formation of initially tiny and long processes which over- and underlie the adjacent acinar cells. This event results in the gradual separation of affected acinar cells from the glandular lumen and vasculature, resulting in their gradual degeneration and their subsequent replacement by altered centroacinar cells. Some degenerated acinar cells seem to be expelled into the lumen. Others obviously undergo autophagic processes. Whether or not some of these cells are phagocytized by altered centroacinar cells, as suggested,[8] is a subject for further study.

The formation of long cytoplasmic processes has been observed, thus far, to be a unique characteristic of pancreatic polypeptide cells,[27] the processes of which have also been found to be associated with acinar cells.[27] Since it is believed that endocrine cells derive from ductular, particularly from centroacinar, cells,[10,26,28-30] the formation of cytoplasmic processes seems to be inherent in centroacinar cells. Considering the known pluripotent nature of centroacinar cells, it can be assumed that the fate of these cells, in terms of their differentiation, depends on the nature of stimuli; they form either phenotypical benign endocrine cells or poorly differentiated, elongated cells with the tendency to damage and to replace acinar cells. This ability is consistent with the malignant nature of these cells, although, in most cases, pseudoductules represent a stationary structure with only a slight tendency for malignant progression. However, the expression of blood group antigens, not present in normal centroacinar cells, clearly reflects fundamental changes in their genotype. The question as to why some of the pseudoductular cells remain dormant and others undergo rapid malignant transformation cannot be answered by the present study.

Our observationss coincide with those of others[7,8] who could not demonstrate dedifferentiation of acinar cells to ductular-like cell elements. We also could not see a gradual transformation of acinar cells, nor were there any cell types which could represent intermediary cells between acinar and ductular cells.

Clearly there is a discrepancy in findings relative to the formation of pseudoductules or tubular structures in some pancreatic diseases. Bockman et al. have studied rats in which the pancreas was exposed to polycyclic aromatic hydrocarbons,[12,13] as well as human pancreatitis cases.[31,32] They have found that under both conditions the tubular structure develops by the gradual atrophy of acinar cells, which lose their apical cytoplasmic portion and zymogen granules and take on the appearance of ductal cells. However, a complete dedifferentiation of these acinar cells to ductular cells has not been clearly shown. Studies by Flaks et al.[14-17] in hamsters exposed to pancreatic carcinogen repeatedly describe the formation of pseudoductules (tubular complexes) from acinar cells, without, however, detailing the possible intermediate cells between acinar cells and ductular cells. Remarkably enough, these authors (contrary to many others working with this model) have not observed alteration of ductal and ductular epithelium during carcinogenesis, although several photomicrographs in their publications clearly depict long cytoplasmic processes covering the acinar cell surface.[15,16] Nevertheless, similar studies employing the same experimental schedule by other groups of researchers led to the opposite view, i.e., that ductal/ductular cells are the primary origin of induced pancreatic exocrine tumors.[7,8]

The reason for the differing findings and concepts is difficult to explain. We entertain the notion that there are possibly two different mechanisms involved in pseudoductular (tubular) formations. Dilatation of acini with flattening of acinar cells and loss of their apical cell portions have been observed in a variety of benign conditions, such as ductal occlusion,[33,34] uremia,[33,35,36] and inflammation.[31,32] This type of acinar cell change most probably represents a reactive process, which can also occur in other conditions, such as in tissue culture preparations,[37] most probably as a reflection of altered environment. Implantation of local active carcinogens, such as polycyclic aromatic hydrocarbons, a method used in studies of Bockman et al., causes, besides the neoplastic process, reactive inflammatory and reparative events and consequently could lead to formation of dilated acini comparable to tubular structures. On the other hand, we[10,22,23] and others[38] have shown that similar degenerative, reparative processes occur in the hamster pancreas by repeated injection of high (toxic) doses of the carcinogen, a schedule which was used in studies by Flaks et al.[14-17] We have shown that a single treatment scheme was more appropriate for the study of tumor histogenesis than chronic carcinogen administration, since repeated toxic insult to pancreatic cells can be avoided by this type of experiment.[10] By using such a treatment schedule, primary alterations during carcinogenesis are clearly shown to affect ductal and ductular cells. Hence, we believe that the mechanisms of pseudoductular (tubular) formation in benign and malignant processes are different. Tubular complexes may well represent an adaptive (reactive) response of acinar cells, whereas the formation of pseudoductules by altered centroacinar cells during carcinogenesis represents, in our view, a neoplastic response. This includes formation of benign-appearing pseudoductules, because some of these structures have been shown to be precursors of either microcytic adenomas or carcinomas.[9,10]

The neoplastic process is generally accepted to be initiated by ''germinal'' or ''stem cells'' of a given tissue, at least in gastrointestinal cancer. Formation of a variety of cells, including endocrine cells, found in experimental and human gastrointestinal tumors (see References 10, 39, and 40), points to the pluripotent nature of the tumor precursor cells, an ability which is reserved for ''stem cells''. Considering the embryogenesis of the pancreas in numerous species, ductal/ductular cells are a precursor of two major types of pancreatic cells, i.e., acinar and islet cells. The development of endocrine cells from ductal and ductular cells has been shown by numerous studies to be retained in adult life in man and in animals (see References 10, 11, 28-30, and 41-48). We have also shown that during the carcinogenesis process endocrine cells take part in the proliferation by their obvious overproduction from hyperplastic ductal/ductular (centroacinar) cells.[10,11,26] Endocrine cells of different types have been demonstrated in experimental and human pancreatic ductal tumors (see References 10, 11, and 26). Our recent study also demonstrated formation of a few acinar cells during carcinogenesis.[49] A similar observation has been made in human neoplasms. Our findings, thus, correlate with those of other gastrointestinal carcinomas, in which a variety of poorly differentiated to highly specialized cell elements was observed.

The presented results and above-cited studies, nevertheless, strongly support the view that in hamsters the centroacinar cells are the primary progenitor cells of induced pseudoductules and of tumors arising from them.

V. SUMMARY

Electron microscopic and immunohistochemical studies were performed to elucidate the histogenesis of pseudoductular structures, which represent early changes during pancreatic carcinogenesis in hamsters. It was found that hypertrophy and hyperplasia of centroacinar cells are the earliest changes occurring in the pancreas. These altered centroacinar cells differentiate toward either endocrine-type cells or to elongated cells with remarkably long and slender cytoplasmic processes. These processes gradually over- and under-lie the adjacent acinar cells and result in progressive degeneration and loss of acinar cells which subsequently are replaced by altered centroacinar cells. Formation of cytoplasmic processes, which are not visible under light microscopy, are associated with the expression of blood group substances, which are also found on the surface of altered ductal cells. Since these antigens cannot be demonstrated in normal pancreatic cells, they seem to represent specific markers for altered ductal/ductular (centroacinar) cells. In no instance was there evidence of dedifferentiation of acinar cells into duct-like cells. The present data, along with our previous findings, demonstrate that centroacinar cells are the foundation for pseudoductular structures and the progenitor cells of tumors arising from them.

ACKNOWLEDGMENTS

This study was supported, in part, by core grant CA36727 from NCI/NIH and SIG-16 from the American Cancer Society.

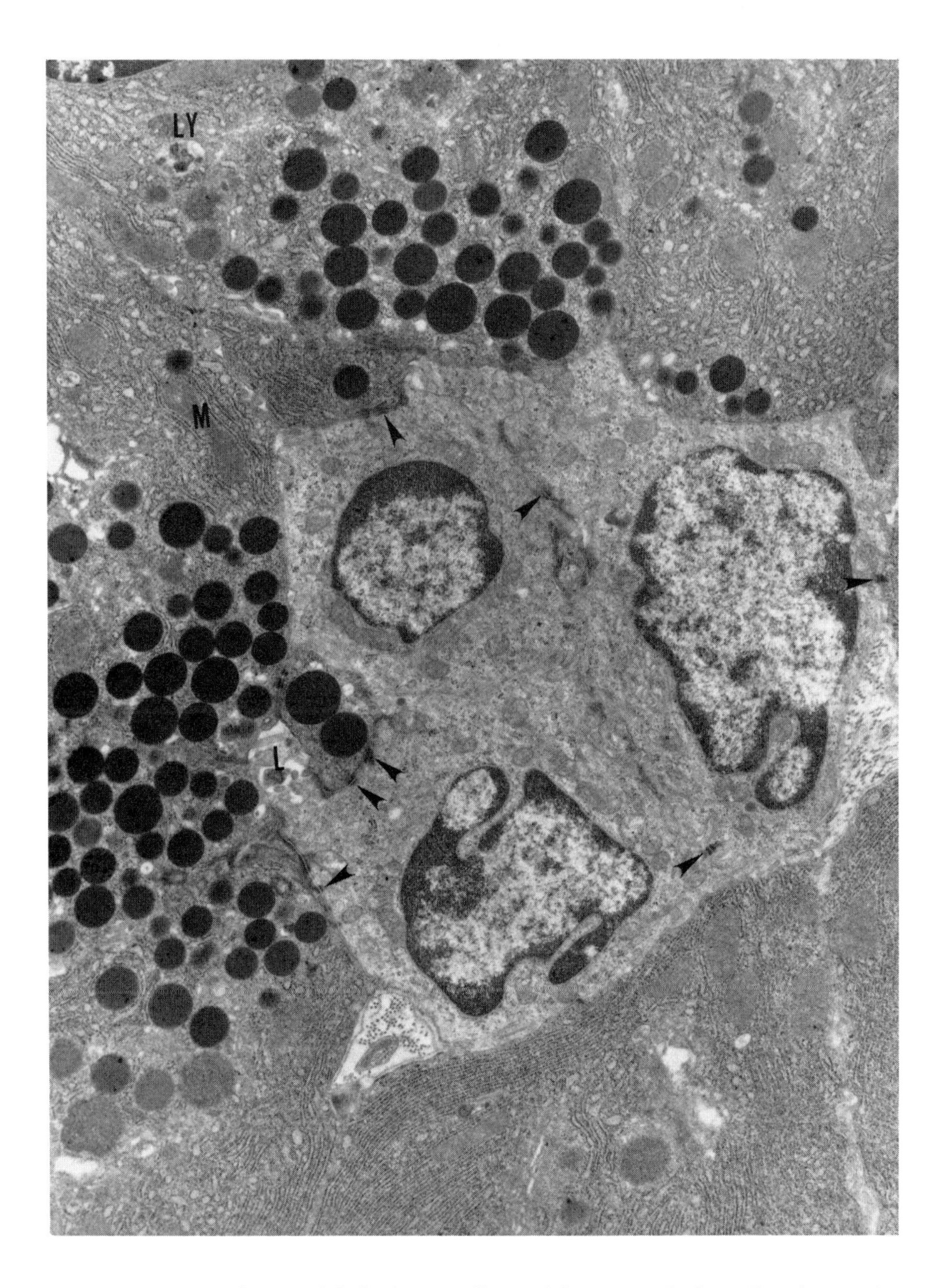

FIGURE 1. Pancreas of untreated Syrian hamster. Characteristic patterns of acinar cells and centroacinar cells. Acinar cells have abundant granular endoplasmic reticulum, electron-dense zymogen granules of various sizes, usually oval-shaped mitochondria (M) and a few lysosomes (LY). Note the sparsity of organelles in centroacinar cells and numerous desmosomes, some indicated by arrowheads. Portion of glandular lumen (L) is seen. (Magnification × 7000.)

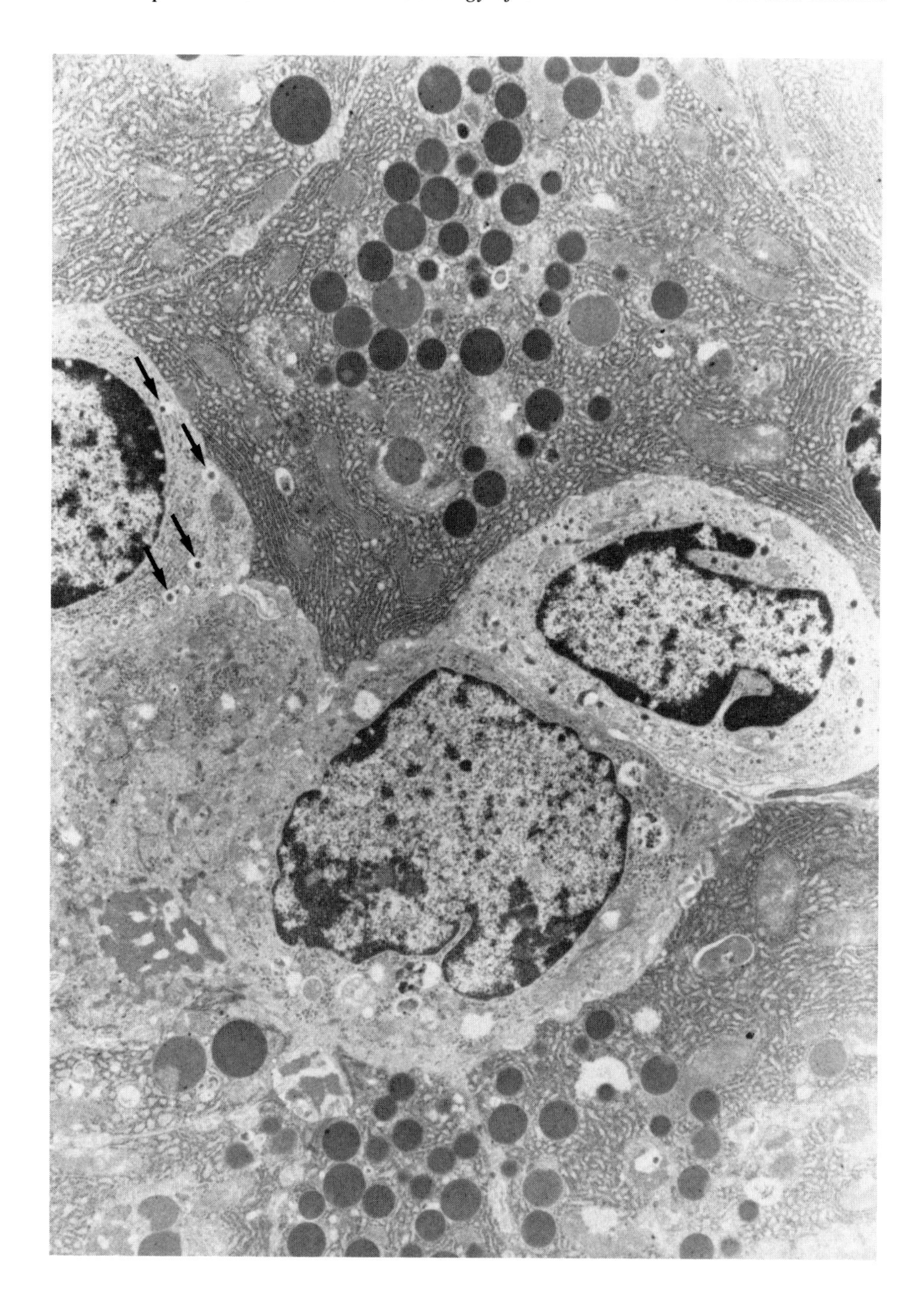

FIGURE 2. Pancreas of a Syrian hamster treated with the pancreatic carcinogen *N*-nitrosobis(2-oxopropyl)amine (BOP). Note mild dilatation of endoplasmic reticulum, vacuolization of cytoplasm, irregular size and electron-density of zymogens and mild swelling of mitochondria in acinar cells. Centroacinar cells show cytoplasmic vacuoles, numerous lysosomes and a few granules of an endocrine type (arrowheads). (Magnification × 9000.)

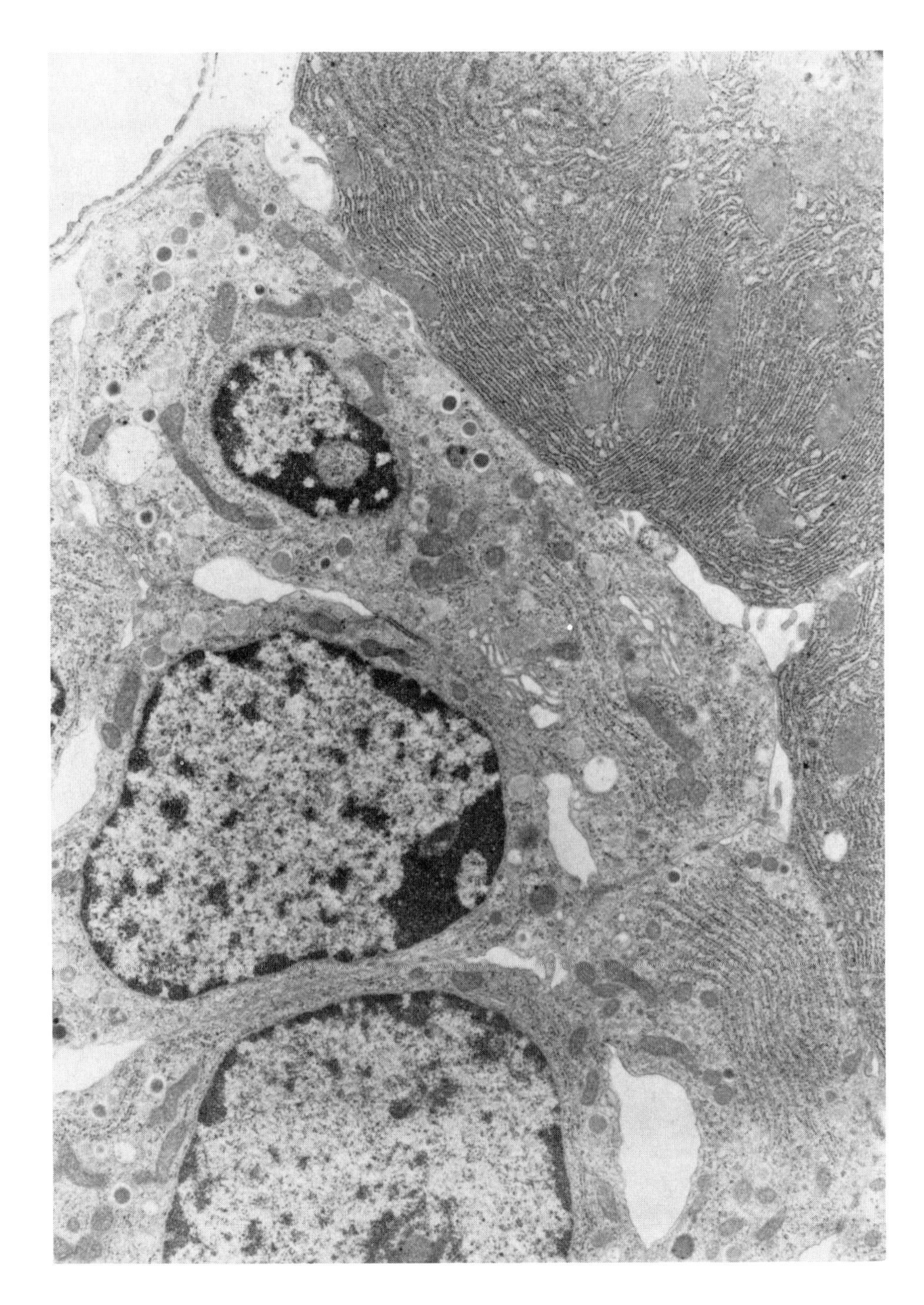

FIGURE 3. Pancreas of a BOP-treated hamster depicting centroacinar cells, two of which show elongated cytoplasm containing an endocrine type of granules Golgi apparatus and numerous mitochondria. (Magnification × 9000.)

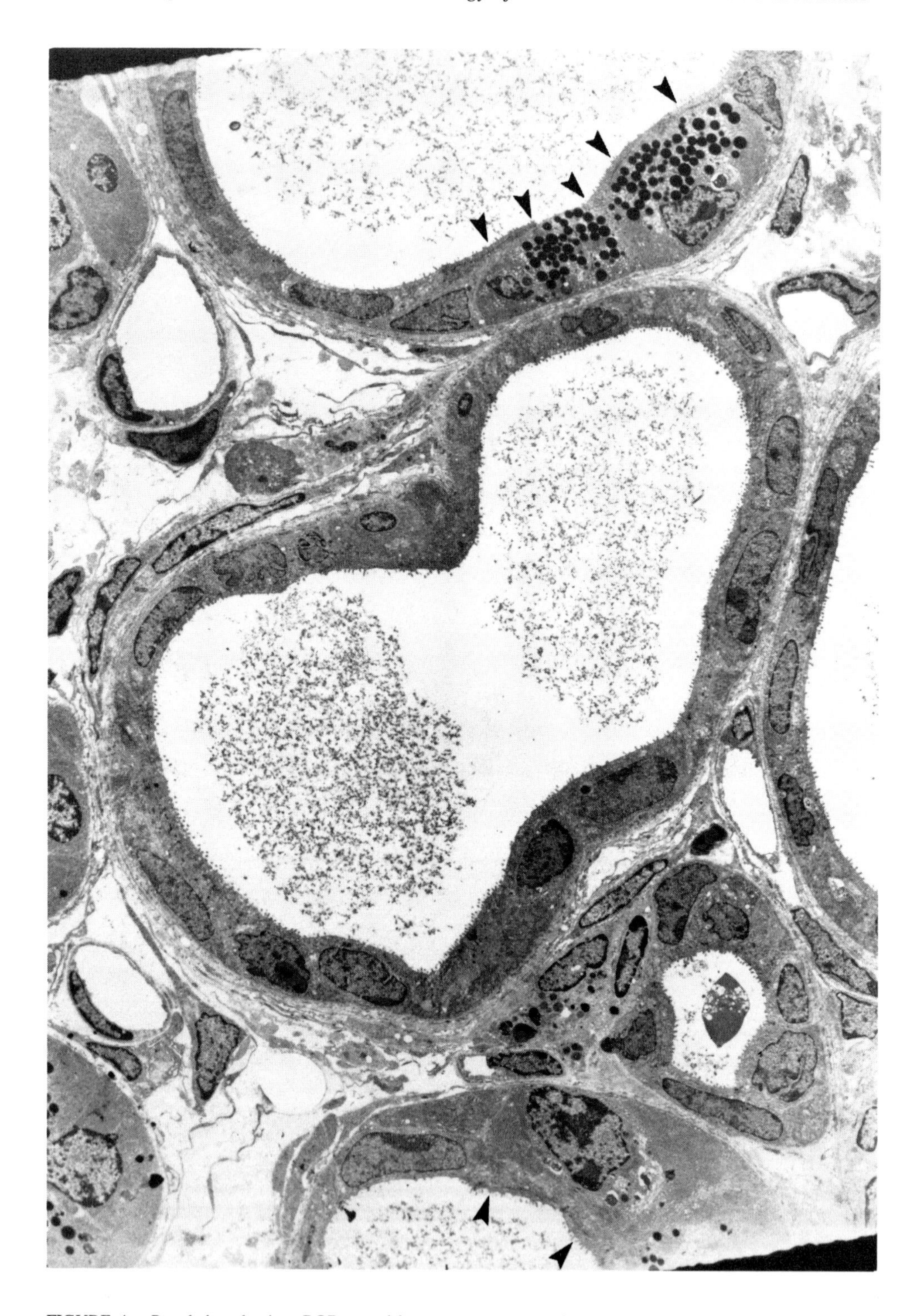

FIGURE 4. Pseudoductules in a BOP-treated hamster pancreas are lined by flat, slightly pleomorphic (altered centroacinar) cells with fine microvilli. Note the position of cell nuclei to the lumen, i.e., their axis runs parallel to that of the basal membrane. Between the flat cells there are a few acinar cells, which show signs of degeneration (bottom). Note the long cytoplasmic processes of flat cells covering the surface of two acinar cells (arrowheads). See also Figure 5. (Magnification × 1200.)

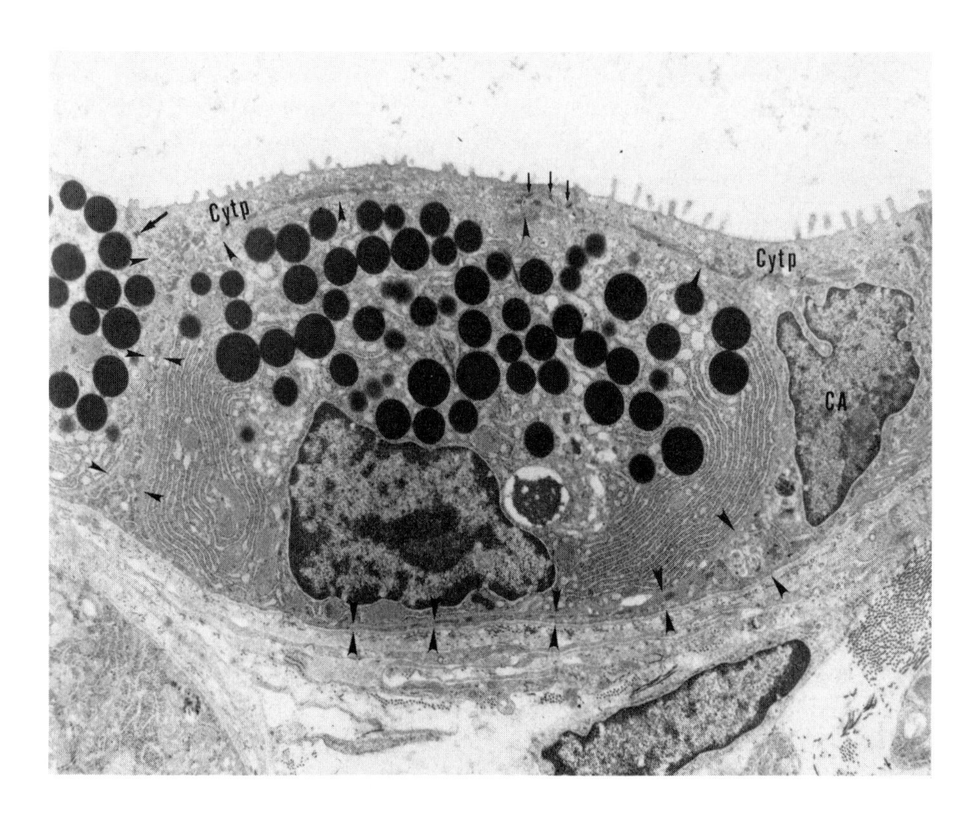

FIGURE 5. High-powered view of Figure 4 depicting an acinar cell, the surface of which is covered by cytoplasmic processes (cytp) of centroacinar cells (small arrowheads). Note also the long and tiny processes of a centroacinar cell (CA) cell beneath the acinar cell (large arrowheads). There are desmosomes between cytp (small arrows) and between these and acinar cell (large arrows). (Magnification × 8000.)

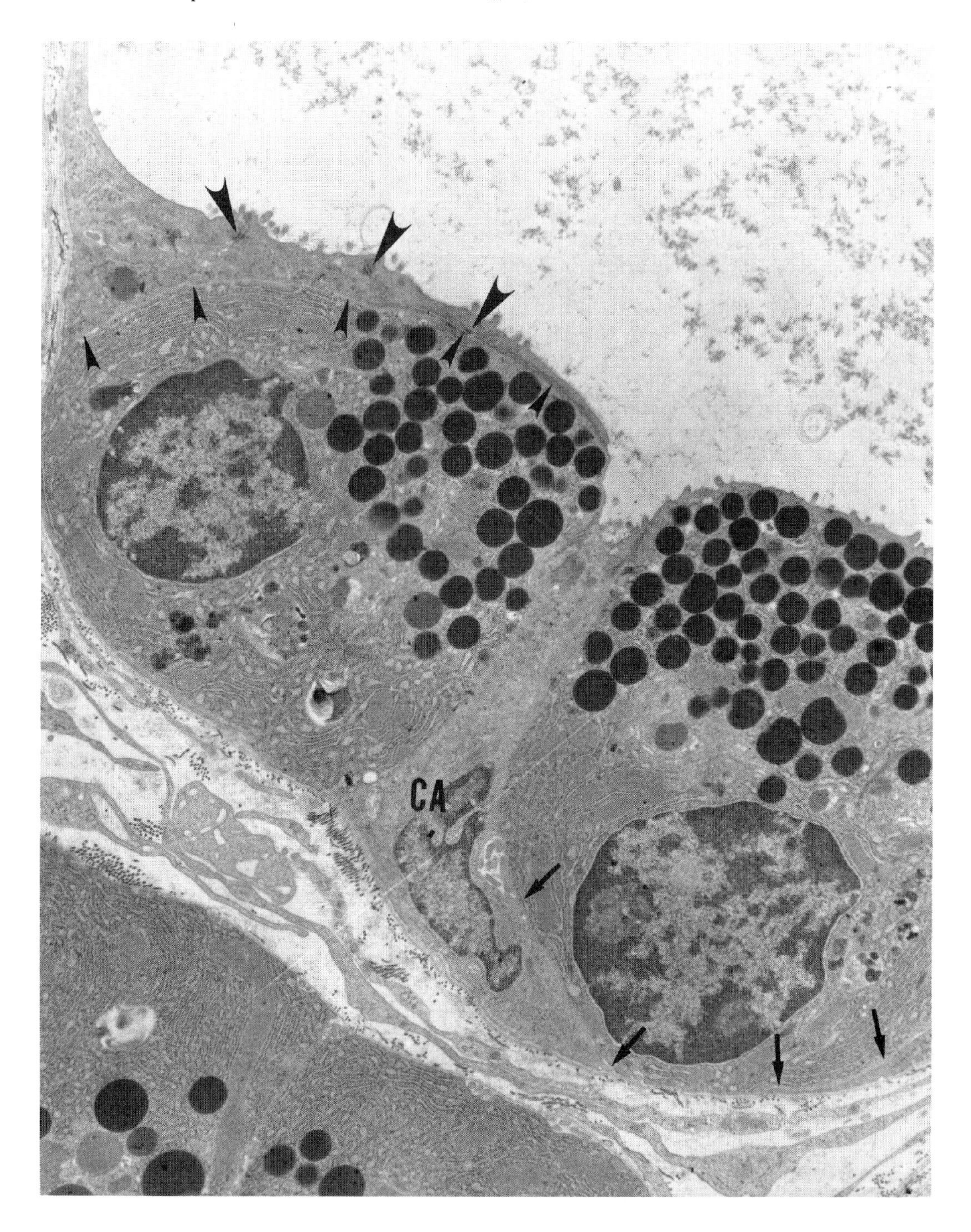

FIGURE 6. Acinar cells with numerous autophagic vacuoles. A greater portion of the acinar cell surface is covered by cytoplasmic processes (small arrowheads) exhibiting several desmosomes (large arrowheads). The cytoplasm of an elongated centroacinar cell (CA) extends beneath an acinar cell (arrows). Zymogen granules face toward the remaining free acinar cell surface. (Magnification × 6000.)

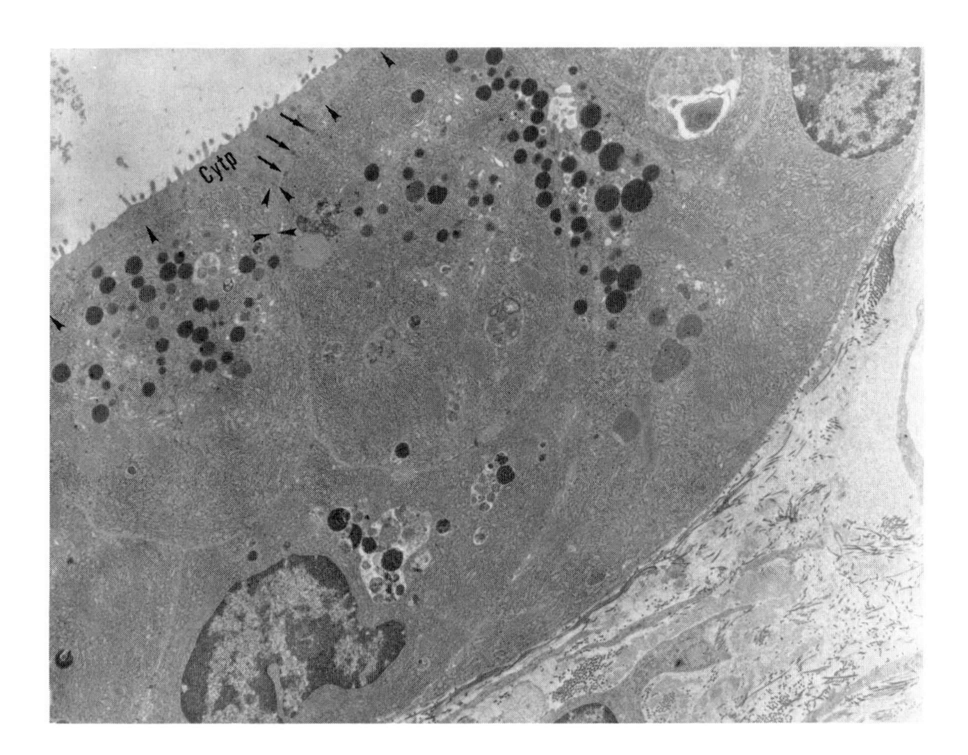

FIGURE 7. Numerous lysosomes and autophagic vacuoles in acinar cells, the surface of which is covered by cytoplasmic processes (cytp; arrowheads). Note presence of desmosomes between cytp and between these and acinar cells (arrows), (magnification × 3000.)

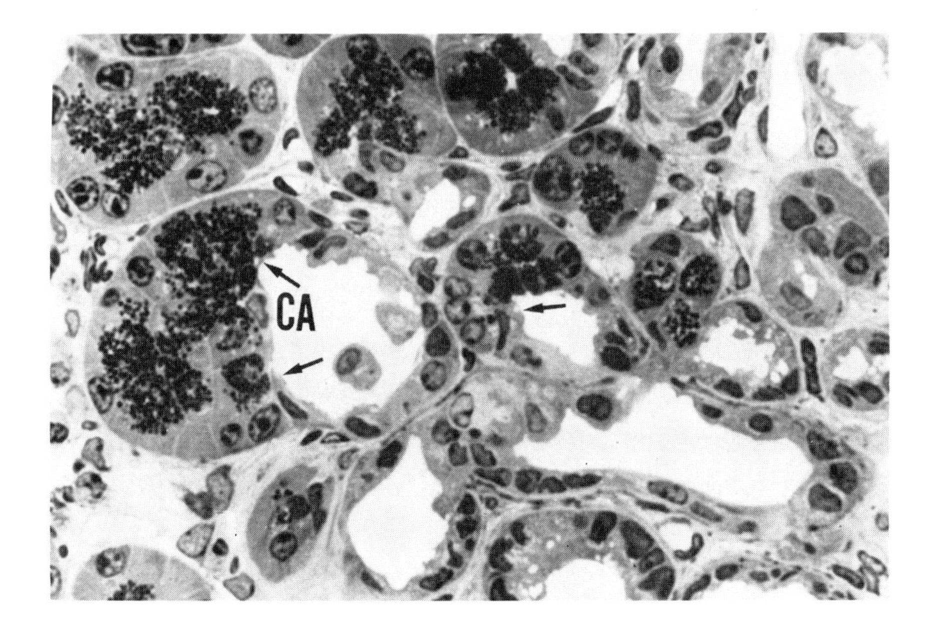

FIGURE 8. Beginning of pseudoductular formation during early stages of carcinogenesis. Several acini are partly or totally replaced by centroacinar cells of various sizes and shapes, some showing vacuolization of their cytoplasm. Note cytoplasmic processes of centroacinar cells (CA) over acinar cell surface (arrows). These processes are hardly seen in routine histologic sections. (Epon, Giemsa; Magnification × 420.)

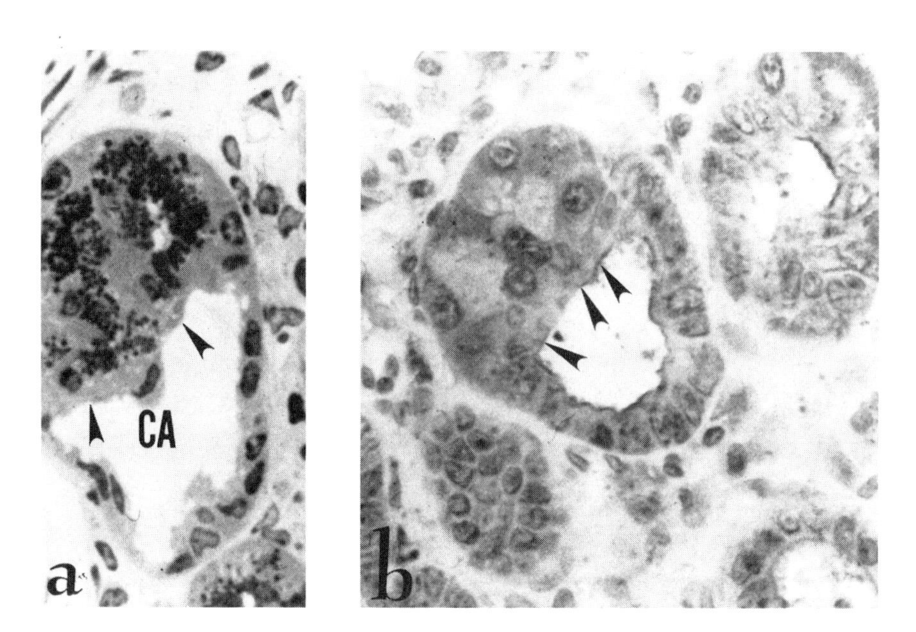

FIGURE 9. Metaplastic changes in acini. In (a) part of the gland is replaced by centroacinar cells (CA), cytoplasmic processes of which are seen covering acinar cell surface (arrowheads). In (b) a similar lesion was processed by immunohistochemistry using Anti-A antibody, which reacts with the luminal surface of centroacinar cells (in photo black). Note the reaction over the acinar cell surface (arrow), corresponding to the cytoplasmic processes of centroacinar cells. (Magnification × 420.)

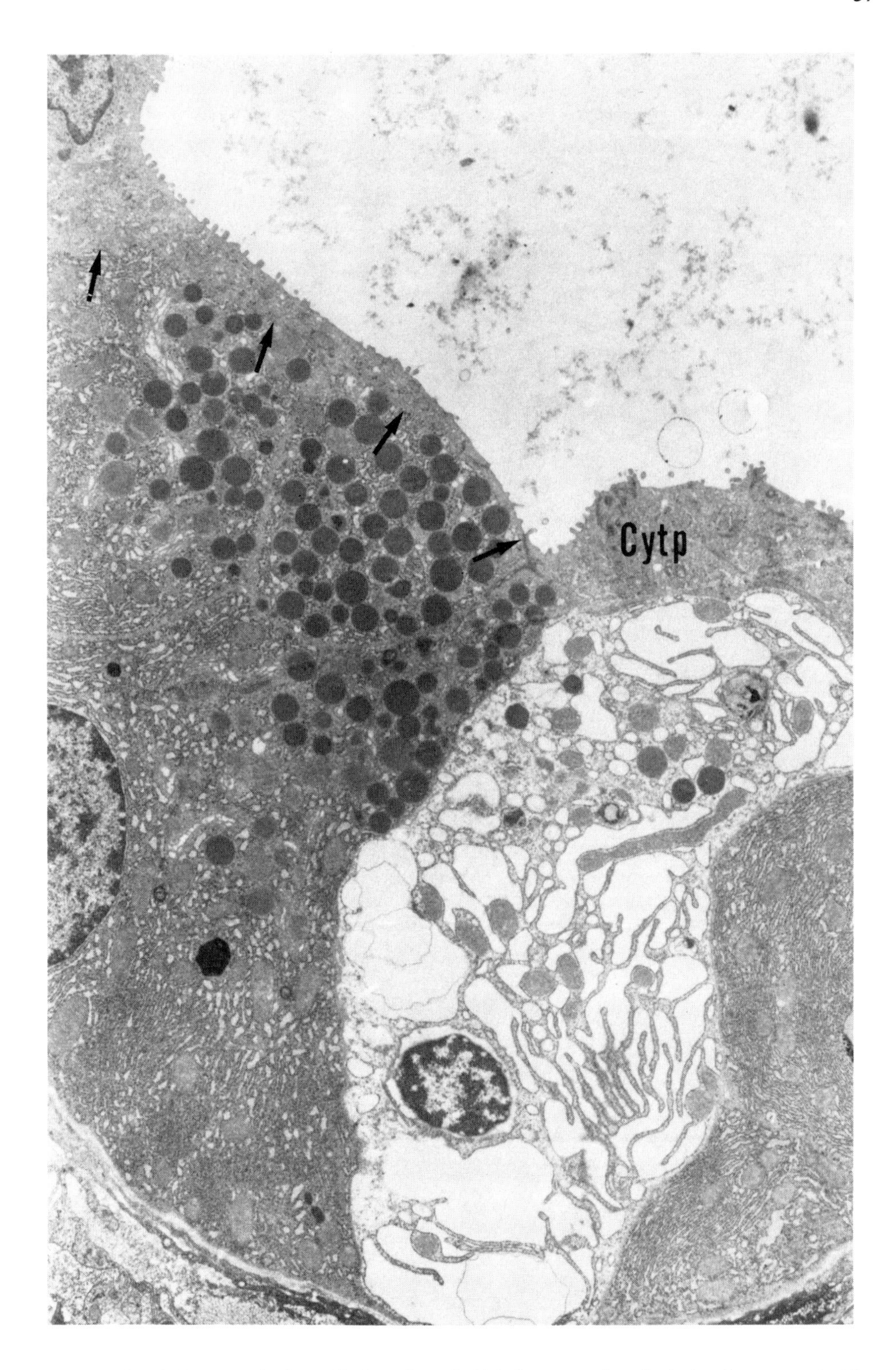

FIGURE 10. A degenerated acinar cell, the surface of which shows a portion of cytoplasmic processes (cytp). Note the presence of tiny processes of a centroacinar cell over the neighboring acinar cell (arrows). (Magnification × 6000.)

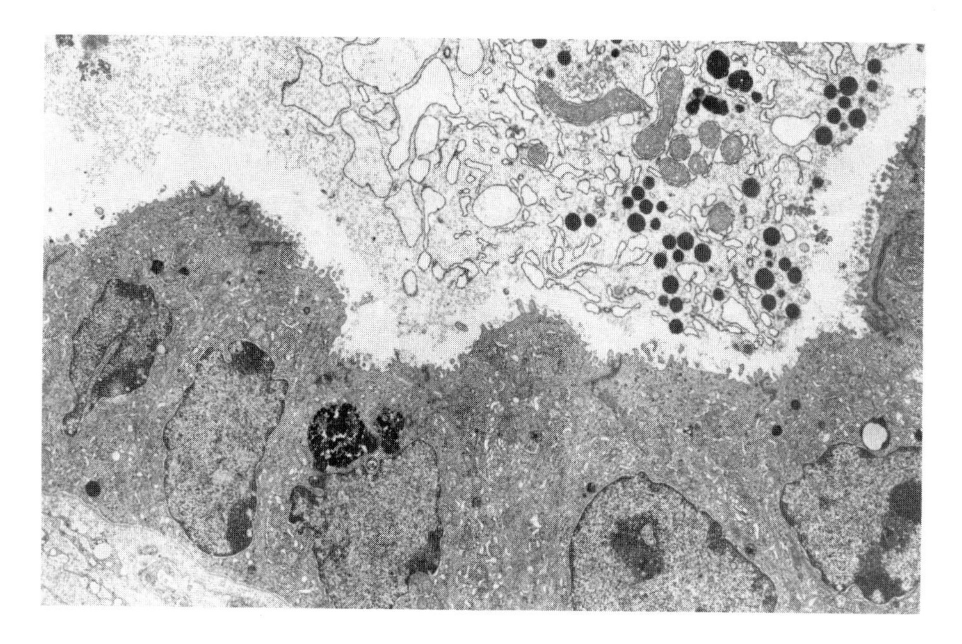

FIGURE 11. A pseudoductule composed of atypical cells showing numerous lysosomes and plump and irregular microvilli. Note the vertical position of the cells to the lumen, which contains a portion of a degenerated acinar cell. (Magnification × 3600.)

REFERENCES

1. **Dissin, J., Mills, L. R., Mains, D. L., Black, O., Jr., and Webster, P. D., III,** Experimental induction of pancreatic adenocarcinoma in rats, *J. Natl. Cancer Inst.*, 55, 857, 1975.
2. **Kirev, T., Toshkov, I., and Mladenov, Z.,** Epithelial tumors of the exocrine pancreas in Guinea fowls treated with virus strain Pts-56. Proliferation of endocrine cells in the neoplastic foci, *Gen. Comp. Pathol.*, 13, 3, 1981.
3. **Longnecker, D. S. and Curphy, T. J.,** Adenocarcinoma of the pancreas in azaserine-treated rats, *Cancer Res.*, 35, 2249, 1975.
4. **Pour, P. M., Krueger, F. W., Althoff, J., Cardesa, A., and Mohr, U.,** Cancer of the pancreas induced in the Syrian golden hamsters, *Am. J. Pathol.*, 76, 349, 1974.
5. **Reddy, J. K., Svoboda, D. J., and Roa, M. S.,** Susceptibility of an inbred strain of guinea pigs to the induction of pancreatic adenocarcinoma by *N*-methyl-*n*-nitrosourea, *J. Natl. Cancer Inst.*, 52, 991, 1974.
6. **Pour, P. M., Mohr, U., Cardesa, A., Althoff, J., and Krueger, F. W.,** Pancreatic neoplasms in an animal model: morphologic, biological and comparative studies, *Cancer*, 35, 379, 1975.
7. **Moore, M. A., Takahashi, M., Ito, N., Bannascsh, P.,** Early lesions during pancreatic carcinogenesis induced in Syrian hamster by DHPN or DOPN. I. Histologic, histochemical and radioautographic findings, *Carcinogenesis*, 4, 431, 1983.
8. **Moore, M. A., Takahashi, M., Ito, N., Bannasch, P.,** Early lesions during pancreatic carcinogenesis induced in Syrian hamster by DHPN or DOPN. II. Ultrastructural findings, *Carcinogenesis*, 4, 439, 1983.
9. **Pour, P. M.,** Experimental pancreatic ductal (ductular) tumors, in *The Pancreas*, Fitzgerald, P. J. and Morrison, A. B., Eds., Williams & Wilkins, Baltimore, 1980, 111.
10. **Pour, P. M. and Wilson, R. B.,** Experimental pancreas tumors, in *Cancer of the Pancreas*, Moossa, A. R., Ed., Williams & Wilkins, Baltimore, 1980, 37.
11. **Pour, P. M.,** Histogenesis of exocrine pancreatic cancer in the hamster model, *Environ. Health Perspect.*, 56, 229, 1984.
12. **Bockman, D. E.,** Cells of origin of pancreatic cancer: experimental animal tumors related to human pancreas, *Cancer*, 47, 1528, 1981.
13. **Bockman, D. E., Black, O., Jr., Mills, L. R., and Webster, P. D., III,** Origin of tubular complexes developing during induction of pancreatic adenocarcinoma by 7,12-dimethylbenz(a)anthracene, *Am. J. Pathol.*, 90, 645, 1978.

14. **Flaks, B., Moore, M. A., and Flaks, A.,** Ultrastructural analysis of pancreatic carcinogenesis: morphological characterization of *N*-nitrosobis(2-hydroxypropyl)amine-induced neoplasms in the Syrian hamster, *Carcinogenesis,* 1, 423, 1980.
15. **Flaks, A., Moore, M. A., and Flaks, B.,** Ultrastructural analysis of pancreatic carcinogenesis. III. Multifocal cystic lesions induced by *N*-nitrosobis(2-hydroxypropyl)amine in the hamster exocrine pancreas, *Carcinogenesis,* 1, 693, 1981.
16. **Flaks, B., Moore, M. A., and Flaks, A.,** Ultrastructural analysis of pancreatic carcinogenesis. IV. Pseudoductular transformation of acini in the hamster pancreas during *N*-nitrosobis(2-hydroxypropyl)amine carcinogenesis, *Carcinogenesis,* 2, 1241, 1981.
17. **Flaks, B., Moore, M. A., and Flaks, A.,** Ultrastructural analysis of pancreatic carcinogenesis. V. Changes in differentiation of acinar cells during chronic treatment with *N*-nitrosobis(2-hydroxypropyl)amine, *Carcinogenesis,* 3, 485, 1982.
18. **Reddy, J. K., Sarpelli, D. G., and Rao, M. S.,** Experimental pancreatic carcinogenesis, in *Digestive Cancer,* Vol. 9, Thatcher, N., Ed., Pergamon Press, Oxford, 1979, 99.
19. **Scarpelli, D. G., Rao, M. S., and Subbarao, V.,** Augmentation of carcinogenesis by *N*-nitrosobis(2-oxopropyl)amine administered during S phase of the cycle in regenerating hamster pancreas, *Cancer Res.,* 43, 611, 1983.
20. **Nagel, D. and Kupper, R.,** Synthesis of *N*-nitrosobis(2-oxopropyl)amine and C^{14} analogue, *J. Labelled Compd.,* 18, 1081, 1981.
21. **Pour, P. M.,** Induction of unusual pancreatic neoplasms with morphologic similarity to human tumors and evidence for their ductal/ductular cell origin, *Cancer,* 55, 2411, 1985.
22. **Pour, P. M., Uchida, E., Burnett, D. A., and Steplewski, Z.,** Blood group antigens expression during pancreatic carcinogenesis in hamsters, *Int. J. Pancreatol.,* 1, 327, 1986.
23. **Althoff, J., Pour, P. M., Malick, L., and Wilson, R. B.,** Pancreatic neoplasms induced in Syrian golden hamsters, *Am. J. Pathol.,* 83, 517, 1976.
24. **Althoff, J., Wilson, R. B., Ogrowsky, D., and Pour, P. M.,** The fine structure of pancreatic duct neoplasias in Syrian golden hamsters, *Prog. Exp. Tumor Res.,* 24, 397, 1979.
25. **Ogrowsky, D., Fawcett, J., Althoff, J., Wilson, R. B., and Pour, P. M.,** Structure of the pancreas in Syrian hamsters, *Acta Anat.,* 107, 121, 1980.
26. **Pour, P. M.,** Islet cells as a component of pancreatic ductal neoplasms. I. Experimental study. Ductular cells, including islet cell precursors are primary progenitor cells of tumors, *Am. J. Pathol.,* 90, 295, 1978.
27. **Rahier, J. and Wallon, J.,** Long cytoplasmic processes in pancreatic polypeptide cells, *Cell Tissue Res.,* 209, 365, 1980.
28. **Komai, Y., Murakami, Y., and Morii, S.,** Immunohistochemical localization of insulin in nesidioblastosis of rat pancreas, *Acta Histochem. Cytochem.,* 14, 261, 1981.
29. **Sacchi, T. B., Bartolini, G., Biliotti, G., and Allara, E.,** Behavior of the intercalated ducts of the pancreas in patients with functioning insulinomas, *J. Submicrosc. Cytol.,* 11, 243, 1979.
30. **Sacchi, T. B., Romagnoli, P., and Biliotti, G. C.,** The effects of chronic hyperglastrinemia on human pancreas, *J. Submicrosc. Cytol.,* 15, 1073, 1983.
31. **Bockman, D. E.,** Pathomorphology of pancreatitis: regressive changes in an acutely or chronically damaged epithelial organ, in *Pancreatitis — Concepts and Classification,* Gyr, D. E., Singer, M. V., and Sarles, H., Eds., Excerpta Medica, Amsterdam, 1984, 11.
32. **Bockman, D. E., Boydston, W. R., and Anderson, M. C.,** Origin of tubular complexes in human chronic pancreatitis, *Am. J. Surg.,* 144, 243, 1982.
33. **Becker, V.,** *Sekretionsstudien am Pancreas,* George Thieme Verlag, Stuttgart, 1957.
34. **Pound, A. W. and Walker, N. I.,** Involution of the pancreas after ligation of the pancreatic ducts. I. A histological study, *Br. J. Exp. Pathol.,* 62, 547, 1981.
35. **Baggenstoss, A. H.,** The pancreas in uremia: a histopathologic study, *Am. J. Pathol.,* 24, 1003, 1948.
36. **Baggenstoss, A. J.,** Dilatation of the acini of the pancreas, *Arch. Pathol.,* 45, 463, 1948.
37. **Resau, J. H., Cottrell, J. R., Hudson, E. A., Trump, B. F., and Jones, R. T.,** Studies on the mechanisms of altered exocrine acinar cell differentiation and ductal metaplasia following nitrosamine exposure using hamster pancreatic explant organ culture, *Carcinogenesis,* 6, 29, 1985.
38. **Reznik-Schüller, H.,** Ultrastructure of pancreatic tumors induced in Syrian hamsters by *N*-nitroso-2,6-dimethylmorpholine, *Vet. Pathol.,* 17, 352, 1980.
39. **Höefler, H., Klöppel, G., and Heitz, Ph. U.,** Combined production of mucus, amines and peptides by goblet-cell carcinoids of the appendix and ileum, *Pathol. Res. Pract.,* 178, 555, 1984.
40. **Klappenbach, R. S., Kurman, R. J., Sinclair, C. F., and James, L. P.,** Composite carcinoma-carcinoid tumors of the gastrointestinal tract. A morphologic, histochemical, and immunocytochemical study, *Am. J. Clin. Pathol.,* 84, 137, 1985.
41. **Bani, D., Bani, T., and Biliotti, G.,** Nesidioblastosis and intermediate cells in the pancreas of patients with hyperinsulinemic hypoglycemia, *Virchows Arch. B.,* 48, 19, 1985.

42. **Bartow, S. A., Mukai, K., and Rosai, J.,** Pseudoneoplastic proliferation of endocrine cells in pancreatic fibrosis, *Cancer,* 47, 2627, 1981.
43. **Buchino, J. J., Castello, F. M., and Nagaraj, H. S.,** Pancreatoblastoma: a histochemical and ultrastructural analysis, *Cancer,* 53, 963, 1984.
44. **Jaffe, R., Hashida, Y., and Yunis, E. J.,** Pancreatic pathology in hyperinsulinemic hypoglycemia of infancy, *Lab. Invest.,* 42, 356, 1980.
45. **Klöppel, G. and Heitz, Ph. U.,** Nesidioblastosis: a clinical entity with heterogeneous lesions of the pancreas, in *Evolution and Tumor Pathology of the Neuroendocrine System,* Falkmer, S., Hakanson, R., and Sundler, F., Eds., Elsevier, Amsterdam, 1984, 349.
46. **Morohoshi, T., Kanda, M., and Klöppel, G.,** On the histogenesis of experimental pancreatic endocrine tumors. An immunocytochemical and electron microscopy study, *Acta Pathol. Jpn.,* 34, 271, 1984.
47. **Bani, T. S., Cortesini, G., and Domenici, L.,** The endocrine pancreas of the rat following portacaval shunt, *J. Submicrosc. Cytol.,* 14, 655, 1982.
48. **Suda, K., Komatsu, K., and Hashimoto, K.,** A histopathological study on the islets of Langerhans and ductal epithelial metaplasia in atrophic lobuli of pancreas, *Acta Pathol. Jpn.,* 26, 561, 1976.
49. **Pour, P. M., Parsa, I., and Hauser, R.,** Evidence of partial acinar differentiation in induced pancreatic tumors in hamsters, *Int. J. Pancreatol.,* 2, 47, 1987.

Chapter 3

PATHOLOGICAL FEATURES OF PRENEOPLASTIC AND NEOPLASTIC LIVER LESIONS IN RODENTS AND HUMANS

Hideki Mori, Takafumi Ichida, Takuji Tanaka, and Gary M. Williams

TABLE OF CONTENTS

I. INTRODUCTION

Numerous experimental studies of hepatocarcinogenesis have been performed since the demonstration by Sasaki and Yoshida in 1935 of chemical induction of liver cancer.[1] Much interest has focused on the properties of the presumed precursor lesions for hepatocellular carcinoma and their role in a multistep progression to cancer. In this chapter, the cellular changes in the pathogenesis of chemically induced hepatocellular neoplasms in rodents represented by rat, mouse, and hamster, and comparable lesions in humans are reviewed with emphasis on the histochemical and ultrastructural features of preneoplastic cell populations.

II. HEPATOCELLULAR LESIONS

In rats, which are the most studied experimental species, altered foci and nodules (adenomas, hyperplastic nodules, neoplastic nodules) are two types of hepatocellular lesions that have been defined and accepted as being related to cancer development.[2,3] Altered foci are small lesions of hepatocytes that do not disrupt the parenchymal architecture. They are equivalent to the lesions that also have been called hyperplastic foci.[1,4-6] Nodules are larger lesions that disrupt and compress the surrounding parenchyma. They have been designated as adenomas,[1] hyperplastic nodules,[4-9] or neoplastic nodules.[2,3] In other rodents such as mice or hamsters, and in humans, similar lesions have been also reported and regarded as possible precursor lesions for hepatocellular malignancies, although their significance has not always been as evident as in rats.[10-16] In this chapter, both foci and nodules in rodents and humans are discussed.

A. ALTERED FOCI

1. Foci in Rats

a. Histology and Cytochemistry

Foci are lesions which are smaller than the size of a lobule.[2,3] Three distinctive types of foci have been described; those composed of cells with eosinophilic, basophilic, or clear cytoplasms (Figures 1, 2, and 3).[3] Each type of focus is generally uniform in its cellular composition, but may contain cells of different types. The plates of foci cells are in continuity with those of the surrounding parenchyma although they are sharply demarcated from hepatocytes due to their altered morphology and staining properties.[17]

In general, foci, particularly early in their development, are more easily identified by their cytochemical abnormalities, which in fact provided important evidence that these lesions were involved in the pathogenesis of liver cancer.[17-22] A variety of phenotypic abnormalities have been reported for rat altered foci (Tables 1 and 2). Cytochemically, foci are recognized as groups of cells which display altered enzymatic activities and other abnormal cellular constituents or functions, which may be decreased (negative markers) or increased (positive

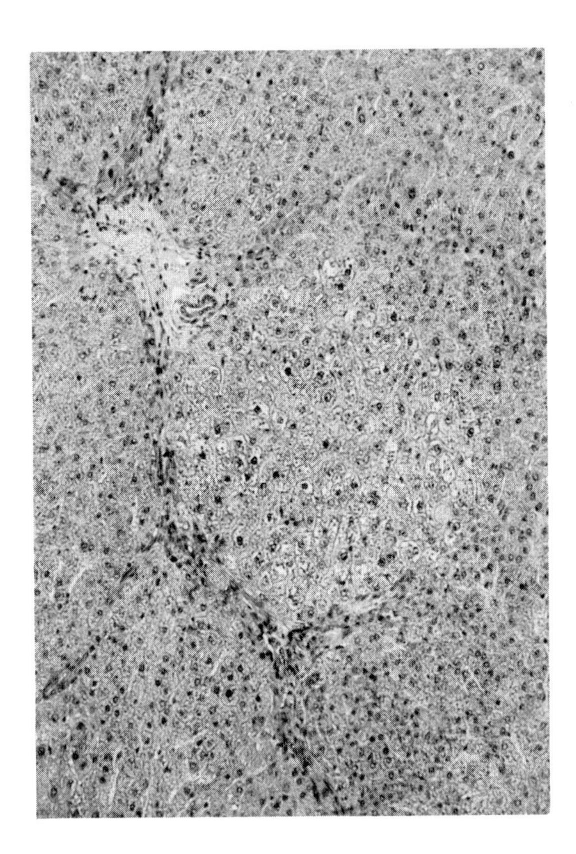

FIGURE 1. An eosinophilic cell focus in the liver ACI rat given 2-acetylaminofluorene. (H and E stain; magnification ×130.)

markers) compared to normal hepatocytes. Negative cytochemical markers such as glucose-6-phosphatase (G6Pase), adenosine triphosphatase (ATPase), or iron exclusion and several positive markers such as glycogen storage, gamma-glutamyltranspeptidase (GGTase), or D-T diaphorase frequently have been used for identification of foci. The abnormalities are generally expressed in most or all cells within a lesion, although some foci can be heterogeneous.[19] Phenotypic properties are often variable between foci.[23-28]

GGTase is one of the most common foci-associated enzyme markers (Figure 4). It characterizes approximately 90% of foci induced by several carcinogens.[21,23] GGTase is a fetal liver enzyme whose activity is quite low in normal adult rat liver, but which is elevated in liver cell lesions.[29-31] Care is required in the use of GGTase as a marker since its activity increases in periportal hepatocytes with age,[32-35] and is induced by xenobiotics including phenobarbital[34-40] and antioxidants[33,40] which also induce drug-metabolizing enzymes. Inhibition of GGTase activity and enhancement of alkaline phosphatase (ALPase) activity by peroxisome proliferators have also been reported.[41,42] It has been reported that glycogen-rich foci display a positive activity for GGTase whereas glycogen-poor foci which were hyperbasophilic (Figure 5) lacked activity.[43]

Besides the enzymatic abnormalities of altered foci, the resistance to accumulation of iron in the foci when the liver is rendered siderotic by iron overload (Figure 6) was developed by Williams and associates.[26,44-46] A number of advantages of iron exclusion as a marker for liver carcinogenesis were suggested,[17,47] including ultrastructural identification of foci as will be discussed. Iron exclusion characterizes foci in other experimental species, as well as humans.

FIGURE 2. A clear cell focus in rat liver. (H and E stain; magnification ×130.)

Increased serum alpha-fetoprotein (AFP) has been associated with the development of foci and proliferation of oval cells,[48,49] which appear to originate in the periportal region and are suggested to be related to hepatocytes, possibly as stem cells.[50,51] Immunofluorescent and peroxidase-antiperoxidase staining have confirmed the localization of AFP to these cells.[52,53] No evidence that foci cells elaborate AFP has yet been found.

An altered pattern of biotransformation enzymes in foci cells which may underlie their resistance to cytotoxic effects of chemicals[22] has been found.[54-58] The purification of certain drug-metabolizing enzymes and the preparation of antibodies has provided the possibility of immunohistochemical detection of these enzymes. Thus, elevated uridine diphosphate glucuronyl transferase and glutathione transferase have been confirmed by immunofluorescent or immunoperoxidase staining.[55,59-65] Recently, a placental form of glutathione transferase has been developed as a possible new marker for foci.[58,66] Hepatocytes in two cell types of foci (clear cell and eosinophilic cell foci) contain considerable amounts of glycogen as demonstrated by the PAS-reaction.[18,67-69] In such foci, the activities of glycogen phosphorylase and G6Pase were decreased while those of G6P dehydrogenase and glyceraldehyde-3-phosphate dehydrogenase were increased. In contrast to those foci, basophilic cell foci were poor in glycogen. Recently, new techniques such as peroxidase-antiperoxidase method and monoclonal antibody technique have been developed for identification of preneoplastic hepatocellular lesions.[55,60,70]

b. Ultrastructure

Electron microscopic observations on altered foci of rats have been made by several investigators. Bruni[71] reported subcellular properties of cells from focal lesions appearing

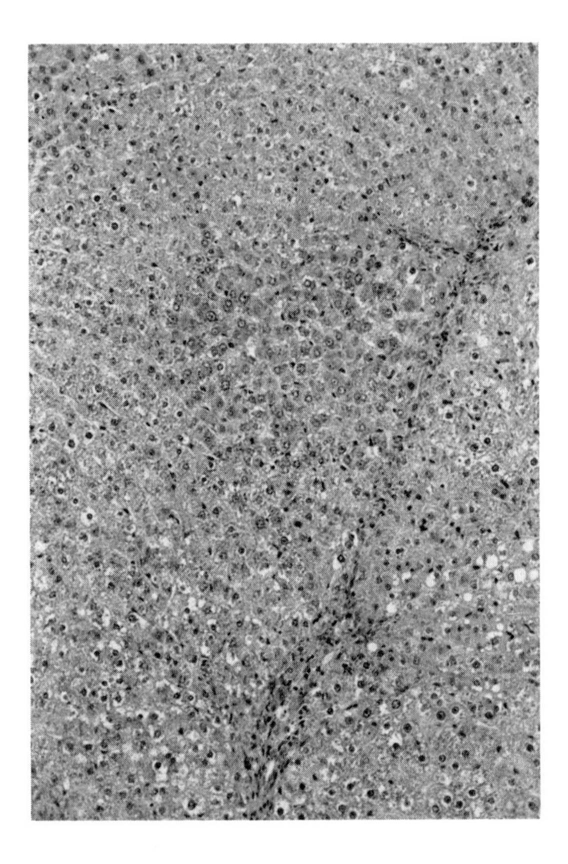

FIGURE 3. A basophilic cell focus in rat liver. (H and E stain; magnification ×130.)

TABLE 1
Negative Markers for Preneoplastic and Neoplastic Liver Lesions

Markers	Ref.
Glucose-6-phosphatase	66, 164—166
Adenosine triphosphatase	131, 167, 168
Beta-glucuronidase	166
Serine dehydratase	166
Glycogen phosphorylase	69, 97, 169
Acid or alkaline nucleases	170
Ornithine carbamyl transferase	171
Ribonucleases and deoxyribonucleases	163, 172
Exclusion of cellular iron	26, 44, 45, 80, 173
Cytochrome P-450	61, 64

in an early stage of diethylnitrosamine-induced liver carcinogenesis. These cells were highly pleomorphic. The majority of the cells were hepatocellular in character and had greatly reduced volume of smooth endoplasmic reticulum (SER). Ogawa et al.[72] also studied the fine structure of altered foci induced by diethylnitrosamine. Subcellular characteristics of the earliest altered cells identified were an increase of SER and glycogen, and nucleolar hypertrophy. Late-appearing altered cells had abnormal rough endoplasmic reticulum (RER), increases of pericanalicular vesicles, polysomes, and microfilaments around bile canaliculi.

TABLE 2
Positive Markers for Preneoplastic and Neoplastic Liver Lesions

Markers	Ref.
Alpha-fetoprotein	51
Gamma-glutamyl transpeptidase (GGTase)	174—176
D-T diaphorase	177
Epoxide hydrolase	102, 178
Esterase	179, 180
Cytosolic glutathione transferase	60, 64
Glutathione transferase, placental form	58, 66
UDP-glucuronyltransferase	55, 59, 62, 65
Aldehyde dehydrogenase	181
Glycogen storage	67—69, 97, 169, 182
Glucose-6-phosphate dehydrogenase	69, 169
Glyceraldehyde-3-phosphate dehydrogenase	69
Preneoplastic antigen	183
Normal rat hepatocyte cell membrane antigens	70

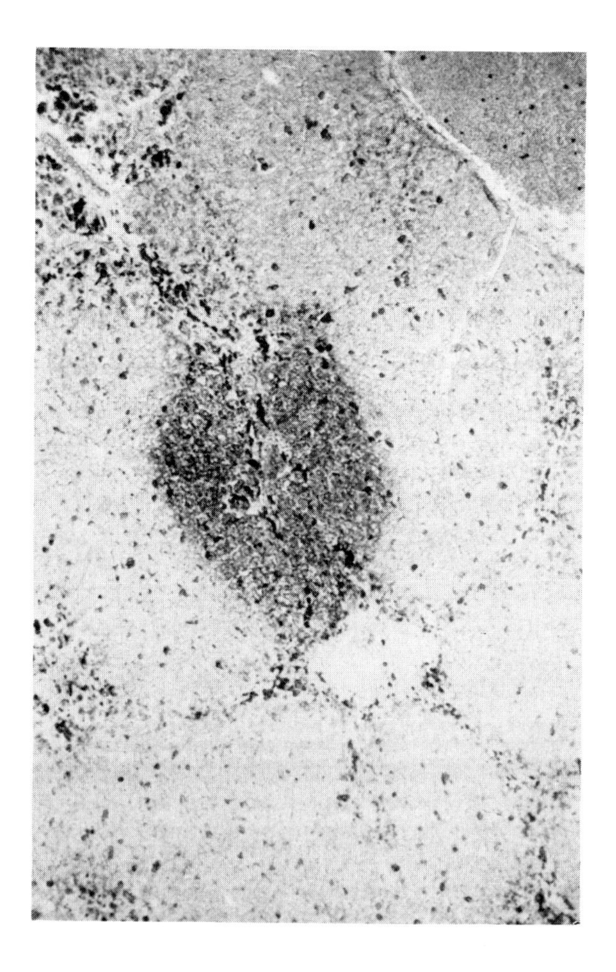

FIGURE 4. A focus in rat liver with positive reaction of GGTase. GGTase reaction. (Magnification ×45.)

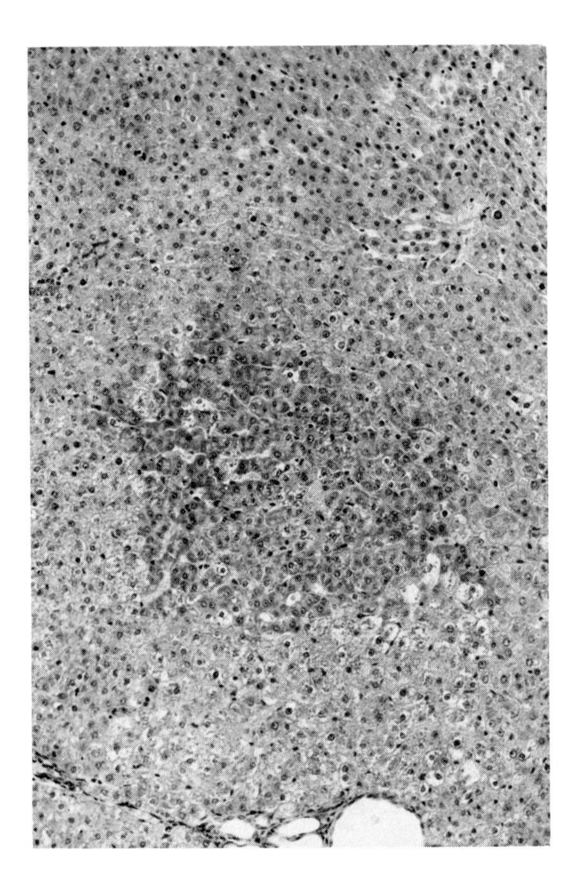

FIGURE 5. A hyperbasophilic cell focus in rat liver. (H and E stain; magnification ×130.)

These altered cells possessed the characteristics of hepatocytes. A similar study on early-appearing altered foci cells induced by diethylnitrosamine by Zaki[73] revealed that the cells exhibited an excessive storage of glycogen and increased coated pinocytotic vesicles. Subsequent cellular changes were disorganization of the RER, dispersal of ribosomes, proliferation of SER, predominance of annulate lamellae, depletion of glycogen, and enlargement of the Golgi apparatuses. These cells were hepatocellular in character. Timme[74] studied altered foci induced by 4-dimethylaminoazobenzene using lack of iron accumulation for identification. Foci showed well developed RER, ribosomes, peroxisomes, and bile canaliculi, and had small stores of glycogen. The subcellular properties of foci were basically similar to those of normal hepatocytes. Hirota and Williams[75] also studied the fine structure of altered foci cells in the siderotic livers of rats given 2-acetylaminofluorene. Cells in all types of foci were hepatocellular in character. Those in eosinophilic foci had increased RER, SER, and Golgi apparatus. The bile canaliculi of cells of the foci were abnormal in form and increased in number. Cells in basophilic foci were characterized by the presence of numerous polyribosomes, distended RER, hypertrophied Golgi apparatuses, and prominent nucleoli. Similar observations have been made by the present authors in the siderotic liver of a different strain rats given 2-acetylaminofluorene (Figures 7 and 8). The cells in foci displayed organelles characteristic of hepatocytes. Reznik-Schüller and Gregg[76] reported ultrastructure of methapyrilene-induced altered foci. One of the characteristic features of altered cells in their study was a marked proliferation of mitochondria. Bannasch and associates[18,67,68] have studied in detail the subcellular properties of cells from altered foci

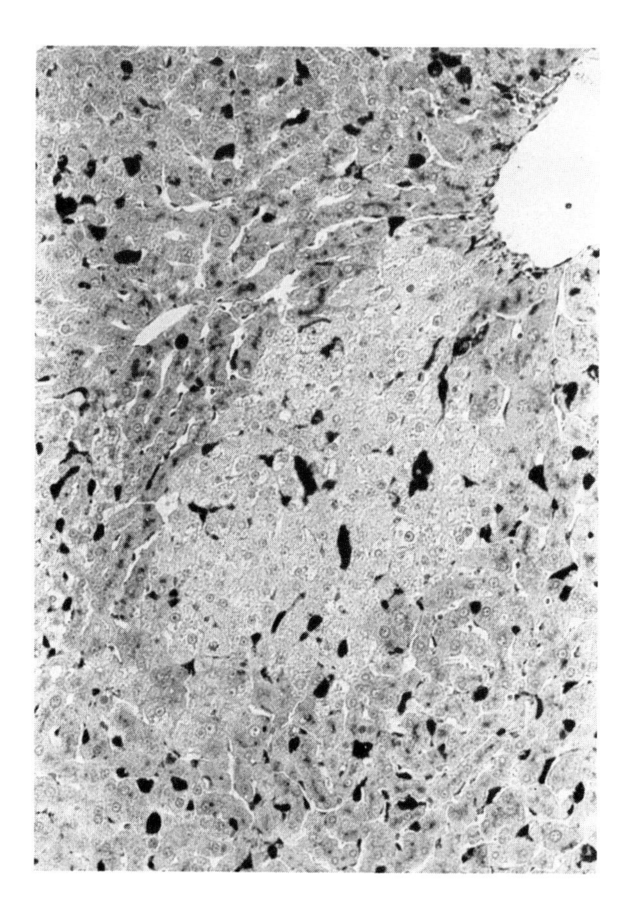

FIGURE 6. An iron-excluding focus in rat liver. (Prussian blue reaction; magnification ×130.)

induced by *N*-nitrosomorpholine, finding that an excess of cellular glycogen is a consistent change. They postulated that a gradual shift of cellular carbohydrate metabolism occurs during liver tumorigenesis.[17,18,67-69]

2. Altered Foci in Mice

a. Histology and Cytochemistry

Altered foci in mouse liver are not as conspicuous as those in the rat, although they have similar morphological appearances.[10,77] Thus, acidophilic (Figure 9), basophilic (Figure 10), vacuolated, clear cytoplasms (Figure 11), and mixed cell types are recognized.[10] Usually, the most conspicuous type of focus is that composed of basophilic cells (Figure 10). Mouse foci also display a variety of phenotypic abnormalities, but these are more variable than in the rat.[46,78-83] Iron exclusion is a fairly consistent property,[46,78,83] as in the rat. A decreased activity of G6Pase in mouse foci has been reported.[79,82,84] In most foci, at later stages, a decrease in cytoplasmic ATPase and an increase of membrane activity have been observed. Decreased acid phosphatase has also been seen. In general, GGTase activity in mouse foci is not increased[82,84] except in foci induced by safrole,[79] and by dimethylnitrosamine in newborn mice.[83] Goldfarb et al.[11] reported an increased activity of alkaline phosphatase in type 1 foci induced by griseofulvin. Williams and Numoto[82] recently reported that in altered foci induced by diethylnitrosamine in male B6C3F1 hybrid mice, either increased or decreased G6Pase was present in the greatest number of foci. These were also iron excluding.

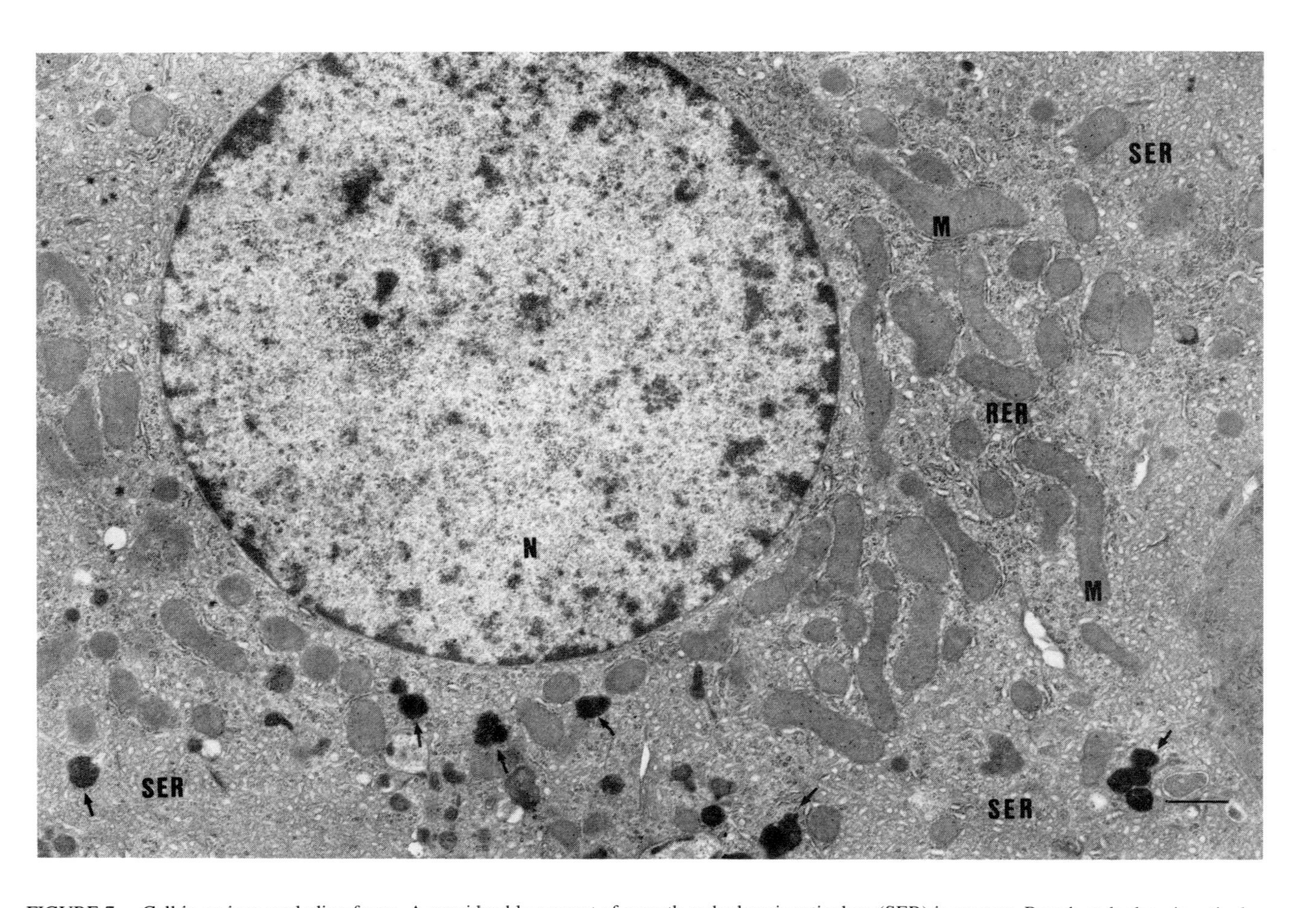

FIGURE 7. Cell in an iron-excluding focus. A considerable amount of smooth endoplasmic reticulum (SER) is present. Rough endoplasmic reticulum (RER) with irregular arrangement is relatively increased. Mitochondria (M) appear to be normal. A few ferritin-laden lysosomes (arrow) are present in the lower part. (Magnification $\times 8400$.)

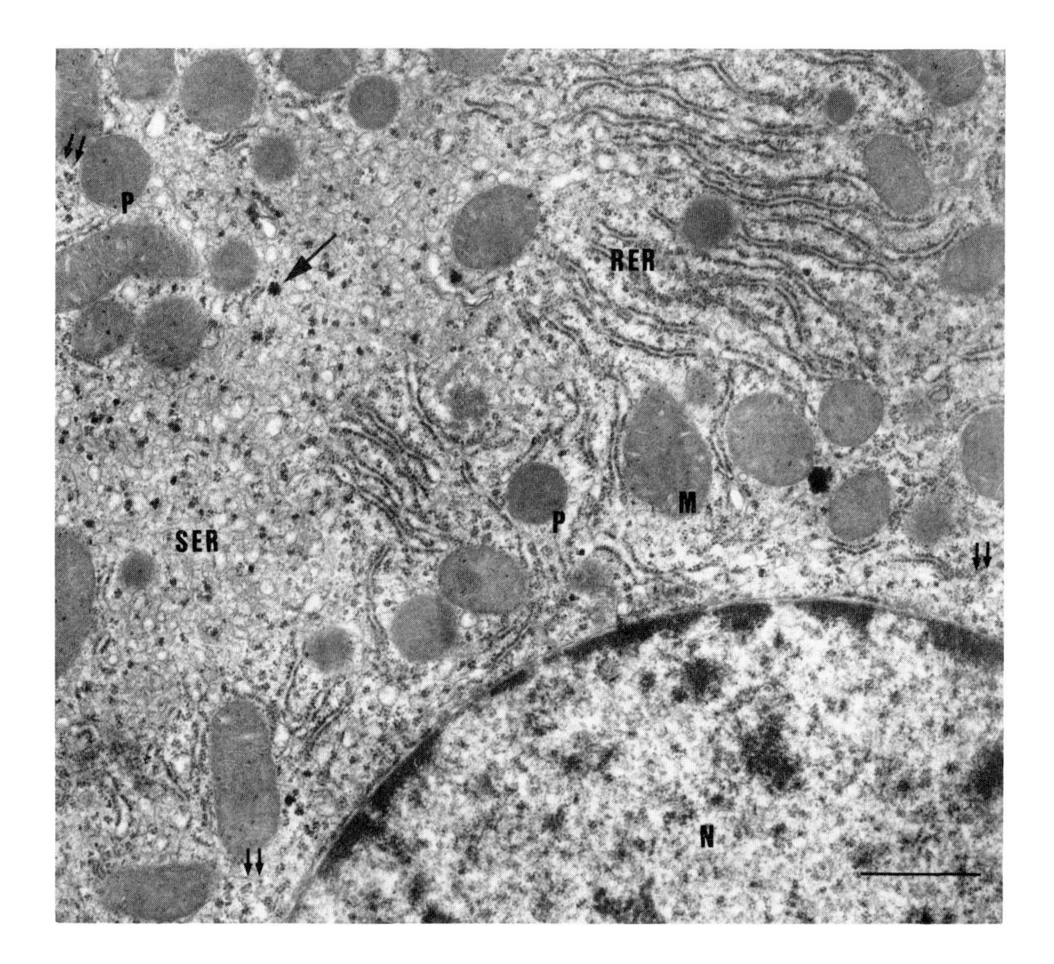

FIGURE 8. A portion of a cell in an iron-excluding focus. RER is partly oriented into small parallel arrays. A small number of glycogen particles in the form of five rosettes (arrow) are present in the vicinity of SER and focal polyribosomal particles (double arrows) are present. N: nucleus, M: mitochondria, P: peroxisome. (Magnification × 16,800.)

They also reported that increased G6Pase of altered liver cell foci induced by diethylnitrosamine disappeared with the discontinuation of the carcinogen exposure.[85] Lipsky et al.[79] reported that foci induced by safrole displayed a decrease in G6Pase and succinate dehydrogenase activities. Foci induced by safrole were also iron excluding. Koen et al.[12] reported that AFP-positive hepatocytes were present in about 20% of foci with diameters between 1.5 and 0.24 mm induced by diethylnitrosamine in infant mice. However, all foci smaller than 0.24 mm in diameter were negative for AFP. This property in mice was different from in rats.

b. Ultrastructure

No specific description of the ultrastructural features of mouse liver cell foci was found.

3. Altered Foci in Hamsters

a. Histology and Cytochemistry

Only a few studies on altered foci in hamster liver have been made.[14,66] Foci induced by dimethylnitrosamine were composed mainly of clear cells (Figure 12) with PAS-positive reaction. These foci were iron excluding (Figure 13) but not GGTase positive.[14] Moore et

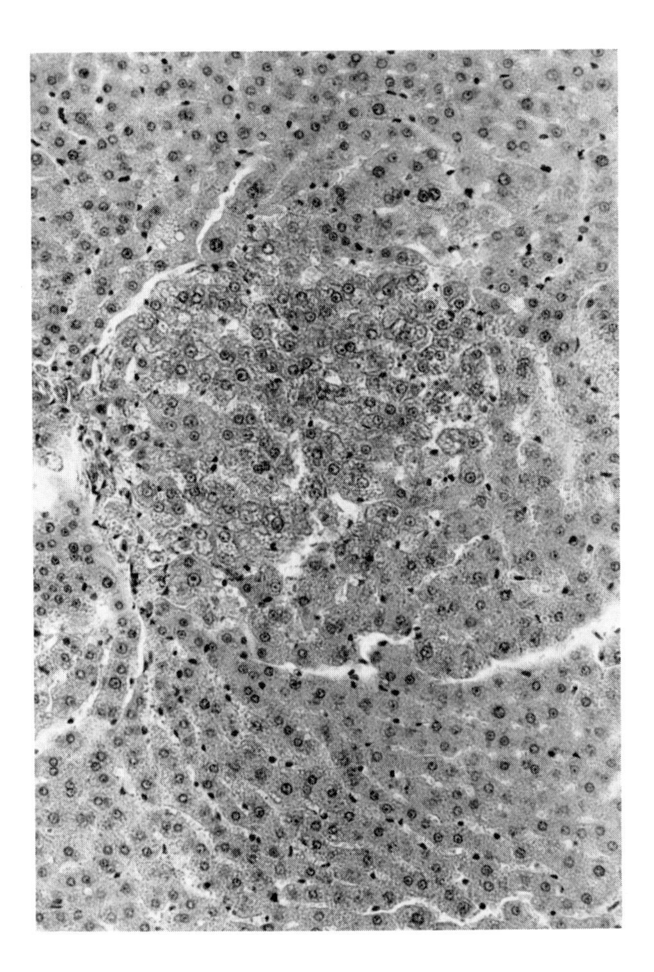

FIGURE 9. An eosinophilic cell focus induced by diethylnitrosamine in mouse liver. (H and E stain; magnification ×145.)

al.[66] reported that cells of foci induced by diethyl-di-*n*-propylnitrosamine in male Syrian hamsters expressed the placental form of glutathione transferase.

b. Ultrastructure

No ultrastructural studies on hamster liver cell foci were found.

4. Altered Foci in Humans

a. Histology and Cytochemistry

Altered hepatocellular foci (Figure 14) in human liver resembling those in rodents have been reported by several authors.[15,16,86,87] Such foci cells were rich in glycogen[16,86] and resistant to iron accumulation,[87] and had lowered activity of ATPase and G6Pase.[86] They have higher nuclear DNA content than the surrounding hepatocytes.[15] The development of these lesions may be related to aging,[86] hepatitis,[87] or the long-term use of drugs.[15,16]

Liver cell dysplasia,[88-90] which was suggested as a premalignant lesion, displays an increase of alkaline phosphatase.[91] However, the pattern of other enzymes including GGTase was not similar to that of hepatocellular carcinoma. Dysplastic cells occasionally have GGTase activity along the whole cytoplasmic membrane of some single cells.[91]

b. Ultrastructure

Ultrastructural observations on altered foci in human liver were not confirmed. However,

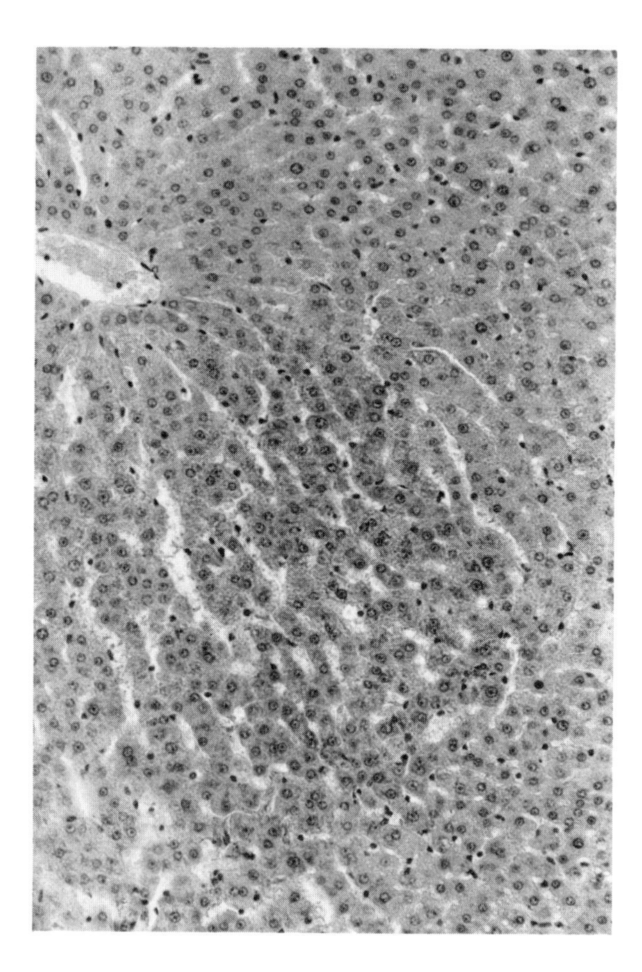

FIGURE 10. A basophilic cell focus in mouse liver. (H and E stain; magnification × 145.)

Watanabe and co-workers[92] classified dysplastic cells into two types and suggested that small dysplastic cells were the more important candidate for the precancerous cell and that the large dysplastic cells of the type described by Anthony[88,89] possessed some features of regenerative cells.

B. ADENOMAS

Because of the evidence in both rats[17,93] and mice[46] that lesions formerly referred to as nodules are actually neoplasms, they will be called adenomas in the review.

1. Adenomas in Rats

a. Histology and Cytochemistry

Adenomas are roughly round and generally as large or larger than the area of a lobule.[2,3] An important diagnostic feature of nodules is their compression of the surrounding liver tissue (Figure 15). At their periphery, adenomas are discontinuous with the cell plates of the adjacent tissue.

Adenoma cells display many of the same functional abnormalities as foci. They are reduced in iron accumulation,[44,94,95] deficient in G6Pase,[96,97] and ATPase.[17] GGTase was elevated in some adenomas[34,98] but was not increased in other circumstances.[43,93] Some

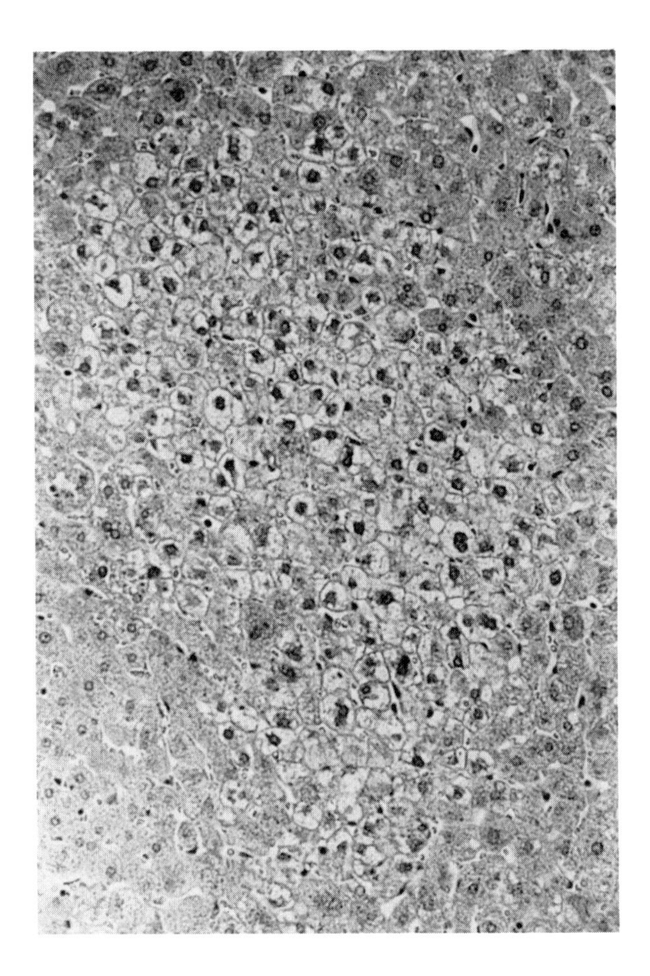

FIGURE 11. A clear cell focus in mouse liver. (H and E stain; magnification × 145.)

basophilic nodules were negative for GGTase activity. Adenomas are not usually associated with elevated serum AFP,[48-53] whereas carcinomas are. Reduced cytochrome P-450 has been reported in nodules,[64,99-101] although increased activities of various types of drug-metabolizing enzymes such as epoxide hydrolase, UDP-glucuronyl transferase, and glutathione transferase have been found.[59,82,101]

There is a type of nodular lesion that develops early during exposure to carcinogens and is apparently reversible.[103] Similar lesions have been induced by a protocol of partial hepatectomy during 2-acetylaminofluorene administration.[104] These actually may be exaggerated foci which assume a nodular configuration as a result of intense selective pressure.

b. Ultrastructure

Electron microscopic studies of adenomas have been performed by many investigators. Merkow et al.[105,106] described the fine structure of 2-acetylaminofluorene-induced nodules. In their studies, nodule cells were characterized by an increase of annulate lamellae and SER. They postulated that basic fine structural properties were similar to those of hepatocellular carcinomas. Bannasch[107] classified nitrosomorpholine-induced nodule cells into five types according to their electron microscopic properties. Clear cells were abundant in glycogen particles. RER was reduced per unit volume. Acidophilic cells had hypertrophic SER

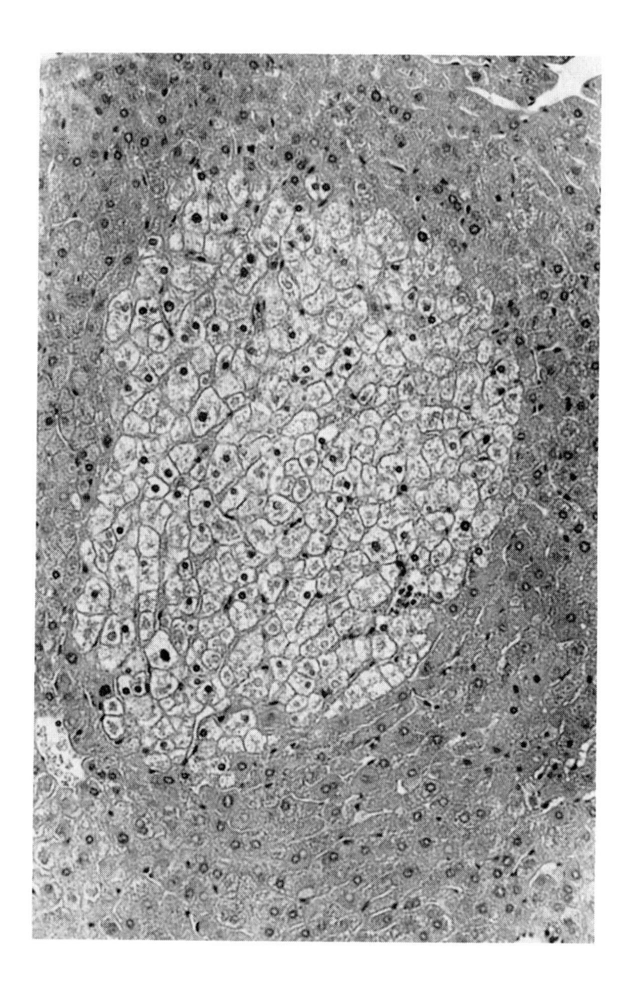

FIGURE 12. A clear cell focus in the liver of hamster given dimethylnitrosamine. (H and E stain; magnification ×145.)

and a considerable amount of glycogen. SER in the cells of this type frequently formed lamellar complexes. Vacuolated cells contained many lipid droplets and liposomes were frequently present in the cisternae of SER. Basophilic cells had abundant free or membrane-bound ribosomes. Intermediate cells possessed large glycogen areas. Many ribosomes were located preferentially in areas where the RER formed pocketing. Ogawa et al.[72] described the fine structure of diethylnitrosamine-induced nodules. Prominent features of the nodule cells were seen at cell surfaces, i.e., widening of intercellular space, irregularly dilated bile canaliculi with invagination into the cytoplasm, and formation of basement membrane-like structures. Feldman et al.[108] reported electron microscopic features of phenobarbital-induced benign hepatocellular neoplasms. Outstanding features of the neoplastic cells were the existence of voluminous whorls, phagosomes filled with whorls, and myelin figures.

Our electron microscopic studies of adenomas induced by 2-acetylaminofluorene revealed that the cells had the basic structures of hepatocytes. Their nuclei were enlarged with an irregular oval shape and indentation. The inner layer of the nuclear envelope was electron dense and thicker than the outer layer. In the nucleolus, an increase of the granular component was recognized. In general, the RER was moderately increased in amount, displaying a lessening of parallel-arrayed arrangement and increased irregular contours. On occasion, vesiculation with degranulation of RER was remarkable. The fingerprints and smooth mem-

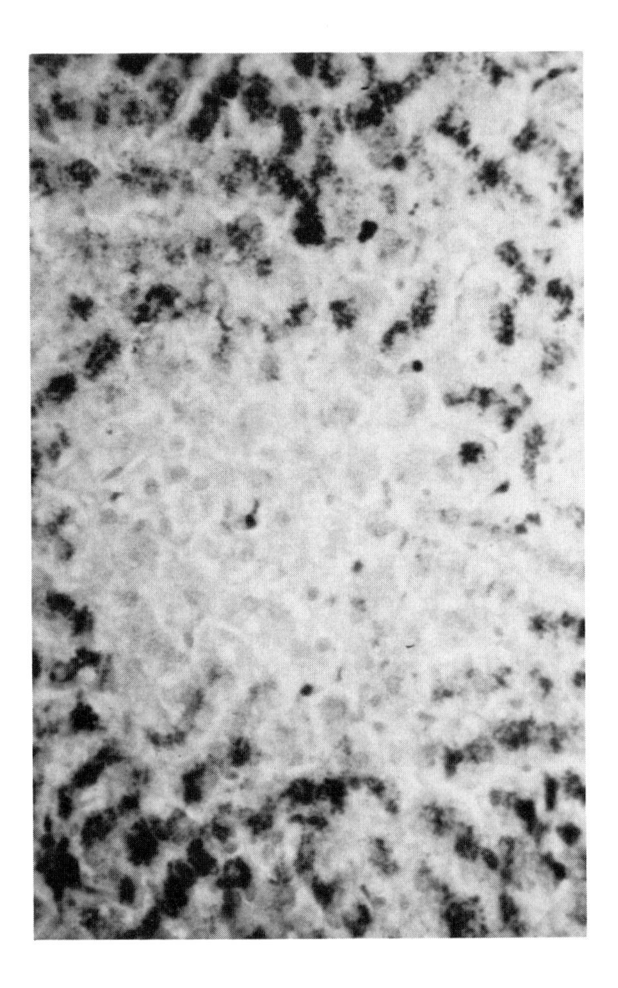

FIGURE 13. An iron-excluding focus in hamster liver. (Prussian blue reaction; magnification ×290.)

brane lamellar bodies which are a common response of endoplasmic reticulum to several agents were apparent in some of the adenoma cells. A striking alteration was the proliferation of SER which was closely associated with considerable amounts of glycogen, and the increase of polyribosomes scattered throughout the cytoplasm (Figure 16).

The mitochondria showed some variability in size, shape, and matrix density. Small and narrow mitochondria, with frequent dilatation of cristae and few matrix granules were striking features, although these alterations have been recognized in hepatocytes in some toxic conditions.[109,110] In some of the cells which showed marked changes of RER, occasionally the mitochondrial abnormalities were more severe with an increased degree of swelling, elongation of shape, absence of matrix dense granules, and reduction in the number of cristae (Figure 17).

Peroxisomes, with small, eccentric nucleoids and variable dense matrix, were slightly increased in number. In some of the adenoma cells, clusters of peroxisomes which had a smooth outer limiting membrane enveloping a homogeneous matrix containing a distinct internal core, were present. Most of the peroxisomes were surrounded by slightly degranulated RER. These alterations of peroxisomes may represent either a pathological process or an adaptive pharmacological response.[111,112]

The Golgi apparatuses, which were slightly reduced in number, slightly hypotrophic,

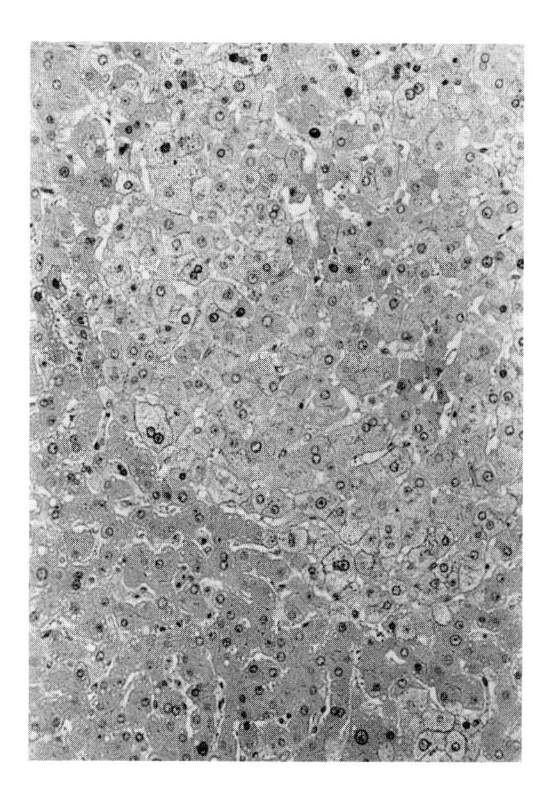

FIGURE 14. A hyperplastic focus in human liver. (H and E stain; magnification ×130.)

and moderately distorted, were distributed in pericanalicular regions and perinuclear sites, resembling the findings with exposure to some carcinogens.

Lysosomes apparently converging on the cytoplasmic zones bordering on the bile canaliculus were decreased in number. Occasionally, large vacuoles were also present, containing membrane materials.

Some of the bile canaliculi formed between several adenoma cells were dilated and distorted in shape, together with a tendency toward either diminution of microvilli or alteration of pericanalicular junctional complexes (Figure 16). In the pericanalicular regions, numerous microfilaments, a small number of free ribosomes, and Golgi apparatuses were often observed.

The microvilli on the sinusoidal surfaces were occasionally decreased in number, together with elongated or short and blunt formations. The intercellular spaces were rather widened. At some of the lateral surfaces, the inner surfaces of the membranes showed abnormal extensive areas of the cytoplasmic condensation and invagination (Figure 17).

In some areas of adenomas, dilated spaces of Disse and collagen fiber concentration were striking features. These alterations in the space of Disse seem to indicate some relationship between fat-storing cells and collagen synthesis. On occasion, basement membrane-like structures were detected beneath endothelial cells in the space of Disse.[113,114] In the sinusoids, enlarged Kupffer cells containing numerous lysosomes and normal shaped endothelial cells were present.

2. Adenomas in Mice

a. Histology and Cytochemistry

Nodular lesions in mouse liver have been referred to as type A and B nodules, adenomatous nodules, adenomas and hepatomas.[10,77] In general, adenomas are composed of

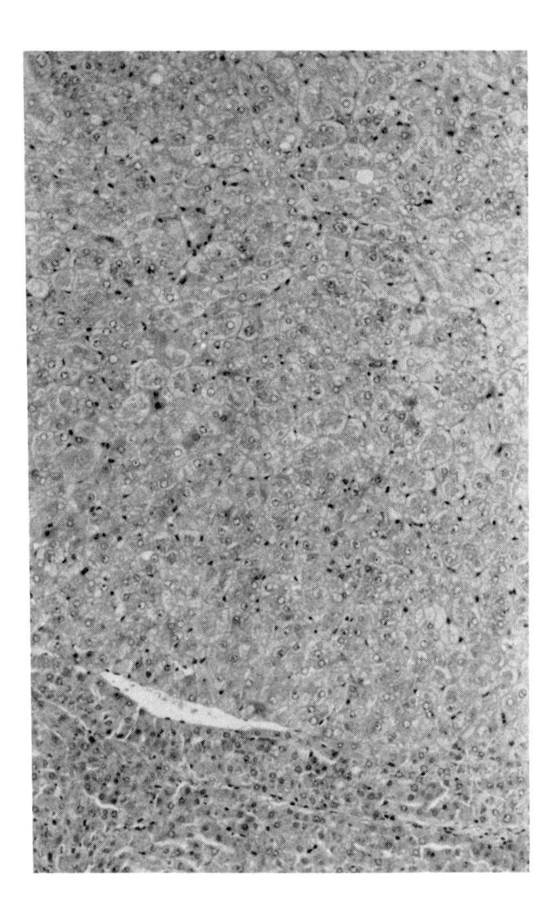

FIGURE 15. A neoplastic nodule induced by 2-acetylamino-fluorene in rat liver. The cells in the lesion are sharply demonstrated from the adjacent parenchyma, which is compressed. (H and E stain; magnification ×130.)

regular solid cords one or two cells thick of closely packed cells which differ little from normal hepatocytes in their morphology and staining characteristics. The adenomas usually compress the surrounding parenchyma, but may merge gradually with it in areas. The nuclei of cells in both spontaneous and induced lesions often display variation in size and shape.

Cytochemically, G6Pase activity is usually reduced in mouse liver neoplasms.[79,81,115,116] Alkaline phosphatase (ALPase) is unchanged or has marked variation.[77,79] ALPase is reported to be increased in adenomas induced by griseofulvin.[11] But the enzyme activity was decreased or unchanged in another report.[116] Usually, the majority of adenomas have similar enzymatic reactions to those of adjacent normal liver cells. The property of iron exclusion is present in both spontaneous and induced mouse liver neoplasms.[46,78,79,83] Benign nodular lesions are not associated with elevated serum AFP, whereas its level is increased with liver cell carcinomas.[117,118] Recently, Numoto et al.[85] reported a relationship between gross appearance or histological types and cytochemical properties of mouse neoplasms induced in B6C3F1 mice by diethylnitrosamine and organochloride pesticides. In their study, hepatocellular adenomas did not differ from hepatocellular carcinomas in their pattern or intensity of enzymatic activity of G6Pase, ALPase, and ATPase. This was similar to the findings of Lipsky et al.[79] but differed from those by Butler and Hempsall,[115] Essigmann and Newberne,[116] and Ruebner.[119] In general, hepatocellular neoplasms with acidophilic cytoplasms more often displayed increased activity of ALPase than did basophilic neoplasms. The

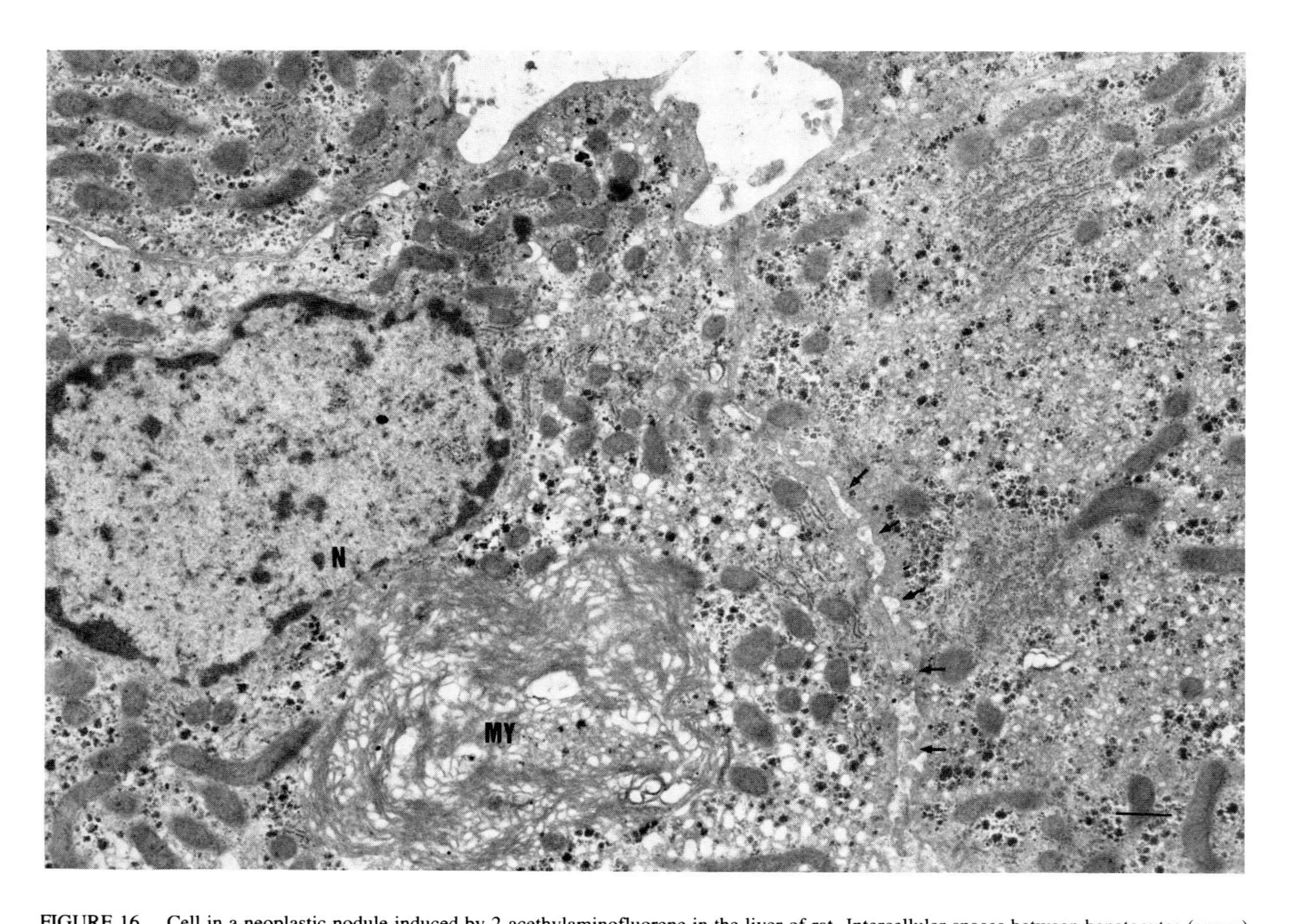

FIGURE 16. Cell in a neoplastic nodule induced by 2-acethylaminofluorene in the liver of rat. Intercellular spaces between hepatocytes (arrow) are irregularly widened. In the cytoplasm, a myelin figure (MY) is in continuity with the RER. Glycogen particles are frequent. (Magnification ×8400.)

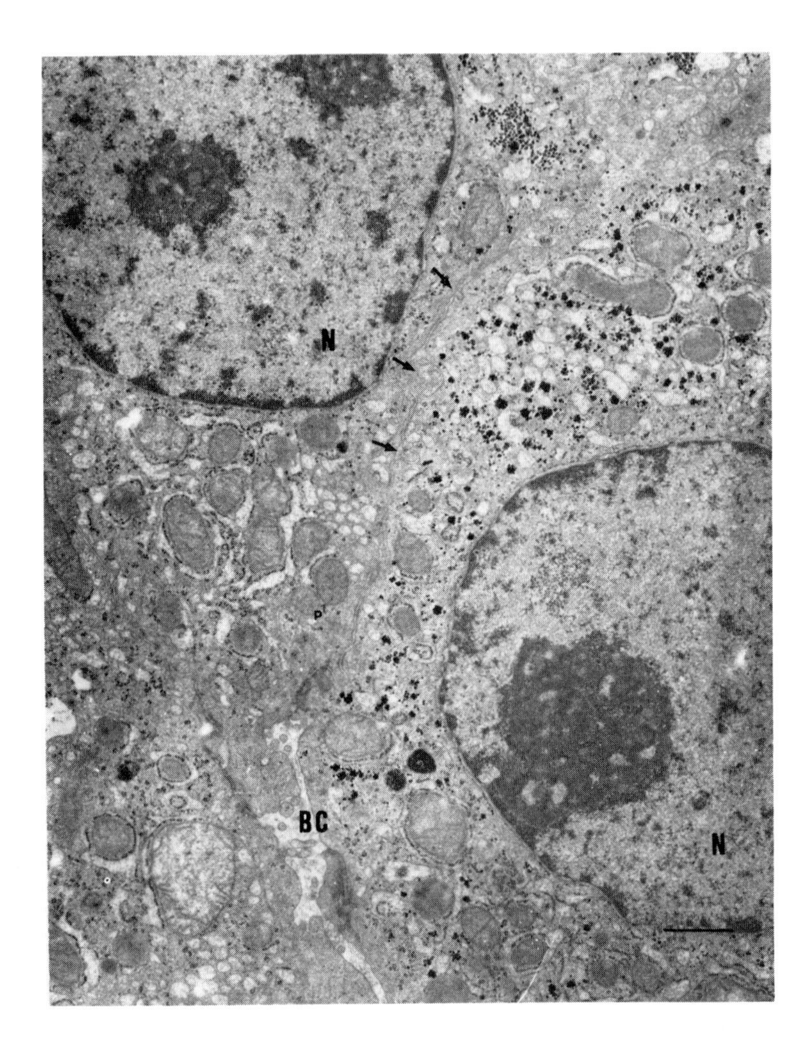

FIGURE 17. Lateral surfaces of two adjacent hepatocytes in a nodule; Invaginations (arrow) of the plasma membrane are apparent. The bile canaliculus (BC) is dilated or distorted in shape with a tendency toward either smaller or decreased number of microvilli. N: nucleus. (Magnification ×8000.)

activities of G6Pase and ATPase were decreased in the benign nodules with acidophilic or basophilic cytoplasms. No difference of GGTase activity between adenomas and carcinomas was seen. The great majority of neoplasms were resistant to iron accumulation.

In mice, several xenobiotics exert a modulating influence on the enzyme activity of liver neoplasms. For examples, nodules induced by 3-5-dichloror(N-1, 1-dimethyl-2-propynyl)benzamide displayed a high ALPase activity regardless of variability of types and induction methods.[11,116] Williams et al.[120] described enhancement of GGTase activity and reduced G6Pase and ATPase activities by phenobarbital in mouse liver neoplasms. Furthermore, chlordane or heptachlor exposure increased the alkaline phosphatase activity in adenomas.[85] Taking all these findings into account, it seems that the results of enzyme studies on mouse hepatocellular adenomas and carcinomas induced in the presence of exposure to chemicals should not be generalized.

b. Ultrastructure

There are several reports on the electron microscopic features of mouse adenomas. Essner[121] performed a fine structural analysis on spontaneously occurring liver nodules of

C3H mice. In that study, alteration of RER such as flattened cisternae containing amorphous, granular, or filamentous materials was characteristic. Confer and Stenger[122] described the subcellular properties of C3H mouse nodules arising after carbon tetrachloride. The cells exhibited enlarged mitochondria, hypertrophied peroxisomes, and abundant dilated RER. Adenomas induced by safrole in BALB/c mice were composed of hepatocytes with considerable heterogeneity of subcellular properties.[123] Cells varied from those with large nuclei with irregular borders and prominent RER to others with abundant SER. Ruebner et al.[124] compared the ultrastructure of induced and spontaneous nodules, finding no definite differences. They also studied the pathogenesis of mouse nodules and postulated that unusual clear hepatocytes which had increased glycogen and lipid droplets and decreased SER were important for nodular transformation.

Our electron microscopic studies on spontaneously occurring adenomas of C3H mice revealed that the cells had a similar morphology to normal hepatocytes except some abnormalities of RER and mitochondria (Figure 18).

3. Adenomas in Hamsters

a. Histology and Cytochemistry

Benign hepatocellular neoplasms of hamsters resembling adenomas in rats have been observed under different experimental conditions.[125] The hamster adenomas differed conspicuously either in size and cytological characteristics from the surrounding liver tissue. Usually, the lesions were composed of cells similar to hepatocytes which were arranged in trabecular fashion. Cytologically, these neoplasms displayed morphology varying from normal appearing hepatocytes to neoplasms composed of cells with multinucleation or enlarged nuclei. The PAS reaction was variable among neoplasms. The nodules were iron excluding, but only rarely positive for GGTase.[14]

b. Ultrastructure

No reports were found on the fine structural features of adenomas in hamsters.

4. Adenomas in Humans

a. Histology and Cytochemistry

Human hepatocellular tumor-like lesions corresponding to adenomas in rodents are infrequently seen. Such lesions have been designated as liver cell adenoma, focal nodular hyperplasia, partial nodular transformation, or regenerative nodular hyperplasia.[126] Among these, liver cell adenoma has been well studied. Use of oral contraceptives and steroids were regarded as etiologic factors.[126]

Gerber and Thung[127] reported that liver cell adenoma related to oral contraceptives displayed significant changes of enzymatic activities of ATPase and G6Pase.

b. Ultrastructure

Electron microscopic studies of human hepatocellular adenoma have been performed.[128,129] The fine structural appearance of neoplastic cells in the adenoma resembled that of normal hepatocytes. Usually, the cells contained relatively large amounts of glycogen granules and lipid droplets. Infrequently, adenoma cells displayed cytoplasmic crystalline structures and have abnormalities of mitochondria.

C. CARCINOMA

Primary malignant liver neoplasms consist of hepatocellular carcinoma, cholangiocellular carcinoma, hepatoblastoma, malignant hemangioendothelioma, and others in both laboratory animals and human. Only hepatocellular carcinoma is considered here.

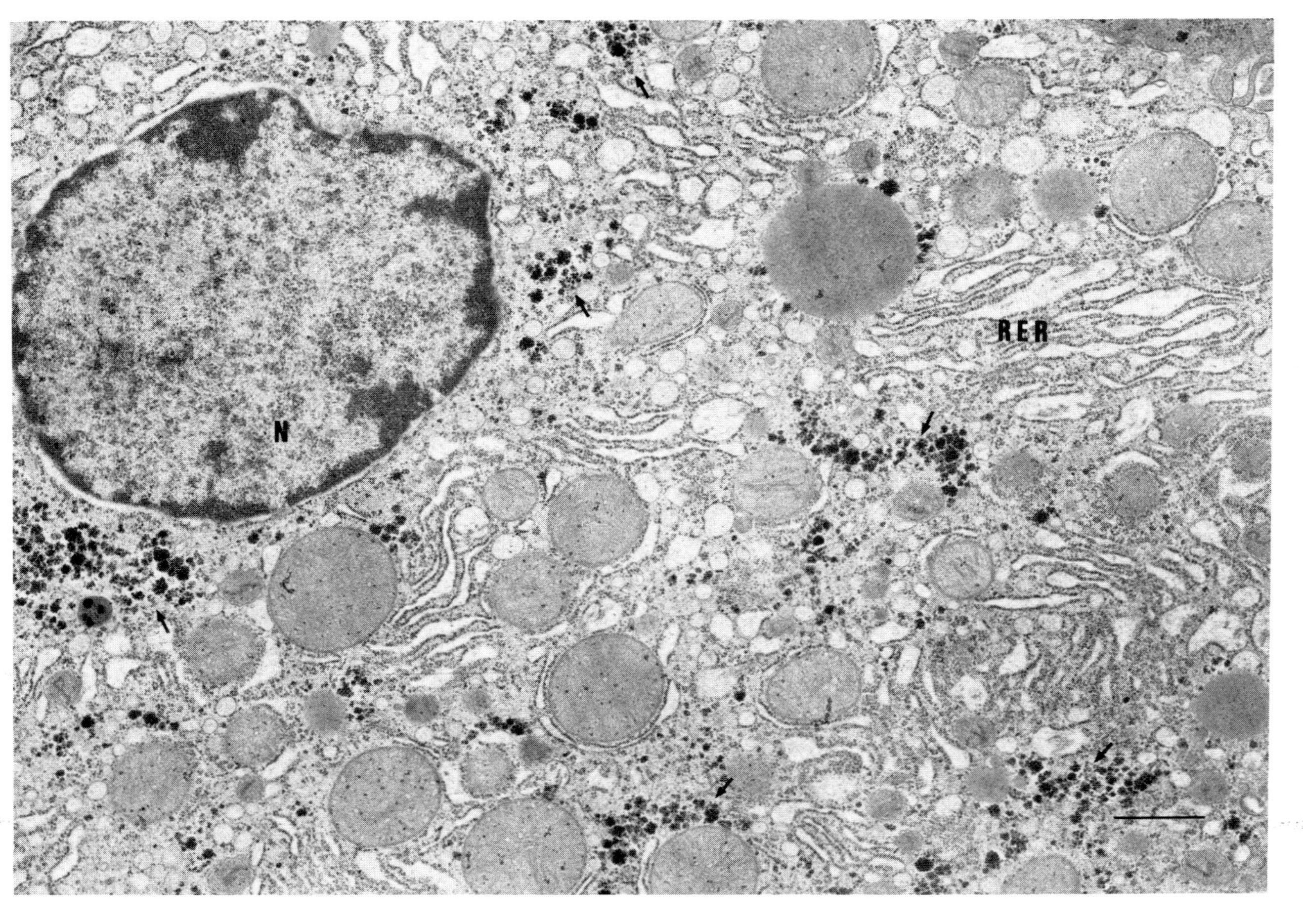

FIGURE 18. Cell in a type A nodule of C3H mouse which spontaneously occurred. In the cytoplasm, large mitochondria are prominent. RER in parallel arrays are well developed. Glycogen particles are clustered (arrow). N: nucleus. (Magnification $\times 12,000$.)

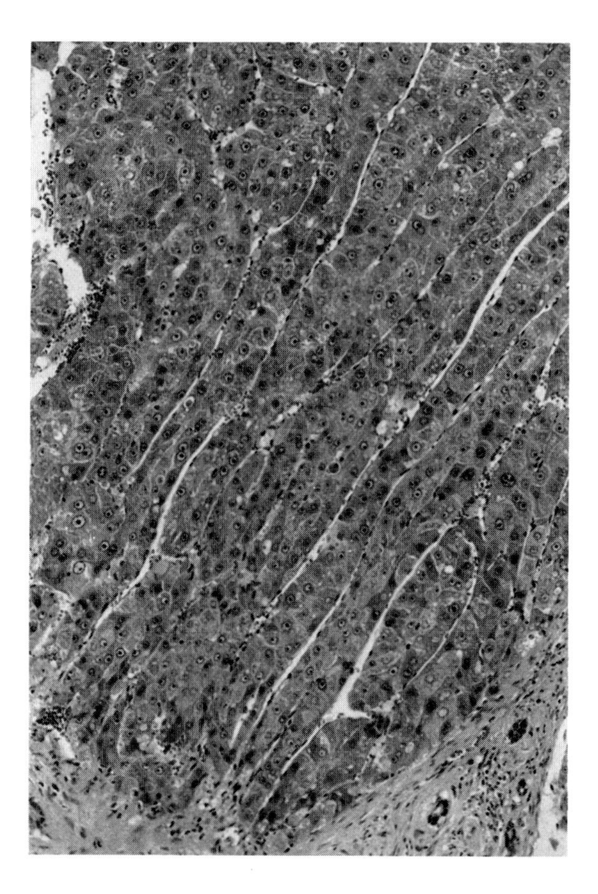

FIGURE 19. A hepatocellular carcinoma with trabecular pattern induced by 2-acetylaminofluorene in rat. (H and E stain; magnification ×130.)

1. Hepatocellular Carcinoma in Rats

a. Histology and Cytochemistry

In general, the architectural appearance of hepatocellular carcinomas consists of three patterns; trabecular (Figure 19), adenocarcinoma and poorly differentiated patterns.[3] Because of the mucus production and an abundant stroma in adenocarcinomas this neoplasm has been considered to be of bile duct cell origin by some investigators.[3,130]

Hepatocellular carcinomas display many of the same phenotypic abnormalities as altered foci and adenomas. They are resistant to iron accumulation.[44,94,95] A number of histochemical abnormalities[131,132] have been found, such as positive activity of GGTase, reported to be present in primary transplantable hepatomas[132] and chemically induced liver cell carcinomas[93,98] and increase of P-450 in response to phenobarbital.[63]

b. Ultrastructure

Electron microscopic studies on hepatocellular carcinoma in the rat have been performed by many investigators.[133-136] In well-differentiated hepatocellular carcinomas, the mitochondria, glycogen particles, Golgi apparatuses, and endoplasmic reticulum system were well developed (Figure 20). In poorly differentiated hepatocellular carcinoma, some of the neoplastic cells revealed irregular arrangement of RER, increased dispersed polyribosomes and annulate lamellae (Figure 21).

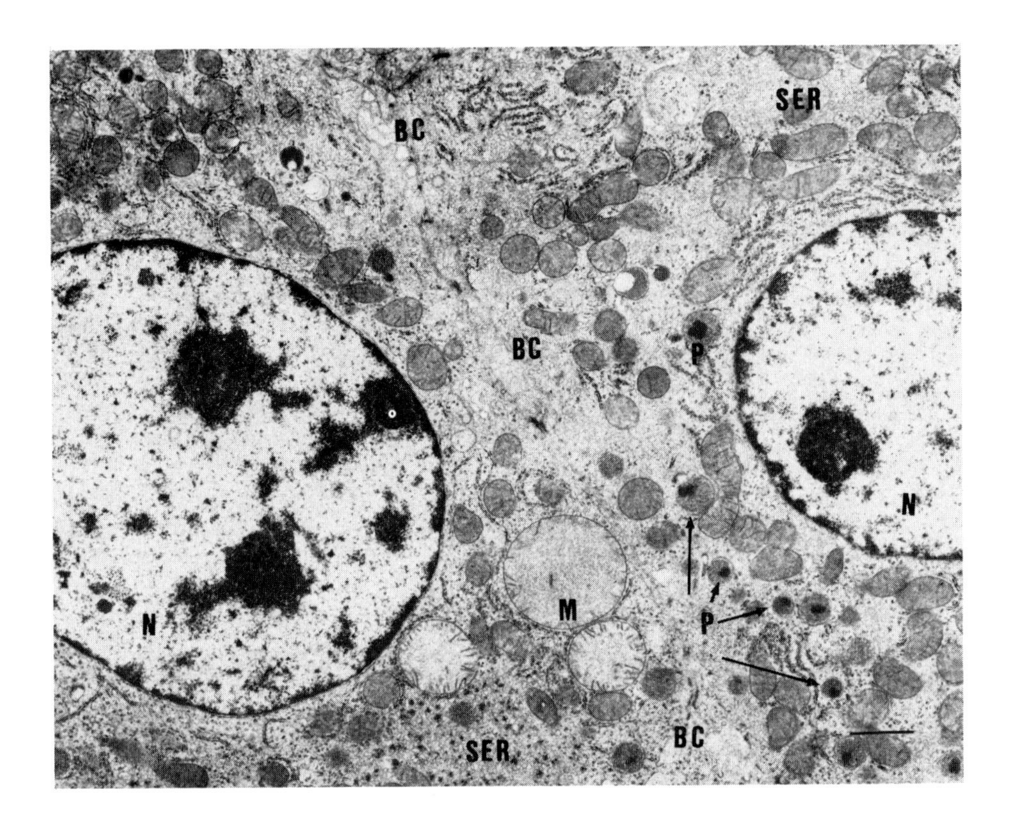

FIGURE 20. Well-differentiated hepatocellular carcinoma in rat; Abundant peroxisomes (P), mitochondria (M) and SER are present. The bile canaliculi are distorted. The nuclei display reduced opacity of the nucleoplasm. (Magnification ×8,000.)

2. Hepatocellular Carcinoma in Mice

a. Histology and Cytochemistry

Hepatocellular carcinomas of mice are often diagnosed on the basis of a distinct trabecular or adenoid pattern as well as the cytologic features characteristic of malignancy (Figure 22).[10] Synonyms for this lesion include type B nodule,[137] type 3 nodule,[138] trabecular carcinoma, and malignant hepatoma. The liver cell plates are more than one cell layer thick, irregular, and composed of well- to poorly differentiated hepatocytes.

Hepatocellular carcinomas of the mouse exhibited a marked increase in alkaline phosphatase (ALPase) activity.[116] In the case of safrole-induced hepatocellular carcinomas, the carcinoma cells possessed a similar histochemical profile to those of adenomas.[79] Similar results on enzymatic activities of G6Pase, ALPase, and ATPase were obtained by Numoto et al.[85] Hepatocellular carcinomas with acidophilic cytoplasms more often displayed increased activity of ALPase than did basophilic neoplasms. (Figure 23) The activities of G6Pase and ATPase were decreased in hepatocellular carcinomas with acidophilic or basophilic cytoplasms. No difference has been reported in the activities of GGTase between carcinomas and adenomas. The property of exclusion of cellular iron is regularly seen in both spontaneous and induced mouse liver carcinomas.[46,78-83,85] Liver cell carcinomas in mice have been associated with elevated serum AFP.[117,118]

Several xenobiotics such as phenobarbital, chlordane, or heptachlor exert some modulating influence on enzyme activities of mouse liver cell carcinomas.[11,85,116,120] Therefore, the results of enzyme studies on mouse hepatocellular neoplasms induced by various chemicals might be influenced by those chemicals and should not be generalized, as mentioned for adenomas.

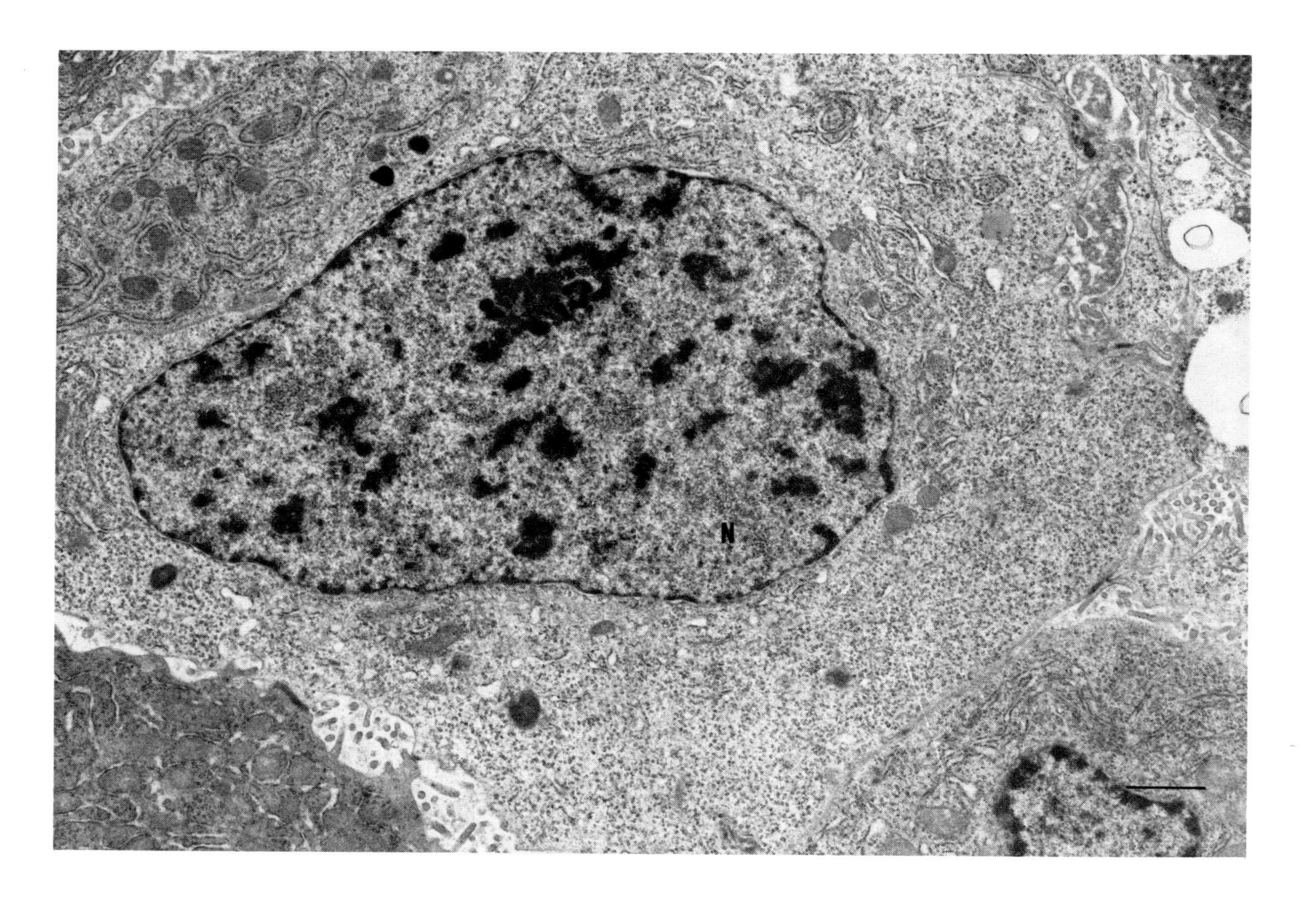

FIGURE 21. Cell in poorly differentiated hepatocellular carcinoma of rat; The nucleus (N) is irregularly shaped. In the cytoplasm, numerous polyribosomes are present, but other organelles are poorly developed. (Magnification $\times$ 10,500.)

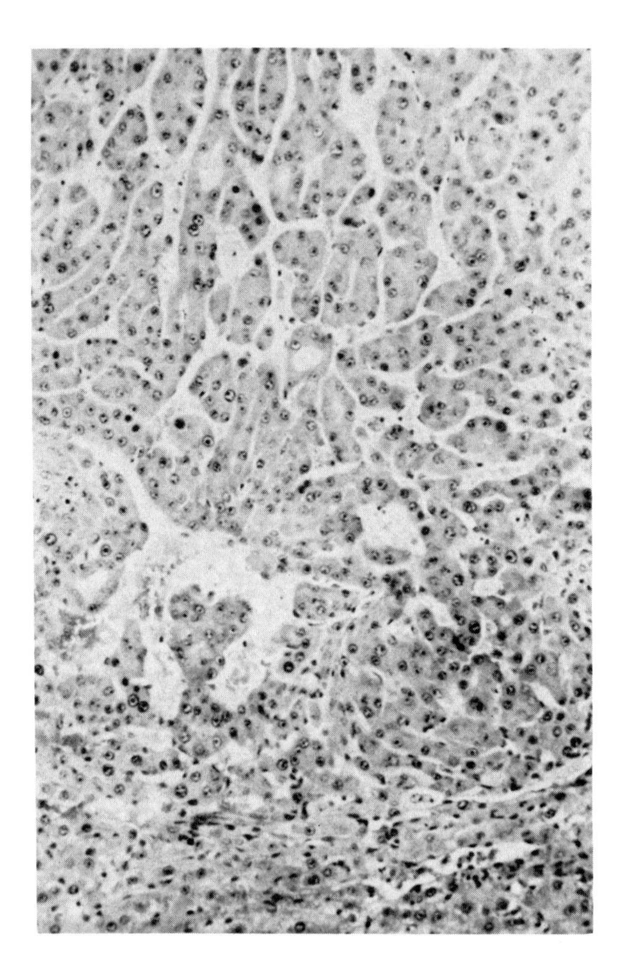

FIGURE 22. An hepatocellular carcinoma with trabecular pattern of mouse. (H and E stain; magnification ×145.)

b. Ultrastructure

Electron microscopic features of mouse hepatocellular carcinomas have been described by several workers. Although mouse carcinomas have some similarities of their fine structures to adenomas, the subcellular properties are known to be different sometimes in several aspects such as greater secretory activity of RER, development of basement membrane-like structures, and nuclear structures.[123,124,139]

3. Hepatocellular Carcinoma in Hamsters

a. Histology and Cytochemistry

Hepatocellular carcinomas in hamsters have been referred to by different names such as malignant hepatoma, adenocarcinoma, carcinoma of the liver, trabecular carcinoma, hepatoma, and liver cell carcinoma. Most carcinomas are relatively well-differentiated neoplasms of the trabecular type. Anaplastic carcinoma and other unusual histological variants are rare.[125]

Hepatocellular carcinomas induced by diethylnitrosamine were iron excluding, like foci and adenomas, but were not positive for GGTase.[14]

b. Ultrastructure

Fine structural observation of hepatocellular carcinomas in hamsters have been infrequently reported. Mori et al. reported subcellular properties of transplanted adenocarcinomas

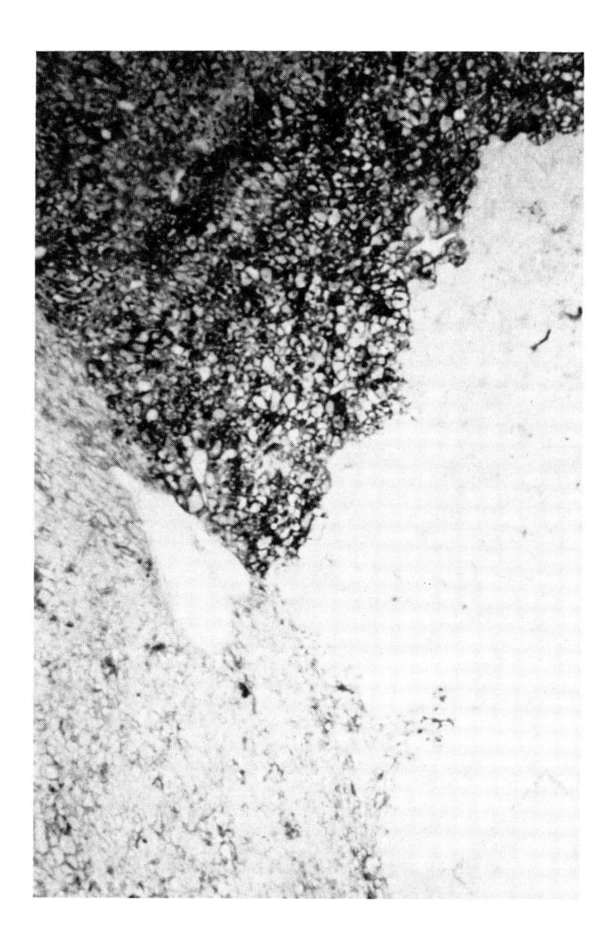

FIGURE 23. An hepatocellular carcinoma having both sites with increased and decreased activities of mouse ALPase reaction. (Magnification ×45.)

and anaplastic carcinomas.[140,141] The fine structure of some adenocarcinomas were similar to those of bile duct epithelium, but the structures of most adenocarcinomas, which included abundant glycogen, well-developed mitochondria, and RER, suggested hepatocellular origin. One of the characteristics of these hamster carcinomas was the existence of various forms of nuclear bodies.

4. Hepatocellular Carcinoma in Humans

a. Histology and Cytochemistry

Human hepatocellular carcinomas show a variety of histological patterns. The four principal histological patterns are trabecular, pseudoglandular, scirrhous, and compact patterns.[142] The variety and degree of morphological alterations depends on the differentiation of the neoplasm. Concerning the differentiation of neoplastic cells, there are four major groups as follows:[143]

Group 1 — similar to adenoma.
Group 2 — a representative of well-differentiated cells.
Group 3 — includes marked giant cells.
Group 4 — characterized by poorly differentiated neoplastic cells.

In light microscopy, clear cytoplasms, multinucleation, and cytoplasmic inclusions such as globular hyaline bodies and Mallory's hyaline are characteristic features in human hepatocellular carcinomas.[142]

In human liver, iron exclusion was also observed in carcinomas in hematochromatotic livers.[87] GGTase activity was positive in human hepatocellular carcinoma.[91,144] No definite decrease of G6Pase, alkaline phosphatase, acid phosphatase, and nonspecific esterase activities has been reported.[91,145,146] Cytochemical changes ranged from alteration of all three markers (GGTase positive; ATPase negative; G6Pase negative) to only one marker (GGTase positive; ATPase positive; G6Pase positive) among liver cell carcinomas.[127] Uchida et al.[91] also reported phenotypic heterogeneity of G6Pase, acid phosphatase and nonspecific esterase activities. Other reports[144] describe a similar heterogeneity of alkaline phosphatase within hepatocellular carcinoma and markedly heterogeneous expression of antigens such as alpha-1-antitrypsin, alpha-fetoprotein, carcinoembryonic antigen, and hepatitis B surface antigen.[147]

b. Ultrastructure

Ultrastructure studies of human hepatocellular carcinomas have been numerous.[148-159] Between the sinusoidal space and tumor, an absence of Kupffer cells, polylayered endothelial cells and formation of basement membrane-like structures are striking features in human hepatocellular carcinomas (Figure 24).[114]

Multiple nucleoli and pseudoinclusions were present in nucleoplasm. Bile canaliculi and glycogen particles, which are important evidences for hepatocellular origin, were detected in the neoplastic cells. In well-differentiated neoplastic cells, well developed RER, round-shaped mitochondria, and moderately developed Golgi apparatuses were distributed throughout the cytoplasm. A few organelles and abundant free polyribosomes were displayed generally in the cytoplasm of poorly differentiated hepatocellular carcinoma. Mallory bodies which consist of fibril components were detected in the cytoplasm together with other cellular depositions (Figure 25).[158,159]

III. PATHOGENESIS

The histogenesis of liver cell carcinoma is becoming increasingly clear, particularly in the rat. The morphological features of foci reviewed here strongly suggest that they are derived from hepatocytes or from an hepatocyte precursor, such as oval cells.[50,51] Foci increase in size and number with continued carcinogen administration,[45,93] and display increased DNA synthesis and mitoses,[160] indicating that they develop by proliferation. Although foci can undergo phenotypic reversion following cessation of carcinogen administration,[45] they are often persistent[14,21] and their nuclear abnormalities[15] strongly suggest that they are a permanently altered population.[17]

Adenomas, as reviewed here, possess many of the features of foci, which suggested that some foci develop into adenomas.[5,6] It is noteworthy that eosinophilic and basophilic foci in rat liver appear to have the same DNA pattern as nodules.[15] Under most circumstances, adenomas are progressively growing[93] and for this reason they have been considered to be neoplasms.[2,3] Nevertheless, there is evidence for regression of some nodules[9,103] and they do not form tumors upon transplantation,[120,161,162] in contrast to carcinomas. In the mouse, the behavior of well-differentiated nodules, now usually called adenomas[77] is different from carcinomas.[46] Thus, adenomas are considered to be a different type of neoplasm from carcinomas, probably a benign neoplasm.[17,93]

For many years, it was postulated that carcinomas were derived from nodules.[4,5,7] Some lines of evidence,[17,163] however, including the deviant DNA pattern in hyperbasophilic foci,[15] suggest that foci may develop directly into carcinomas. This requires further study, but in any event, foci and adenomas are part of the spectrum of lesions that appear prior to the development of hepatocellular carcinomas.

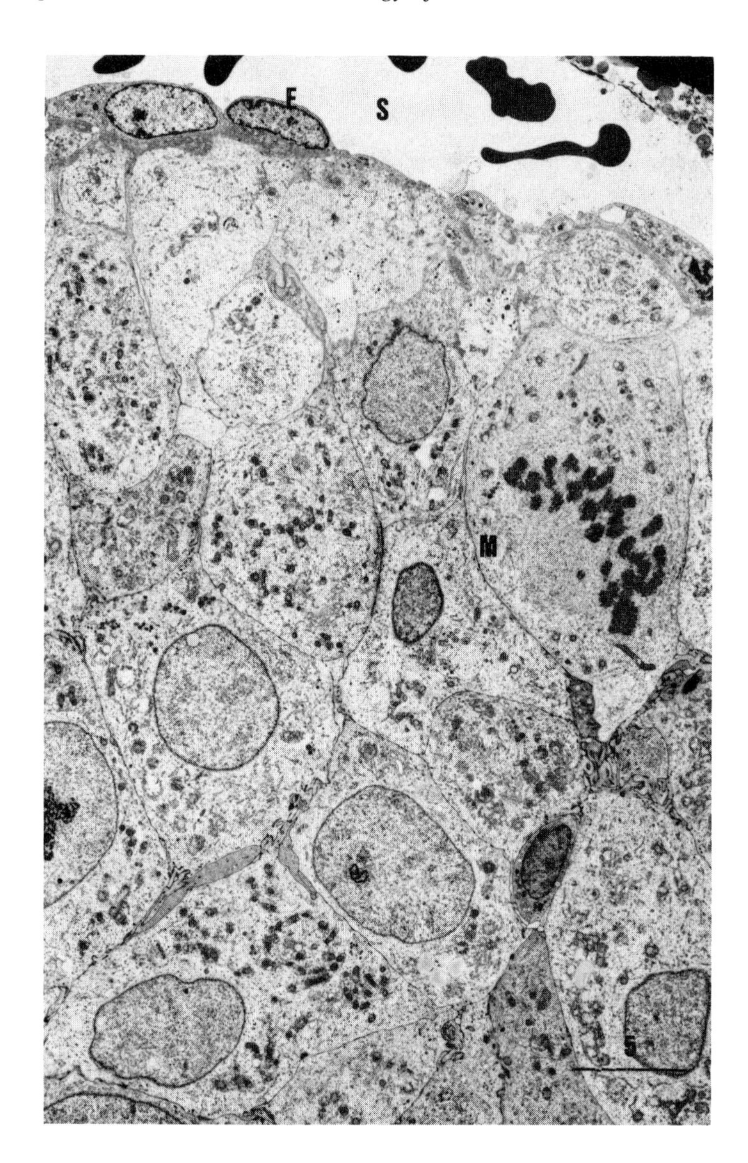

FIGURE 24. Trabecular pattern of poorly differentiated human hepatocellular carcinoma; Between sinusoidal space (S) and tumor, endothelial cells (E) and amorphous substances in the space of Disse are arranged. Mitotic figure (M) can be detected in the tumor. (Magnification × 2800.)

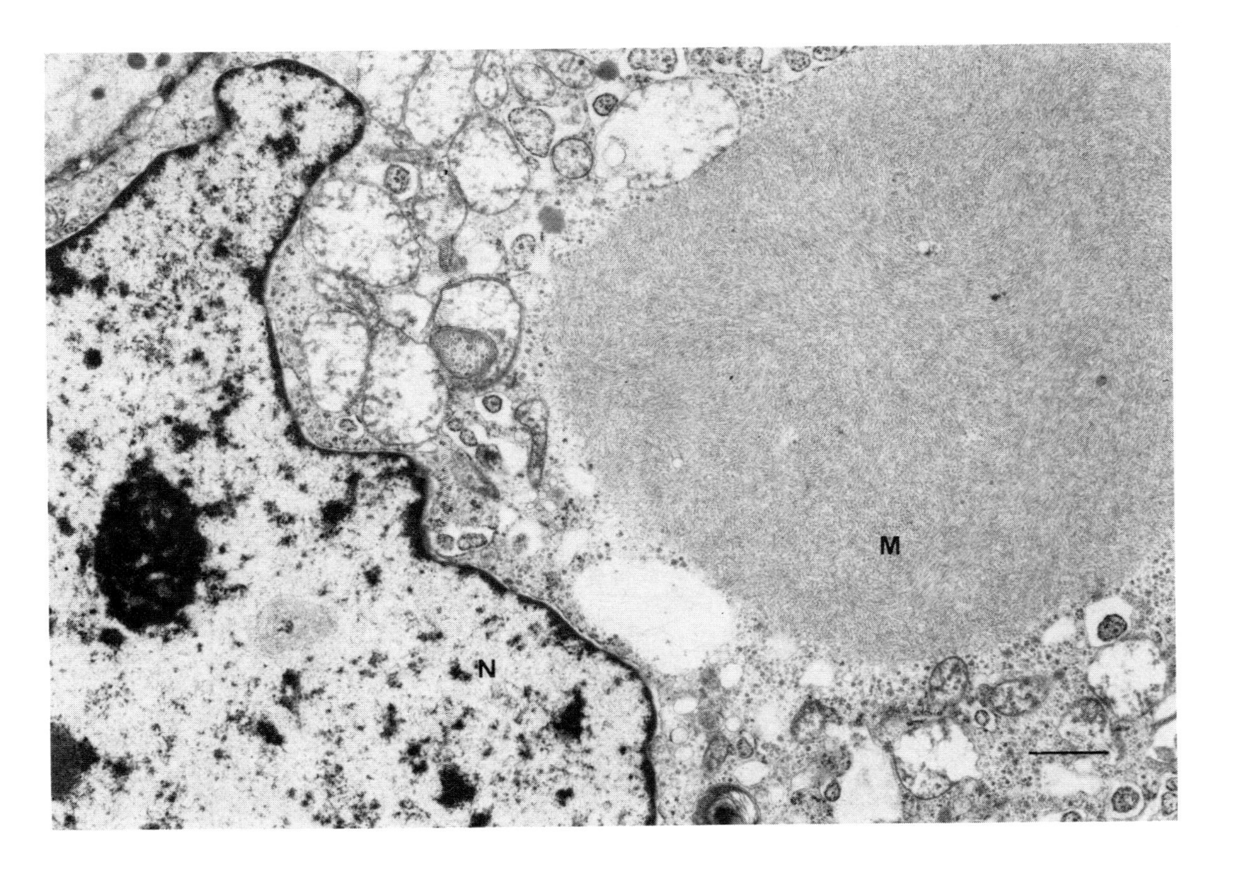

FIGURE 25. Mallory body (M) in the well-differentiated human hepatocellular carcinoma. N: nucleus. (Magnification $\times 8000$.)

REFERENCES

1. **Sasaki, T. and Yoshida, T.,** Experimentelle Erzeugung des Leberkarzinomas durch Fütterung mit *o*-Amidoazotoluol, *Virchows Arch. Pathol. Anat.*, 95, 175, 1935.
2. **Squire, R. A. and Levitt, M. H.,** Report of a workshop on classification of specific hepatocellular lesions in rats, *Cancer Res.*, 35, 3214, 1975.
3. **Stewart, H. L., Williams, G. M., Keysser, C. H., Lombard, L. S., and Montali, R. J.,** Histologic typing of liver tumors of the rat, *J. Natl. Cancer Inst.*, 64, 179, 1980.
4. **Firminger, H. I.,** Histopathology of carcinogenesis and tumors of liver in rats, *J. Natl. Cancer Inst.*, 15, 1427, 1955.
5. **Reuber, M. D.,** Development of preneoplastic and neoplastic lesions of the liver in male rats given 0.025 percent N-2-fluorenylacetamide, *J. Natl. Cancer Inst.*, 34, 697, 1965.
6. **Newberne, P. M. and Wogan, G. N.,** Sequential morphologic changes in aflatoxin B1 carcinogenesis in the rat, *Cancer Res.*, 28, 770, 1968.
7. **Farber, E. and Ichinose, H.,** On the origin of malignant cells in experimental liver cancer, *Acta Unio. Int. Contra Cancrum.*, 15, 152, 1959.
8. **Popper, H., Sternberg, S. S., Oser, B. L., and Oser, H.,** The carcinogenic effect of aramite in rats: a study of hepatic nodules, *Cancer*, 13, 1035, 1960.
9. **Farber, E.,** Hyperplastic liver nodules, *Methods Cancer Res.*, 7, 345, 1973.
10. **Frith, C. H. and Ward, J. M.,** A morphologic classification of proliferative and neoplastic lesions in mice, *J. Environ. Pathol. Toxicol.*, 3, 329, 1980.
11. **Goldfarb, S., Pugh, T. D., and Cripps, D. J.,** Increased alkaline phosphatase activity — a positive histochemical marker for griseofulvin-induced mouse hepatocellular nodules, *J. Natl. Cancer Inst.*, 64, 1427, 1980.
12. **Koen, H., Pugh, T. D., Nychka, D., and Goldfarb, S.,** Presence of alpha-fetoprotein positive cells in hepatocellular foci and microcarcinomas induced by single injection of diethylnitrosamine in infant mice, *Cancer Res.*, 43, 702, 1983.
13. **Moore, M. R., Drinkwater, N. R., Miller, E. C., Miller, J. A., and Pitot, H. C.,** Quantitative analysis of the time-dependent development of glucose-6-phosphatase-deficient foci in the livers of mice treated neonatally with diethylnitrosamine, *Cancer Res.*, 41, 1585, 1981.
14. **Stenbäck, F., Mori, H., Furuya, K., and Williams, G. M.,** Pathogenesis of dimethylnitrosamine-induced hepatocellular cancer in hamster liver and lack of enhancement by phenobarbital, *J. Natl. Cancer Inst.*, 76, 327, 1986.
15. **Mori, H., Tanaka, T., Sugie, S., Takahashi, M., and Williams, G. M.,** DNA content of liver cell nuclei of N-2-fluorenylacetamide-induced altered foci and neoplasms in rats and human hyperplastic foci, *J. Natl. Cancer Inst.*, 69, 1277, 1982.
16. **Mori, H., Tanaka, T., Sugie, S., and Takahashi, M.,** Adenoma with multiple hyperplastic nodules in noncirrhotic liver possibly related to long-term administration of a hypolipidemic drug, *Acta Pathol. Jpn.*, 32, 671, 1982.
17. **Williams, G. M.,** The pathogenesis of rat liver cancer caused by chemical carcinogens, *Biochim. Biophys. Acta*, 605, 167, 1980.
18. **Bannasch, P., Mayer, D., and Hacker, H-J.,** Hepatocellular glycogenosis and hepatocarcinogenesis, *Biochim. Biophys. Acta*, 605, 217, 1980.
19. **Emmelot, P. and Scherer, E.,** The first relevant cell stage in rat liver carcinogenesis. A quantitative approach, *Biochim. Biophys. Acta*, 605, 247, 1980.
20. **Farber, E.,** The sequential analysis of liver cancer induction, *Biochim. Biophys. Acta*, 605, 149, 1980.
21. **Pitot, H. C. and Sirica, A. E.,** The stages of initiation and promotion in hepatocarcinogenesis, *Biochim. Biophys. Acta*, 605, 191, 1980.
22. **Farber, E.,** The multistep nature of cancer development, *Cancer Res.*, 44, 4217, 1984.
23. **Pugh, T. D. and Goldfarb, S.,** Quantitative histochemical and autoradiographic studies of hepatocarcinogenesis in rats fed 2-acetylaminofluorene followed by phenobarbital, *Cancer Res.*, 38, 4450, 1978.
24. **Kitagawa, T.,** Histochemical analysis of hyperplastic lesions and hepatomas of the liver of rats fed 2-fluorenylacetamide, *Gann*, 62, 207, 1971.
25. **Pitot, H. C., Barsness, L., Goldworthy, T., and Kitagawa, T.,** Biochemical characterization of stages of hepatocarcinogenesis after a single dose of diethylnitrosamine, *Nature*, 271, 456, 1978.
26. **Hirota, N. and Williams, G. M.,** The sensitivity and heterogeneity of histochemical markers for altered foci involved in liver carcinogenesis, *Am. J. Pathol.*, 95, 317, 1979.
27. **Ogawa, K., Solt, D. B., and Farber, E.,** Phenotypic diversity as an early property of putative preneoplastic hepatocyte populations in liver carcinogenesis, *Cancer Res.*, 40, 725, 1980.
28. **Estadella, M. D., Pujol, M. J., and Domingo, J.,** Enzyme pattern and growth rate of liver preneoplastic clones during carcinogenesis by diethylnitrosamine, *Oncology*, 41, 276, 1984.

29. **Fiala, S., Mohindru, A., and Kettering, W. G.,** Glutathione and gamma-glutamyl transpeptidase in rat liver during chemical carcinogenesis, *J. Natl. Cancer Inst.*, 57, 591, 1976.
30. **Iannoccone, P. M. and Koizumi, J.,** Pattern and rate of disappearance of gamma-glutamyl transpeptidase activity in fetal and neonatal rat liver, *J. Histochem. Cytochem.*, 31, 1312, 1983.
31. **Hanigan, M. H. and Pitot, H. C.,** Gamma-glutamyl transpeptidase — its role in hepatocarcinogenesis, *Carcinogenesis*, 6, 165, 1985.
32. **Kitagawa, T., Imai, F., and Sato, K.,** Re-evaluation of gamma-glutamyl transpeptidase activity in periportal hepatocytes of rats with age, *Gann*, 71, 362, 1980.
33. **Ogawa, K., Onoe, T., and Takeuchi, M.,** Spontaneous occurrence of gamma-glutamyl transpeptidase-positive hepatocytic foci in 105-week-old Wistar and 72-week-old Fischer 344 male rats, *J. Natl. Cancer Inst.*, 67, 407, 1981.
34. **Furukawa, K., Maeura, Y., Furukawa, N. T., and Williams, G. M.,** Induction by butylated hydroxytoluene of rat liver gamma-glutamyl transpeptidase activity in comparison to expression in carcinogen-induced altered lesions, *Chem.-Biol. Interactions*, 48, 43, 1983.
35. **Ward, J. M.,** Increased susceptibility of livers of aged F344/NCr rats to the effects of phenobarbital on the incidence, morphology, and histochemistry of hepatocellular foci and neoplasms, *J. Natl. Cancer Inst.*, 71, 815, 1983.
36. **Williams, G. M., Tanaka, T., and Maeura, Y.,** Dose-dependent inhibition of aflatoxin B_1 induced hepatocarcinogenesis by the phenolic antioxidants, butylated hydroxyanisole and butylated hydroxytoluene, *Carcinogenesis*, 7, 1043, 1986.
37. **Ratanasavanh, D., Tazi, A., Galteau, M. M., and Seist, G.,** Localization of gamma-glutamyltransferase in subcellular fractions of rat and rabbit liver: effect of phenobarbital, *Biochem. Pharmacol.*, 28, 1363, 1979.
38. **Mochizuki, Y. and Furukawa, K.,** Histochemical investigation of hepatic gamma-glutamyl transpeptidase of rats treated with high dose of phenobarbital, *Acta Histochem. Cytochem.*, 16, 155, 1983.
39. **Moore, M. A., Hacker, H.-J., Kunz, H. W., and Bannasch, P.,** Enhancement of NNM-induced carcinogenesis in the rat liver by phenobarbital: a combined morphological enzyme histochemical approach, *Carcinogenesis*, 4, 473, 1983.
40. **Sawada, N. and Tsukada, H.,** Evaluation of gamma-glutamyl transpeptidase activity in rat liver induced by feeding of butylated hydroxytoluene, *Gann*, 74, 806, 1983.
41. **Numoto, S, Furukawa, K., Furuya, K., and Williams, G. M.,** Effects of the hepatocarcinogenic peroxisome-proliferating hypolipidemia agents Clofibrate and Nafenopin on the rat liver cell membrane enzymes gamma-glutamyltranspeptidase and alkaline phosphatase and on the early state of liver carcinogenesis, *Carcinogenesis*, 5, 1603, 1984.
42. **Furukawa, K., Numoto, S., Furuya, K., Furukawa, N., and Williams, G. M.,** Effect of the hepatocarcinogen nafenopin, a peroxisome proliferator, on the activities of rat liver glutathione-requiring enzymes and catalase in comparison to the action of phenobarbital, *Cancer Res.*, 45, 5011, 1985.
43. **Hirota, N. and Yokoyama, T.,** Comparative study of abnormality in glycogen storing capacity and other histochemical phenotypic changes in carcinogen-induced hepatocellular preneoplastic lesions in rats, *Acta Pathol. Jpn.*, 35, 1163, 1985.
44. **Williams, G. M. and Yamamoto, R. S.,** Absence of stainable iron from preneoplastic and neoplastic lesions in rat liver with 8-hydroxyquinoline-induced siderosis, *J. Natl. Cancer Inst.*, 48, 685, 1972.
45. **Williams, G. M. and Watanabe, K.,** Quantitative kinetics of development of N-2-fluorenylacetamide-induced, altered (hyperplastic) hepatocellular foci resistant to iron accumulation and of their reversion or persistence following removal of carcinogen, *J. Natl. Cancer Inst.*, 61, 113, 1978.
46. **Williams, G. M., Hirota, N., and Rice, J.,** The resistance of spontaneous mouse hepatocellular tumors to iron accumulation during rapid iron loading by parenteral administration and their transplantability, *Am. J. Pathol.*, 94, 65, 1979.
47. **Williams, G. M.,** Phenotypic properties of preneoplastic rat liver lesions and applications to detection of carcinogens and tumor promoters, *Toxicol. Pathol.*, 10, 3, 1982.
48. **Watabe, H.,** Early appearance of embryonic-globulin in rat serum during carcinogenesis with 4-dimethylazobenzene, *Cancer Res.*, 31, 1192, 1971.
49. **Kroes, R., Williams, G. M., and Weisburger, J. H.,** Early appearance of serum alpha-fetoprotein during hepatocarcinogenesis as a function of age of rats and extent of treatment with 3-methyl-4-dimethylaminoazobenzene, *Cancer Res.*, 32, 1526, 1972.
50. **Inaoka, Y.,** Significance of the so-called oval cell proliferation during azo-dye hepatocarcinogenesis, *Gann*, 58, 355, 1967.
51. **Sell, S., Leffert, H. L., Shinozuka, H., Lombardi, B., and Gochman, N.,** Rapid development of large numbers of alpha-fetoprotein-containing oval cells in the liver of rats fed N-2-fluorenylacetamide in a choline devoid diet, *Gann*, 72, 479, 1981.
52. **Tchipysheva, T. A., Guelstein, V. I., and Bannikov, G. A.,** Alpha-fetoprotein-containing cells in the early stages in liver carcinogenesis induced by 3′-methyl-4-dimethylaminoazobenzene and 2-acetylaminofluorene, *Int. J. Cancer*, 20, 388, 1977.

53. **Kuhlmann, W. D.**, Localization of alpha-fetoprotein and DNA-synthesis in liver cell populations during experimental hepatocarcinogenesis in rats, *Int. J. Cancer,* 21, 368, 1978.
54. **Neal, G. E., Metcalf, S. A., Judah, D. J., and Green, J. A.**, Mechanism of the resistance to cytotoxicity which precedes aflatoxin B1 hepatocarcinogenesis, *Carcinogenesis,* 2, 457, 1981.
55. **Fischer, G. and Schauer, A.**, Immunohistochemical demonstration on increased UDP-glucuronyltransferase in putative preneoplastic foci, *Naturwissenschaften,* 70, 153, 1983.
56. **Kitahara, S., Satoh, K., and Sato, K.**, Properties of the increased glutathione S-transferase A form in rat preneoplastic hepatic lesions induced by chemical carcinogens, *Biochem. Biophys. Res. Commun.,* 112, 20, 1983.
57. **Oesch, F., Vogel-Bindel, U., Guenthner, T. M., Cameron, R., and Farber, E.**, Characterization of microsomal epoxide hydrolase in hyperplastic liver nodules of rats, *Cancer Res.,* 43, 313, 1983.
58. **Sato, K., Kitahara, A., Satoh, K., Ishikawa, T., and Tatematsu, M.**, The placental form of glutathione S-transferase as a new marker protein for preneoplasia in rat chemical hepatocarcinogenesis, *Gann,* 75, 199, 1984.
59. **Bock, K. W., Lilienblum, W., Pfeil, H., and Eriksson, L. C.**, Increased uridine diphosphate-glucuronyltransferase activity in preneoplastic liver nodules and Morris hepatomas, *Cancer Res.,* 42, 3747, 1982.
60. **Aström, A., DePierre, J. W., and Eriksson, L.**, Characterization of drug-metabolizing systems in hyperplastic nodules from the livers of rats receiving 2-acetylaminofluorene in their diet, *Carcinogenesis,* 4, 577, 1983.
61. **Eriksson, L., Ahluwalia, M., Spiewak, J., Lee, G., Sarma, D. S. R., Rommi, M. J., and Farber, E.**, Distinctive biochemical pattern associated with resistance of hepatocytes in hepatocytes nodules during liver carcinogenesis, *Environ. Health Perspect.,* 49, 171, 1983.
62. **Sato, K., Kitahara, A., Yin, Z., Ebina, T., Satoh, K., Hatayama, I., Nishimura, L., Yamazaki, T., Tsuda, H., Ito, N., and Dempo, K.**, Molecular forms of glutathione S-transferase and UDP-glucuronyl transferase as hepatic preneoplastic marker enzymes, *Ann. N.Y. Acad. Sci.,* 417, 213, 1984.
63. **Schulte-Hermann, R., Roome, N., Timmermann-Trosiener, I., and Schuppler, J.**, Immunocytochemical demonstration of a phenobarbital-inducible cytochrome P-450 in putative preneoplastic rat liver, *Carcinogenesis,* 5, 149, 1984.
64. **Buckmann, A., Kuhlmann, W., Schwarz, M., Kunz, W., Wolf, C. R., Moll, E., Friedberg, T., and Oesch, F.**, Regulation and expression of four cytochrome P-450 isoenzymes, NADPH-cytochrome P-450 reductase, the glutathione transferases B and C and microsomal epoxide hydrolase in preneoplastic and neoplastic lesions in rat liver, *Carcinogenesis,* 6, 513, 1985.
65. **Fischer, G., Ullrich, D., and Bock, K. W.**, Effects of *N*-nitrosomorphiline and phenobarbital on UDP-glucuronyl-transferase in putative preneoplastic foci of rat liver, *Carcinogenesis,* 6, 605, 1985.
66. **Moore, M. A., Satoh, K., Kitahara, A., Sato, K., and Ito, N.**, A protein cross-reacting immunohistochemically with rat glutathione S-transferase placental form as a marker for preneoplasia in Syrian hamster pancreatic and hepatocarcinogenesis, *Jpn. J. Cancer Res. (Gann),* 76, 1, 1985.
67. **Bannasch, P. and Müller, H. A.**, Lichtmikroskopische Untersuchungen über die Wirkung von N-Nitrosomorpholin auf die Leber von Ratte und Maus, *Arzneim. Forsch.,* 14, 805, 1964.
68. **Bannasch, P.**, The cytoplasm of hepatocytes during carcinogenesis, *Rec. Res. Cancer Res.,* 19, 1, 1968.
69. **Bannasch, P., Hacker, H.-J., Klimek, F., and Mayer, D.**, Hepatocellular glycogenesis and related pattern of enzymatic changes during hepatocarcinogenesis, *Adv. Enzyme Regul.,* 22, 97, 1984.
70. **Holmes, C. H., Autin, E. B., Fisk, A., Gurm, B., and Baldwin, R. W.**, Monoclonal antibodies reacting with normal rat liver cells as probes in hepatocarcinogenesis, *Cancer Res.,* 44, 1161, 1984.
71. **Bruni, C.**, Distinctive cells similar to fetal hepatocytes associated with liver carcinogenesis by diethylnitrosamine. Electron microscopic study, *J. Natl. Cancer Inst.,* 50, 1513, 1973.
72. **Ogawa, K., Medline, A., and Farber, E.**, Sequential analysis of hepatic carcinogenesis: a comparative study of the ultrastructure of preneoplastic, malignant, prenatal, postnatal, and regenerating liver, *Lab. Invest.,* 41, 22, 1979.
73. **Zaki, F. G.**, Ultrastructure of altered hepatocytes induced by diethylnitrosamine (DENA), *Toxicol. Pathol.,* 10, 50, 1982.
74. **Timme, A. H.**, Hyperplastic foci in precancerous rat liver: light microscopic and electron microscopic study, *J. Natl. Cancer Inst.,* 61, 407, 1978.
75. **Hirota, N. and Williams, G. M.**, Ultrastructural abnormalities in carcinogen-induced hepatocellular altered foci identified by resistance to iron accumulation, *Cancer Res.,* 42, 2298, 1982.
76. **Reznik-Schüller, H. M., and Gregg, M.**, Sequential morphologic changes during methapyrilene-induced hepatocellular carcinogenesis in rats, *J. Natl. Cancer Inst.,* 71, 1021, 1983.
77. **Doull, J., Bridges, B. A., Kroe, R., Goldberg, L., Munro, I. C., Paynter, O. E., Pitot, H. C., Squire, R., Williams, G. M., and Darby, W.**, The relevance of mouse liver hepatoma to human carcinogenic risk, A report of the International Expert Advisory Committee to the Nutrition Foundation, The Nutrition Foundation Inc., Washington, D. C., 1983.

78. **Lipsky, M. M., Hinton, D. E., Goldblatt, P. J., Klaunig, J. E., and Trump, B. F.,** Iron-negative foci and nodules in safrole-exposed mouse liver made siderotic by iron-dextran injection, *Pathol. Res. Pract.*, 164, 178, 1979.
79. **Lipsky, M. M., Hinton, D. E., Klaunig, J. E., Goldblatt, P. J., and Trump, B. F.,** Biology of hepatocellular neoplasia in the mouse. II. Sequential enzyme histochemical analysis of BALB/c mouse liver during safrole-induced carcinogenesis, *J. Natl. Cancer Inst.*, 67, 377, 1981.
80. **Nigam, S. K., Babu, K. A., Bhatt, D. K., Karnik, A. B., Thakore, K. N., Lakkad, B. C., Kashyap, S. K., and Chatterjee, S. K.,** Pattern of glycogen and iron accumulation in early appearing BHC-induced liver lesions and liver tumours, *Indian J. Med. Res.*, 74, 289, 1981.
81. **Ohmori, T., Rice, J. M., and Williams, G. M.,** Histochemical characteristics of spontaneous and chemically-induced hepatocellular neoplasms in mice and the development of neoplasms with gamma-glutamyl transpeptidase activity during phenobarbital exposure, *Histochem. J.*, 13, 85, 1981.
82. **Williams, G. M. and Numoto, S.,** Promotion of mouse liver neoplasms by the organochlorine pesticides chlordane and heptachlor in comparison to dichlorodiphenyltrichloroethane, *Carcinogenesis*, 5, 1689, 1984.
83. **Carter, C. A., Gandolfi, A. J., and Sipes, I. G.,** Characterization of dimethylnitrosamine-induced focal and nodular lesions in the livers of newborn mice, *Toxicol. Pathol.*, 13, 3, 1985.
84. **Vesselinovitch, S. D., Hacker, H. J., and Bannasch, P.,** Histochemical characterization of focal hepatic lesions induced by single diethylnitrosamine treatment in infant mice, *Cancer Res.*, 45, 2774, 1985.
85. **Numoto, S., Tanaka, T., and Williams, G. M.,** Histochemical and morphological properties of mouse liver neoplasms induced by diethylnitrosamine and promoted by 4,4′-dichlorodiphenyltrichloroethane, chlordane or heptachlor, *Toxicol. Pathol.*, 13, 325, 1985.
86. **Minase, T., Ogawa, K., and Onoe, T.,** Hyperplastic foci in aged human liver, *Proc. Jpn. Cancer Assoc.*, 397, 1981.
87. **Hirota, N., Hamazaki, M., and Williams, G. M.,** Resistance to iron accumulation and presence of hepatitis B surface antigen in preneoplastic and neoplastic lesions in human hemochromatotic livers, *Hepatogastroenterology*, 29, 49, 1982.
88. **Anthony, P. O., Vogel, C. L., and Barker, L. F.,** Liver cell dysplasia: a premalignant condition, *J. Clin. Pathol.*, 26, 217, 1973.
89. **Anthony, P. P.,** Precancerous changes in the human liver, in *Liver Carcinogenesis*, Lapis, K. and Johannessen, J. V., Eds., Hemisphere, Washington, D. C., 1979, 301.
90. **Cohen, C., Berson, S. D., and Geddes, E. W.,** Liver cell dysplasia; association with hepatocellular carcinoma, cirrhosis and hepatitis B antigen carrier status, *Cancer*, 44, 1671, 1971.
91. **Uchida, T., Miyata, H., and Shikata, T.,** Human hepatocellular carcinoma and putative precancerous disorders: their enzyme histochemical study, *Arch. Pathol. Lab. Med.*, 105, 180, 1981.
92. **Watanabe, S., Okita, K., Harada, T., Kodama, T., Numa, Y., Takemoto, T., and Takahashi, T.,** Morphologic studies of the liver cell dysplasia, *Cancer*, 51, 2197, 1983.
93. **Hirota, N. and Williams, G. M.,** Persistence and growth of rat liver neoplastic nodules following cessation of carcinogen exposure, *J. Natl. Cancer Inst.*, 63, 1257, 1979.
94. **Yamamoto, R. S., Williams, G. M., Frankel, H. H., and Weisburger, J. H.,** 8-Hydroxyquinoline: chronic toxicity and inhibitory effect of the carcinogenicity of N-2-fluorenylacetamide, *Toxicol. Appl. Pharmacol.*, 19, 687, 1971.
95. **Furuya, K., Maeura, Y., and Williams, G. M.,** Abnormalities in liver iron accumulation during N-2-fluorenylacetamide hepatocarcinogenesis that are dependent or independent of continued carcinogen action, *Toxicol. Pathol.*, 12, 136, 1984.
96. **Goldfarb, S. and Zak, F. G.,** Role of injury and hyperplasia in the induction of hepatocellular carcinoma, *J. Am. Med. Assoc.*, 178, 729, 1961.
97. **Epstein, S. M., Ito, N., Merkow, L., and Farber, E.,** Cellular analysis of liver carcinogenesis: the induction of large hyperplastic nodules in the liver with 2-fluorenylacetamide or ethionine and some aspects of their morphology and glycogen metabolism, *Cancer Res.*, 27, 1702, 1967.
98. **Harada, M., Okabe, K., Shibata, K., Masuda, H., Miyata, K., and Enomoto, K.,** Histochemical demonstration of increased activity of gamma-glutamyl transpeptidase in rat liver during hepatocarcinogenesis, *Acta Histochem.*, 9, 168, 1976.
99. **Cameron, R., Sweeney, G. D., Jones, K., Lee, G., and Farber, E.,** A relative deficiency of cytochrome P-450 and aryl hydrocarbon(benzo'a'pyrene)hydroxylase in hyperplastic nodules induced by 2-acetylaminofluorene in rat liver, *Cancer Res.*, 36, 3888, 1976.
100. **Okita, K., Noda, K., Fukumoto, Y., and Takemoto, T.,** Cytochrome P-450 in hyperplastic liver nodules during hepatocarcinogenesis with *N*-2-fluorenylacetamide in rats, *Gann*, 67, 899, 1976.
101. **Feo, F., Canuto, R. A., Garcea, R., Brossa, O., and Caselli, G. C.,** Phenobarbital stimulation of cytochrome P-450 and aminopyrine *N*-demethylase in hyperplastic liver nodules during LD-ethionine carcinogenesis, *Cancer Lett.*, 5, 25, 1978.
102. **Kuhlmann, W. D., Krischan, R., Kunz, W., Guenther, T. M., and Oesch, F.,** Focal elevation of liver microsomal epoxide hydrolase in early preneoplastic stages and its behaviour in the further course of hepatocarcinogenesis, *Biochem. Biophys. Res. Commun.*, 98, 417, 1981.

103. **Teebor, G. W. and Becker, F. F.,** Regression and persistence of hyperplastic hepatic nodules induced by *N*-2-fluorenylacetamide and their relationship to hepatocarcinogenesis, *Cancer Res.*, 31, 1, 1971.
104. **Solt, D. B., Medline, A., and Farber, E.,** Rapid emergence of carcinogen-induced hyperplastic lesions in a new model for the sequential analysis of liver carcinogenesis, *Am. J. Pathol.*, 88, 595, 1977.
105. **Merkow, L. P., Epstein, S. M., Caito, B. J., and Bartus, B.,** The cellular analysis of liver carcinogenesis: ultrastructural alterations within hyperplastic liver nodules induced by 2-fluorenylacetamide, *Cancer Res.*, 27, 1712, 1967.
106. **Merkow, I. P., Epstein, S. M., Farber, E., Pardo, M., and Bartus, B.,** Cellular analysis of liver carcinogenesis. III. Comparison of the ultrastructure of hyperplastic liver nodules and hepatocellular carcinomas induced in rat liver by 2-fluorenylacetamide, *J. Natl. Cancer Inst.*, 43, 33, 1969.
107. **Bannasch, P.,** Cytology and cytogenesis of neoplastic (hyperplastic) hepatic nodules, *Cancer Res.*, 36, 2555, 1976.
108. **Feldman, D., Swarm, R. L., and Bocker, J.,** Ultrastructural study of rat liver and liver neoplasms after long-term treatment with phenobarbital, *Cancer Res.*, 41, 2151, 1981.
109. **Flaks, B., Flaks, A., and Milsom, A.,** A comparative study of ultrastructural changes induced by chronic treatment with 3′-methyl-dimethylaminoazobenzene and the non-carcinogen 3′-methyl-4-diethylaminoazobenzene in the rat hepatocyte, *Carcinogenesis,* 3, 767, 1982.
110. **Flaks, B. and Challis, B. C.,** A comparative electron microscope study of early changes in rat liver induced by *N*-nitrosopiperidine and 2,2′,6,6′-tetramethyl-*N*-nitrosopiperidine, *Carcinogenesis,* 2, 385, 1981.
111. **Tsukada, H., Gotoh, H., and Mochizuki, Y.,** Alteration in peroxisomes of hepatomas, in *Morris Hepatomas,* Morris, H. P. and Criss, W. E., Eds., Plenum Press, New York, 1978, 331.
112. **Svoboda, D. and Higginson, J.,** Comparison of ultrastructural changes in rat liver due to chemical carcinogens, *Cancer Res.*, 28, 1703, 1968.
113. **Schaffner, F. and Popper, H.,** Capillarization of hepatic sinusoids in man, *Gastroenterology,* 44, 239, 1963.
114. **Ichida, T., Miyagiwa, M., Miyabayashi, C., and Sasaki, H.,** Electron microscopic study of sinusoidal lining cells in human hepatocellular carcinoma, *J. Clin. Electron Microsc.*, 16, 468, 1983.
115. **Butler, W. H. and Hempsall, V.,** Histochemical observations on nodules induced in the mouse liver by phenobarbitone, *J. Pathol.*, 125, 155, 1978.
116. **Essigmann, E. M. and Newberne, P. M.,** Enzymatic alterations in mouse hepatic nodules induced by a chlorinate hydrocarbon pesticide, *Cancer Res.*, 41, 2823, 1981.
117. **Becker, F. F., Stillman, D., and Sell, S.,** Serum alpha-fetoprotein in a mouse strain (C3H-AvyfB) with spontaneous hepatocellular carcinomas, *Cancer Res.*, 870, 1977.
118. **Becker, F. F. and Sell, S.,** Alpha-fetoprotein levels and hepatic alterations during chemical carcinogenesis in C57BL/6N mice, *Cancer Res.*, 39, 3491, 1979.
119. **Ruebner, B. H., Gershwin, M. E., Meierhenry, E. F., and Dunn, P.,** Enzyme histochemical characteristics of spontaneous and induced hepatocellular neoplasms in mice, *Carcinogenesis,* 3, 899, 1982.
120. **Williams, G. M., Ohmori, T., Katayama, S., and Rice, J. M.,** Alteration by phenobarbital of membrane-associated enzymes inducing gamma-glutamyl transpeptidase in mouse liver neoplasms, *Carcinogenesis,* 1, 813, 1980.
121. **Essner, E.,** Ultrastructure of spontaneous hyperplastic nodules in mouse liver, *Cancer Res.*, 27, 2137, 1967.
122. **Confer, D. B. and Stenger, R. J.,** Nodules in the livers of C3H mice after long-term carbon tetrachloride administration: a light and electron microscopic study, *Cancer Res.*, 26, 834, 1966.
123. **Lipsky, M. M., Hinton, D. E., Klaunig, J. E., and Trump, B. F.,** Biology of hepatocellular neoplasia in the mouse. III. Electron microscopy of safrole-induced hepatocellular adenomas and hepatocellular carcinomas, *J. Natl. Cancer Inst.*, 67, 393, 1981.
124. **Ruebner, B. H., Gershwin, M. E., Hsieh, L., and Dunn, P.,** Ultrastructure of spontaneous neoplasms induced by diethylnitrosamine and dieldrin in the C3H mouse, *J. Environ. Pathol. Toxicol.*, 4, 237, 1980.
125. **Greenblatt, M.,** Tumor of the liver, in *Pathology of Tumours in Laboratory Animals,* Vol. 3, Turusov, V. S., Ed., IARC, Lyon, 1982, 69.
126. **Anthony, P. P.,** Hepatic neoplasms, in *Pathology of the Liver,* MacSween, R., Anthony, P. P., and Scheuer, P. J., Eds., Churchill Livingstone, Edinburgh, 1979, 387.
127. **Gerber, M. A. and Thung, S. N.,** Enzyme patterns in human hepatocellular carcinoma, *Am. J. Pathol.*, 98, 395, 1980.
128. **Kay, S. and Schatzki, P. E.,** Ultrastructure of a benign liver cell adenoma, *Cancer,* 28, 755, 1971.
129. **Phillips, M. J., Langer, B., Stone, R., Fisher, M. M., and Ritchie, S.,** Benign liver cell tumors. Classification and ultrastructural pathology, *Cancer,* 32, 463, 1973.
130. **Firminger, H. I. and Mulay, A. S.,** Histochemical and morphologic differentiation of induced tumors of the livers in rats, *J. Natl. Cancer Inst.*, 13, 19, 1952.
131. **Kitagawa, T.,** Responsiveness of hyperplastic lesions and hepatomas to partial hepatectomy, *Gann,* 62, 217, 1971.

132. **Fiala, S. and Fiala, E. S.,** Activation by chemical carcinogens of gamma-glutamyl transpeptidase in rat and mouse liver, *J. Natl. Cancer Inst.*, 51, 151, 1973.
133. **Svoboda, D. J.,** Fine structure of hepatomas induced in rats by *p*-dimethylaminoazobenzene, *J. Natl. Cancer Inst.*, 33, 315, 1964.
134. **Onoe, T. and Fuse, Y.,** Electron microscopic study on azo-dye carcinogenesis, *Tumor Res.*, 1, 143, 1966.
135. **Merkow, L. P., Epstein, S. M., Slikin, M., Farber, E., and Pardo, M.,** The cellular analysis of liver carcinogenesis. V. Ultrastructural alterations within hepatocellular carcinomas induced by ethionine, *Lab. Invest.*, 26, 300, 1972.
136. **Butler, W. H. and Jones, G.,** Ultrastructure of hepatic neoplasia, in *Rat Hepatic Neoplasia,* Newberne, P. M. and Butler, H. W., Eds., MIT Press, Cambridge, MA, 1978, 142.
137. **Walker, A. I. T., Thorpe, E., and Stevenson, D. E.,** The toxicology of dieldrin (HEOD). I. Long-term oral toxicity studies in mice, *Fed. Cosmet. Toxicol.*, 11, 415, 1972.
138. **Gellatly, J. B. M.,** The natural history of hepatic parenchymal nodule formation in a colony of C57BL mice with reference to the effect of diet, in *Mouse Hepatic Neoplasia,* Butler, W. H. and Newberne, P. W., Eds., Elsevier, Amsterdam, 1985, 77.
139. **Hruban, Z., Kirstev, W. H., and Slesers, A.,** Fine structure of spontaneous hepatic tumors of male C3H/fGs mice, *Lab. Invest.*, 15, 576, 1966.
140. **Mori, H., Hayashi, K., and Kato, T.,** Electron microscopic observation on cholangiocarcinomas and hepatocellular carcinoma of hamsters, *Acta Sch. Med. Univ. Gifu,* 21, 437, 1973.
141. **Mori, H., Hayashi, K., Kato, T., and Hirobo, I.,** Virus-like particles observed in transplantable intrahepatic bile duct carcinomas and hepatoma of Syrian golden hamsters, *Gann,* 64, 79, 1973.
142. **Gibson, J. B.,** Hepatocellular carcinoma, in *Histological Typing of Tumors of the Liver, Biliary Tract and Pancreas,* Gibson, J. B., Ed., WHO, Geneva, 1978, 20.
143. **Edmondson, H. A.,** Tumors of the liver and intrahepatic bile duct, in *Atlas of Tumor Pathology,* Sect. 7, Fasc. 25, Washington, D. C., Armed Forces Institutes of Pathology, 1958, 1.
144. **Ida, M.,** Enzyme histochemical study on hepatoma: the relation between enzyme activity and histological type, *Acta Pathol. Jpn.*, 27, 647, 1977.
145. **Pepler, W. J. and Theron, J. J.,** A histochemical study of some of the hydrolytic enzymes in malignant hepatoma of the South African Bantu, *Cancer,* 19, 939, 1966.
146. **Viranuvatti, V., Haraphongse, M., Stitnimankarn, T., Limwongse, K., and Plengvanit, U.,** Histochemical studies of alkaline phosphatase in carcinoma of the liver, *Am. J. Dig. Dis.*, 14, 625, 1969.
147. **Thung, S. N., Gerber, M. A., Sarno, E., and Popper, H.,** Distribution of five antigens in hepatocellular carcinoma, *Lab. Invest.*, 41, 101, 1979.
148. **Toker, C. and Trevino, N.,** Ultrastructure of human primary hepatic carcinoma, *Cancer,* 19, 1594, 1966.
149. **Ghadially, F. N. and Parry, E. W.,** Ultrastructure of human hepatocellular carcinoma and surrounding non-neoplastic liver, *Cancer,* 19, 1989, 1966.
150. **Creemers, J. and Jardin, J. M.,** Ultrastructure of a human hepatocellular carcinoma, *J. Microscop.*, 7, 257, 1968.
151. **Schaff, Z. S.,** The ultrastructure of primary hepatocellular cancer in man, *Virchows Arch.*, A, 352, 340, 1971.
152. **O'Conor, G. T., Tralka, T. S., Heuson, E., and Vogel, C. L.,** Ultrastructure survey of primary liver cell carcinomas from Uganda, *J. Natl. Cancer Inst.*, 48, 587, 1972.
153. **Ma, M. H. and Blackburn, C. R. B.,** Fine structure of primary liver tumors and tumor-bearing livers in man, *Cancer Res.*, 33, 1766, 1973.
154. **Tanikawa, K.,** Primary carcinoma of the liver, in *Ultrastructural Aspects of the Liver and its Disorders,* Tanikawa, K., Ed., Igaku-Shoin, Tokyo, 1979, 338.
155. **Lapis, K. and Johannessen, J. V.,** Pathology of primary liver cancer, in *Liver Carcinogenesis,* Lapis, K. and Johannessen, J. V., Eds., Hemisphere, Washington, D. C., 1979, 145.
156. **Johannessen, J. V.,** Primary liver tumors, in *Electron Microscopy in Human Medicine,* Johannessen, J. V., Ed., McGraw-Hill, New York, 1979, 188.
157. **Isomura, T. and Nakashima, T.,** Ultrastructure of human hepatocellular carcinomas, *Acta Pathol. Jpn.*, 30, 713, 1980.
158. **Ichida, T., Inoue, K., and Sasaki, H.,** Ultrastructural study of Mallory bodies in human hepatocellular carcinoma, *J. Clin. Electron Microsc.*, 15, 426, 1982
159. **Ichida, T.,** Ultrastructural study of intracytoplasmic deposits in human hepatocellular carcinoma, *Gastroenterol. Jpn.*, 18, 560, 1983.
160. **Williams, G. M., Klaiber, M., Parkers, S. E., and Farber, E.,** Nature of early appearing, carcinogen-induced liver lesions resistant to iron accumulation, *J. Natl. Cancer Inst.*, 57, 157, 1976.
161. **Williams, G. M., Klaiber, M., and Farber, E.,** Differences in growth of transplants of liver, liver hyperplastic nodules and hepatocellular carcinomas in the mammary fat pad, *Am. J. Pathol.*, 89, 379, 1977.

162. **Mori, H., Furuya, K., and Williams, G. M.,** Enhanced survival and absence of progressive growth of transplanted rat liver altered eosinophilic foci and neoplastic nodules in phenobarbital-treated rats, *J. Natl. Cancer Inst.,* 71, 849, 1983.
163. **Daoust, R.,** Cellular populations and nucleic acid metabolism in rat liver parenchyma during azo dye carcinogenesis, *Can. Cancer Conf.,* 5, 225, 1963.
164. **Gössner, W. and Friedrich-Freksa, H.,** Histochemische Untersuchungen über die Glucose-6-phophatase in der Ratten während der kancerisierung durch Nitrosamine, *Z. Naturforsch.,* 196, 862, 1964.
165. **Friedrich-Freksa, H., Papadopulu, G., and Gössner, W.,** Histochemische Uktersuchungen der Kancerogenese in der Rattenleber nach zeitlich begrenzter Verabfolung von Diäthylnitrosamine, *Z. Krebsforsch.,* 72, 240, 1969.
166. **Kitagawa, T. and Pitot, H. C.,** The regulation of serine dehydratase and glucose-6-phosphatase in hyperplastic nodules of rat liver during diethylnitrosamine and *N*-2-fluorenylacetamide feeding, *Cancer Res.,* 35, 1075, 1975.
167. **Rabes, H. M., Scholze, P., and Jantsch, B.,** Growth kinetics of diethylnitrosamine-induced, enzyme-deficient "preneoplastic" liver cell populations in vivo and in vitro, *Cancer Res.,* 32, 2577, 1972.
168. **Scherer, E., Hoffman, M., Emmelot, P., and Friedrich-Freksa, M.,** Quantitative study on foci of altered liver cells induced in the rat by a single dose of diethylnitrosamine and partial hepatectomy, *J. Natl. Cancer Inst.,* 49, 93, 1972.
169. **Klimek, F., Mayer, D., and Bannasch, P.,** Biochemical microanalysis of glycogen content and glucose-6-phosphate dehydrogenase activity in focal lesions of the rat liver induced by *N*-nitrosomorpholine, *Carcinogenesis,* 5, 265, 1984.
170. **Taper, H. S., Fort, L., and Brucher, J.-M.,** Histochemical activity of alkaline and acid nucleases in the rat liver parenchyma during *N*-nitrosomorpholine carcinogenesis, *Cancer Res.,* 31, 913, 1971.
171. **Lian, D., Daoyun, Z., Jinyan, C., Juantan, W., Yuanbu, W., Shaoru, P., and Shuhua, W.,** A histological study of the promotive effect of diethylstilbestrol on diethylnitrosamine initiated carcinogenesis of liver in rat, *Pathol. Res. Pract.,* 178, 339, 1984.
172. **Fontaniere, B. and Daoust, R.,** Histochemical studies on nuclease activity and neoplastic transformation in rat liver during diethylnitrosamine carcinogenesis, *Cancer Res.,* 33, 3108, 1973.
173. **Malvaldi, G.,** Appearance of iron free hyperplastic hepatocytes following ferric nitrilotriacetate and 2-acetaminofluorene sequential administration, *Eur. J. Cancer,* 17, 481, 1981.
174. **Fiala, S., Fiala, A. E., and Dixon, B.,** Gamma-glutamyl transpeptidase in transplantable chemically induced rat hepatomas and "spontaneous" mouse hepatomas, *J. Natl. Cancer Inst.,* 48, 1393, 1972.
175. **Kalengi, M. M. R., Rochi, G., and Desmet, V. J.,** Histochemistry of gamma-glutamyl transpeptidase in rat liver during aflatoxin B1-induced carcinogenesis, *J. Natl. Cancer Inst.,* 55, 579, 1975.
176. **Cameron, R., Kellen, A., Lolin, A., Malkin, A., and Farber, E.,** Gamma-glutamyl transferase in putative premalignant liver cell populations during hepatocarcinogenesis, *Cancer Res.,* 38, 823, 1978.
177. **Schor, N., Ogawa, K., Lee, G., and Farber, E.,** The use of D-T diaphorase for the detection of foci of early neoplastic transformation in rat liver, *Cancer Lett.,* 5, 167, 1978.
178. **Enomoto, K., Ying, T. S., Griffin, M. J., and Farber, E.,** Immunohistochemical study of epoxide hydrolase during experimental liver carcinogenesis, *Cancer Res.,* 41, 3281, 1981.
179. **Goldfarb, S. A.,** Morphological and histochemical study of carcinogenesis of the liver in rats fed 3′-methyl-4-dimethylaminoazobenzene, *Cancer Res.,* 33, 1119, 1973.
180. **Mori, M., Kaku, T., Dempo, K., Satoh, M., Kaneko, A., and Onoe, T.,** Histochemical investigation of precancerous lesions induced by 3′-methyl-4-dimethylaminoazobenzene, *Gann Monogr. Cancer Res.,* 25, 103, 1980.
181. **Jones, D. D., Jr., Evces, S., and Lindahl, R.,** Expression of tumor-associated aldehyde dehydrogenase during rat hepatocarcinogenesis using the resistant hepatocyte model, *Carcinogenesis,* 5, 1679, 1984.
182. **Hacker, H.-J., and Bannasch, P.,** Histochemische Muster einiger Schlüsselenzyme des Kohlenhydratstoffwechsels während der Hepatocarcinogenese, *Verh. Dtsch. Pathol.,* 61, 481, 1977.
183. **Okita, K., Kligman, L. H., and Farber, E.,** A new common marker for premalignant and malignant hepatocytes induced in the rat by chemical carcinogens, *J. Natl. Cancer Inst.,* 54, 199, 1975.

Chapter 4

THE COMPARATIVE PATHOLOGY OF DIMETHYLNITROSAMINE-INDUCED RENAL TUMORS IN THE RAT

Gordon C. Hard and William H. Butler

TABLE OF CONTENTS

I. INTRODUCTION

Animal models in which a high incidence of neoplasia is induced are indespensible for acquiring an understanding of the processes which lead to the development of cancer. Chemical systems are particularly suitable for identifying the target cells of origin for various types of cancer, for studying tumor pathogenesis in terms of the sequence of precursor lesions, for describing models whereby this process can be modified or reversed, and for defining the biochemical and molecular mechanisms involved in neoplastic transformation. The kidney is a complex organ comprising at least 20 different subpopulations of cells in mammalian species and so there is potential for the development of a range of unrelated tumors which are observed in both man and other mammals.

Renal tumors can be categorized according to their derivation from epithelium, connective tissue, or embryonal primordia. Work in our laboratories has endeavored to seek out or devise potent animal models for these various tumor types. In each case, the model sought has a high incidence of neoplasia induced by a single administration of chemical carcinogen in conventional laboratory mammals. Although this is not the mode of human exposure, at the experimental level, resolution of the essential precursor stages of carcinogenesis is best achieved without the perturbing effects incurred by continued or repeated applications of a toxic carcinogen. Thus, single dose systems avoid repeated episodes of cytotoxicity and inflammation as well as the potential for modification of the precursor lesions. The ability to discriminate nonspecific toxicity from the emerging precursor lesions that will become overt cancer permits more accurate establishment of the time of development for the prodromal lesions and thus a more finite delineation of the sequential stages in the carcinogenic process. Single dose systems can have a further advantage in terms of a longer survival span for the animal from initiation of the cancer process to death, providing a more prolonged period available for tumor progression to secondary invasion.

II. RENAL EFFECT OF DMN

Following Magee and Barne's observation in 1962[1] that dimethylnitrosamine (DMN) was sometimes able to induce kidney tumors in rats after only one injection, Swann and McLean[2] demonstrated the enhancing capacity of a "no protein-high carbohydrate" diet for renal tumor induction, leading to the basis of a high frequency system for the rat renal mesenchymal tumor (RMT) described in detail below.[3,4] Pretreatment with the diet decreases microsomal enzyme activity in the liver and increases the amount of DMN reaching the kidney and other extrahepatic organs.[2,5] Under this regime the LD_{50} for DMN is almost doubled[6] enabling a much higher dose to be administered to the animal which results in a substantially increased tumor yield. Thus, a single intraperitoneal injection of 60 mg/kg body weight DMN administered to immature rats 3 to 5 d after commencement of a diet consisting solely of a glucose/sucrose mixture can produce a 100% incidence of renal tumors of which the great majority are RMT.[4] In fact, between 90 and 100% of affected rats develop this tumor type, but 40% or less also develop cortical epithelial tumors. Earlier studies seemed to indicate that the induction of both RMT and cortical epithelial tumors in the rat might be age dependent. For example, in one study[7] using neonatal rats, only RMT were induced while in another,[8] utilizing 5-month-old rats, the tumors were solely of an epithelial type. This discrepancy prompted an investigation into the age dependency of the two tumor types using DMN in the protein-deprived Wistar rat.[9,10] The results clearly demonstrated a biphasic distribution of the two types of renal tumors, with RMT affecting the immature rat and epithelial tumors increasing in frequency and becoming the predominant tumor in the more mature animal. Susceptibility to the chemical induction of RMT was characterized by a peak of predisposition for treatment at 3 to 4 weeks, with a rapid decline beyond 6 weeks

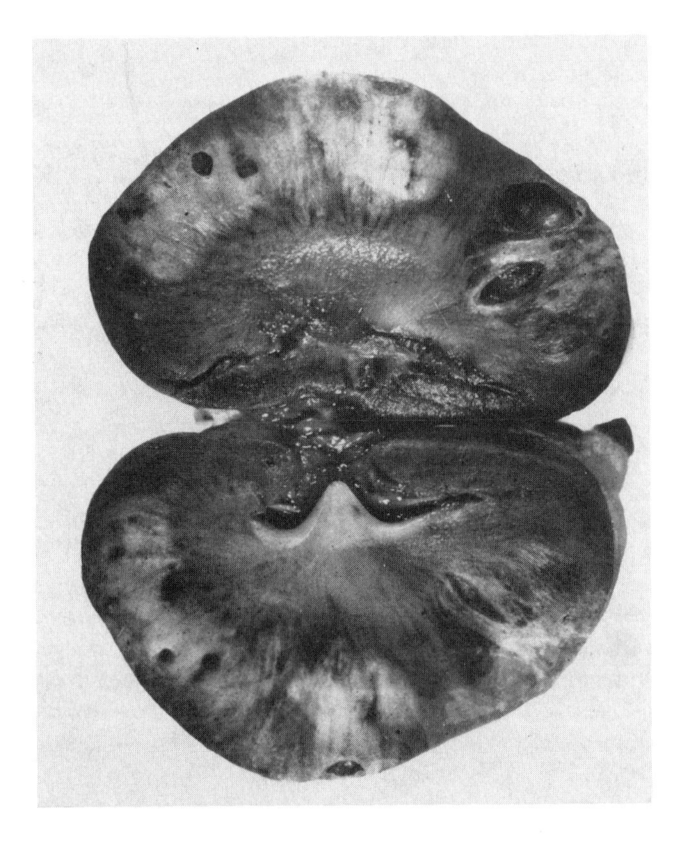

FIGURE 1. Rat kidney showing the gross appearance of a mesenchymal tumor diffusely infiltrating the outer zones. (Magnification ×4.5.) (From Hard, G. C., *Pathology of Laboratory Animals,* Vol. 4, Jones, T. C., Mohr, U., and Hunt, R. D., Eds., Springer-Verlag, N. Y., 1986, 61. With permission.)

of age. Thus, 4- to 6-week old female Porton albino or Charles River Wistars have been the chief animal models for studies on RMT, although inbred strains such as the Fischer-344 and Nb rat[11] are also highly susceptible to RMT induction by DMN at juvenile ages.

III. RENAL MESENCHYMAL TUMOR (RMT)

A. MACROSCOPIC APPEARANCE

Small tumors are visible as a white discoloration on the surface. Such lesions are usually confined to the outer zones of the organ as an ill-defined infiltrative growth following the sagittal plane of the cortex and without the rounded form which typifies epithelial tumors of the kidney (Figure 1). The poorly delineated tumor outline is particularly evident when the kidney is prepared as a histological section. The texture of the small- to intermediate-sized tumors is usually fibrous and they are often cystic (Figure 1). In some cases the affected kidney appears uniformly enlarged, and when sectioned, the tumor tissue is found to be distributed throughout much of the organ. In larger tumors, proliferation occurs beyond the renal capsule to form bulging nodular masses which may fill the abdominal cavity and invade the abdominal wall. Very large tumors are usually multilobular, with prominent areas of hemorrhage, necrosis, and gelatinous tissue. Despite the destruction of kidney substance in these cases, invariably a small rim of normal renal parenchyma remains intact at some point on the surface of the tumor.

The tumor is usually a rapidly growing one often reaching macroscopic dimensions by 5 months after treatment. However, fibromatous variants with a predominance of sclerosing tissue have a slower growth rate. Distant invasion to the lungs and liver does occur but in only a low percentage of cases, probably because the primary neoplasm itself frequently causes death of the animal through hemorrhage, extensive necrosis, or impairment of function. Proliferation into the renal vein, along the length of the vena cava and into the heart is occasionally seen, while peritoneal seeding is a further mode of metastatic spread.

B. HISTOLOGICAL APPEARANCE

The histology of RMT has been described under various synonyms but was reevaluated as a neoplasm of secondary mesenchyme.[4,12] The hallmark of RMT is the heterogeneous range of connective tissue cell types that constitute the differentiative capacity of the cell of origin, and which typically are represented within a single tumor. The basic cell is a spindle form but other neoplastic types which are invariably present within RMT are stellate cells typifying embryonic secondary mesenchyme and smooth muscle fibers. Other elements within the range of connective tissue which may accompany the preceding components but are not always present include neoplastic vascular tissue, striated muscle, cartilage, and osteoid tissue. Extensive deposition of collagen as a product of the tumor cells is also a feature. Characteristically, the tumors have less cellular myxoid areas, as well as more dense cellular areas, and all contain profiles of tubules, glomeruli, or cystic spaces and/or nests of urothelium which represent sequestered remnants of preexisting nephrons or epithelial lining.

Spindle cells are always found at the invading edge with normal kidney, infiltrating between the preexisting renal tubules (Figure 2). In most RMT, spindle cells are aggregated in some part of the tumor, particularly areas which have extended beyond the renal capsule, as sheets of fibrosarcoma typified by a fascicular ("herringbone") cellular pattern (Figure 3). In the larger tumors, condensations of highly basophilic tumor cells are sometimes present as palisades or cords within the fibrosarcomatous sheets. As an additional feature of RMT, spindle cells often condense around sequestered tubule profiles to form stratified whorls (Figure 2). Smooth muscle fibers are usually sparsely distributed throughout the areas of lower cell density (Figure 4), and often partially circumscribe sequestered renal tubules and cystic spaces. Not infrequently, smooth muscle is more profuse, forming sheets with an interlacing fascicular pattern representing leiomyosarcoma. These areas usually merge with areas of fibrosarcoma.

Collagen is demonstrable in almost all parts of every tumor although it is less conspicuous in the dense sarcomatous sheets. Typically, the extent of collagen deposition is unlike reactive fibrocollagenous stroma in that it often forms coarse whorls and tufts of eosinophilic material surrounding solitary or small clumps of tumor cells (Figure 5). On the other hand, in the fibromatous variants of RMT, collagenization is the most predominant feature of the tumor, forming a homogeneous matrix between the sparsely scattered cell clumps and collapsed tubule profiles. All tumors also have a dense reticulin network throughout, but elastin is absent.

In at least one third of RMT, there is a conspicuous development of abnormal vascular structures which can be identified as hemangioma, hemangioendothelioma, hemangiosarcoma (Figure 6), or hemangiopericytoma. Quite frequently a fibrosarcomatous sheet will merge imperceptibly with an area of hemangiosarcoma or hemangiopericytoma indicating that the potential for vascular differentiation is an integral aspect of these tumors. Additionally, mesenchymal tumor cells in denser zones often appear to be circumscribing capillary-like spaces (Figure 4), while a rich vascular network is typical in most tumor areas.

In a small percentage of RMT, large cells with bizarre nuclei, typical of rhabdomyoblasts, or sparse bundles of more mature striated muscle fibers are present. Occasionally, striated

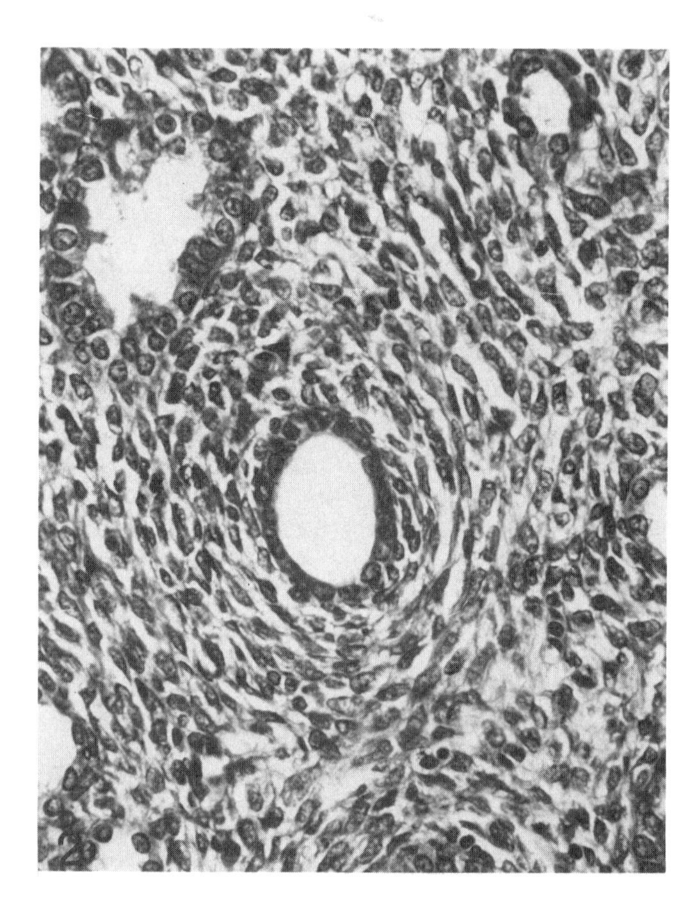

FIGURE 2. Mesenchymal tumor. The predominant cell form is spindle shaped often aligned in layers around pre-existing renal tubules. (Hematoxylin and eosin (H and E); magnification ×300.) (From Hard, G. C. and Butler, W. H., *Cancer Res.*, 30, 2796, 1970. With permission.)

muscle is sufficiently well developed as to be consistent with rhabdomyosarcoma. More infrequently, differentiation of tumor stem cells into nests of cartilage or osteoid tissue are observed.

All tumors contain altered tubules and glomeruli scattered throughout that part of the tissue which has invaded original kidney (Figures 2, 5 and 7). These nonneoplastic elements are derived from the preexisting parenchyma. This comes about by virtue of the infiltrative growth of the tumor cells within the interstitial space and the ability of the engulfed tubules or Bowman's capsules to survive within the tumor tissue. Some of the nephrons are collapsed and become atrophic but many survive as dilated cysts lined by flattened epithelium or as tubular profiles in which the epithelium is hyperplastic (Figure 7). This latter phenomenon does not indicate neoplastic transformation of epithelium nor does it imply bipotential differentiation on the part of the mesenchymal tumor cells. The transition from normal tubules to surviving, but pathologically altered, remnants can be traced from the invading edge of the tumor into the interior of the neoplastic tissue. In large tumors where extension has occurred well beyond the original kidney outline, isolated tubule profiles become progressively less frequent and ultimately absent from the extra-renal tissue. In the frequent cases where invasion occurs through the renal pelvis, the polyp-like extensions retain a covering of urothelium. With tumor expansion, tongues of this epithelial lining become incorporated within the neoplastic tissue. Very occasionally such nests display metaplastic keratinization.

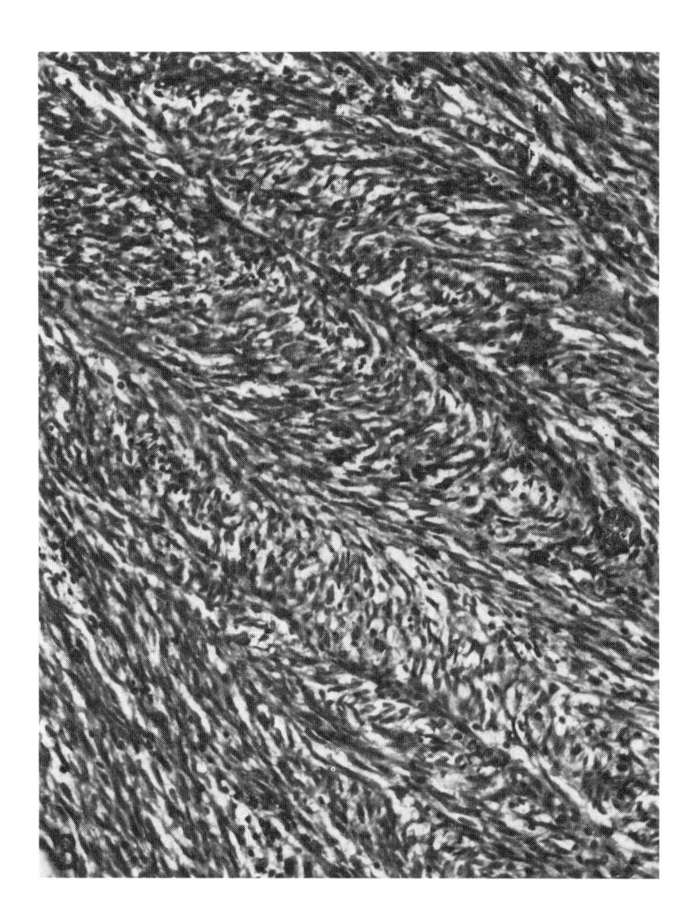

FIGURE 3. Mesenchymal tumor. Cells are arranged in a herring bone pattern characteristic of fibrosarcoma. (H and E; magnification $\times 150$.) (From Hard, G. C. and Butler, W. H., *Cancer Res.*, 30, 2796, 1970. With permission.)

C. ULTRASTRUCTURE

RMT has been studied in detail by electronmicroscopy.[13] Ultrastructurally, the basic spindle cell resembles an active fibroblast being characterized by abundant, anastomosing channels of rough endoplasmic reticulum and bundles of actin-like microfilaments (Figure 8). Stellate cells conform to fibrocytes. There is a close association of both cell types with newly formed collagen. Smooth muscle fibers, rhabdomyoblasts, and striated muscle exhibit the distinctive myofibrillar features which normally characterize these cell types.

One of the outstanding features of RMT at the electron microscopic level is the frequency with which mesenchymal tumor cells are clumped into small annuli enclosing cleft-like lumens suggestive of primitive capillaries (Figure 9). Individual cells within such aggregates are conjoined laterally by intercellular junctions and the structure is usually invested in part by basement membrane (Figure 10). They are found throughout various parts of the tissue which do not necessarily coincide with the histologically obvious areas of vascular neoplasia. In addition, some spindle cells which encircle isolated tubules in stratified layers display the ultrastructural features of vascular pericytes. Electron microscopy therefore emphasizes vasoformative potential as an integral aspect of this tumor.

Ultrastructural examination of the epithelial components confirms their identity as pathologically altered remnants of the preexisting parenchyma and not as neoplastic components. Thus, all epithelial profiles are discretely separated from the surrounding mesenchymal tumor

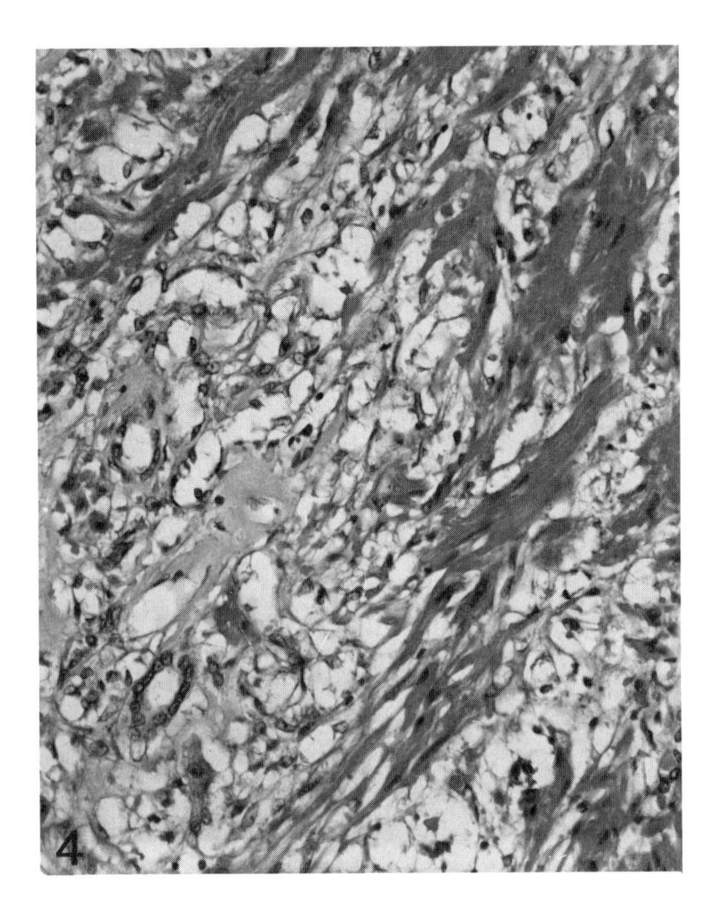

FIGURE 4. Mesenchymal tumor. Area in which the tumor cells circumscribe small vascular spaces sometimes containing erythrocytes. Tracts of smooth muscle fibres are present. (Heme-phloxin-saffron (HPS); magnification ×260.) (From Hard, G. C., *Carcinogenesis,* 6, 1551, 1985. With permission.)

cells by intact basal lamina with no imperceptible merging of transitional cell forms that would be expected in bipotential differentiation. Tubule profiles retain certain features of the mature nephrons but also show aspects consistent with ischemic or atrophic alteration (Figure 11), such as lipid accumulation, prominent autophagic vacuoles, myelin figures, and grossly thickened, detached, and tortuous basal lamina. Hyperplastic lining of isolated tubules recapitulates the typical epithelial cell form without any of the ultrastructural aspects characteristic of neoplastic tubule epithelium.[14] The engulfed tongues and nests of urothelium are also consistent with nonneoplastic transitional epithelium.

D. PATHOGENESIS

To account for the heterogeneous connective tissue constituents of RMT, including vascular types, and location of the earliest lesions, origin must involve a multipotential mesenchymal cell in the cortical interstitium. We have traced the sequential development of chemically induced RMT by DMN.[15,16] Following a short phase of acute toxicity and inflammation in the outer cortex, sporadic hypercellular foci persist within the intertubular space of the cortex during the latent period (Figure 12). These consist of mononuclear inflammatory cells, macrophages, lymphocytes, and plasma cells, but always include occasional and solitary fibroblast-like cells of atypical form with bizarre or enlarged nuclei (Figure 13). Although their relationship to tumor development is not known, these mes-

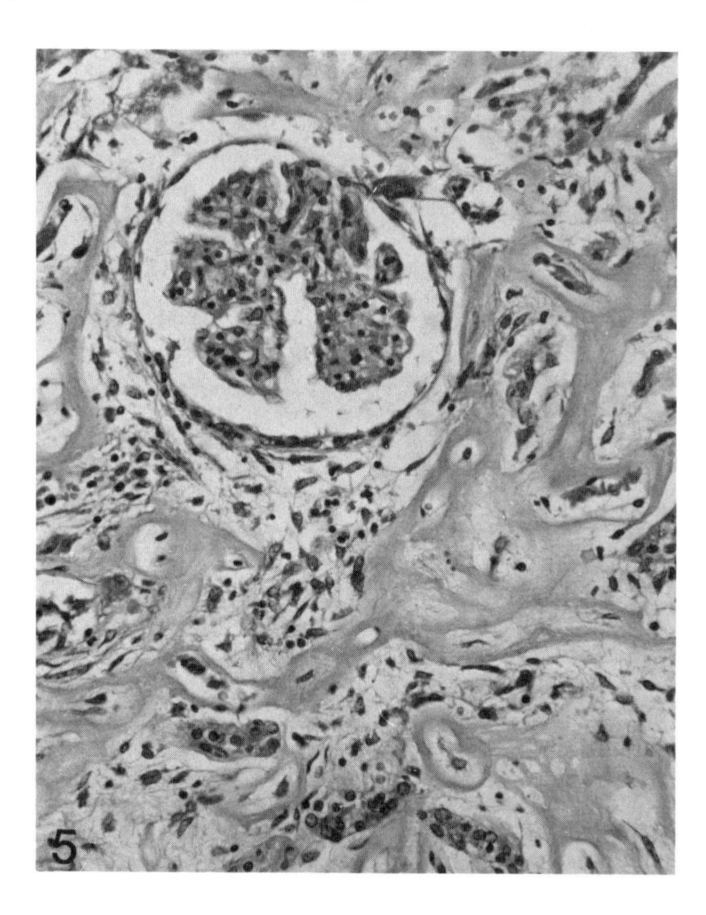

FIGURE 5. Mesenchymal tumor. Area of tumor with residual glomerulus surrounded by an irregular collagenous stroma. (HPS; magnification ×250.) (From Hard, G. C., *Carcinogenesis,* 6, 1551, 1985. With permission.)

enchymal cells represent abnormal constituents of the interstitial space and characterize the early phase of induction. The first unequivocal tumor cell aggregates can be observed microscopically at 12 to 16 weeks after the inciting dose of DMN. Most frequently they are situated in the inner cortex (zone 2) of the rat kidney and always within the intertubular space (Figure 14). Extravasated red cells appear to be constant features of the earliest microscopic neoplasms. The proliferation around and between tubules of the renal parenchyma is particularly evident in these small lesions, and the subsequent and progressive sequestration of the preexisting epithelial elements is distinct. Because of the specific intertubular location of the earliest lesions, RMT most likely takes its origin from the cortical fibrocyte, a cell which resides in the interstitial space of zones 1 and 2 of the rat kidney. This cell is adjacent to the intertubular capillaries and in the young rat possibly may have a broad differentiative potential along the connective tissue pathway including a vasoformative capability. At the same time, origin of the tumor in vascular endothelium of the cortical intertubular capillary network cannot be discounted.

The detailed sequence of ultrastructural changes during the early phase of acute toxicity was also traced in the 6-week-old, DMN-treated rats.[17] The first alterations, moderate lipid droplet formation and proliferation of smooth endoplasmic reticulum, were observed between 2 and 24 h, primarily in the epithelium of the P2 segment of the proximal tubule. There was no accompanying evidence of cytoplasmic degeneration within tubule epithelium at this stage. By 18 to 24 h, cytotoxicity was observed in zone 1 resident cortical fibrocytes (Figure

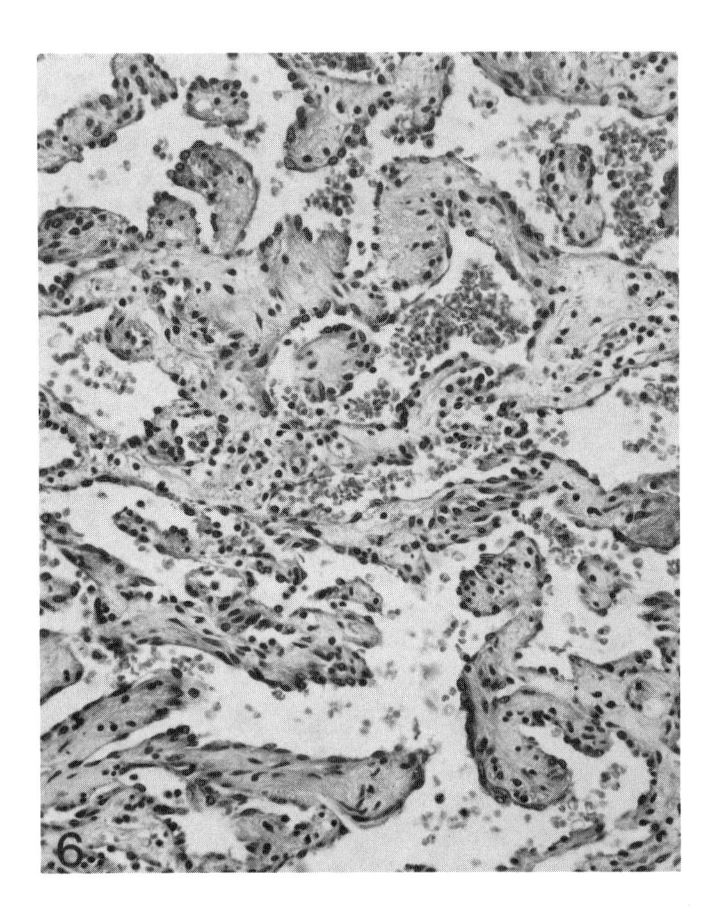

FIGURE 6. Mesenchymal tumor. Area of hemangiosarcomatous differentiation. (H and E; magnification ×250.) (From Hard, G. C., *Carcinogenesis,* 6, 1551, 1985. With permission.)

15), followed by focal necrosis of vascular endothelium of the zone 1 peritubular capillary network at 2 to 4 d (Figure 16). With disruption of the capillary lining, red cells leaked into the interstitial space. At the same time, overt cytotoxicity occurred specifically in the P2 segment of the nephron, with marked lipid accumulation and focal cytoplasmic degradation, leading to scattered single-cell necrosis spatially related to the damaged capillaries by 5 d. From day 4, infiltration of mononuclear phagocytes into the interstitial space of zone 1 was prominent. Macrophages were observed to be involved in erythrophagocytosis during the course of this inflammatory reaction. Capillary repair appeared to be complete by 7 d, and proximal tubule regeneration was present through the 11th d. By 14 d, the inflammatory process was decreasing.

The following hypothesis was presented as an explanatory basis for this pattern of injury acknowledging the pharmacokinetics of DMN in the protein-deprived rat.[17] The results suggest the possibility that DMN might be metabolically activated to a toxic species by the tubular microsomal enzyme system in the P2 segment of the proximal nephron with subsequent diffusion of a reactive intermediate(s), or toxic cell product, to produce cytotoxic injury in the susceptible mesenchymal cell populations, that is the resident cortical fibrocyte, and the adjacent capillary endothelium. The observed time sequence also suggests that the transient circulatory disturbance presumed to result from the focal endothelial necrosis, in turn, causes ischemic injury to the P2 segment of the proximal tubule as a relatively delayed event. Resolution of the renal lesion by 3 weeks is associated with endothelial and tubule

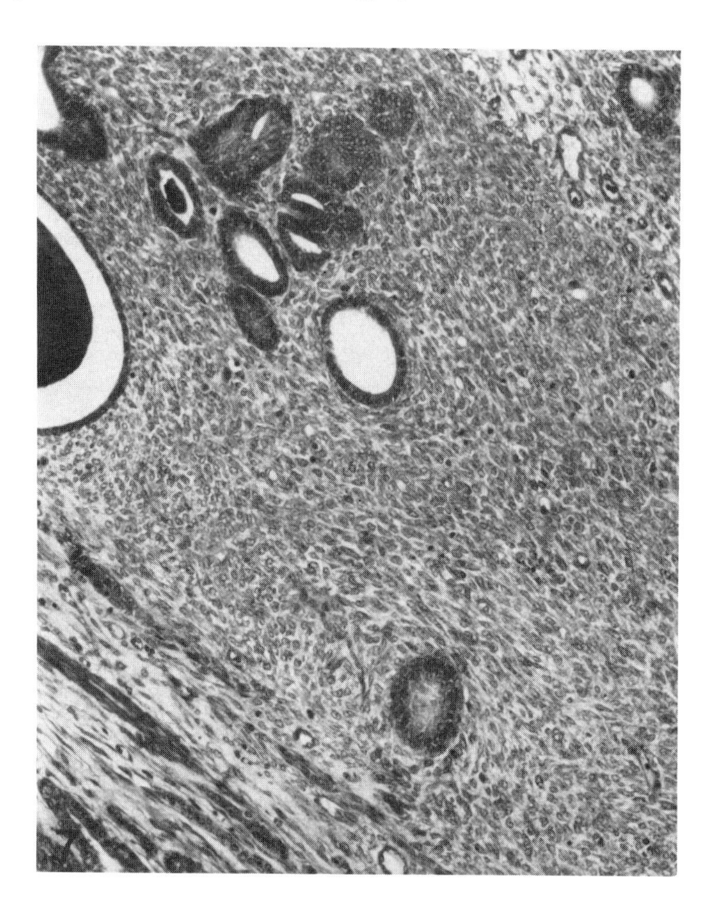

FIGURE 7. Mesenchymal tumor. Cellular area with residual renal tubules which are either flattened and atrophic cysts, or hyperplastic. (H and E; magnification ×150.) (From Hard, G. C., *Pathology of Tumours in Laboratory Animals — Tumours of the Rat,* Vol. 1, Part 2, Turusov, V. S., Ed., JARC, Lyon, 1976, 73. With permission.)

regeneration but in particular by clearance of the extravasated red cells from the interstitial space by macrophages. The cell populations involved as targets in the acute toxic reponse to DMN implicate the proximal convoluted tubule and either the resident cortical fibrocyte or peritubular capillary endothelium as origins for the DMN-induced cortical epithelial and mesenchymal tumors, respectively.

E. DIFFERENTIAL DIAGNOSIS

RMT has been the subject of much confusion with respect to diagnosis and classification, the literature revealing no less than 13 synonyms for this complex neoplasm. Most commonly, RMT is mistaken for an embryonal kidney tumor and misclassified as nephroblastoma or Wilms' tumor. It is, however, a distinct entity unrelated to nephroblastoma. There are several reasons for this confusion, but primarily it is the presence of epithelial profiles in RMT (the isolated remnants of preexisting parenchyma) which are mistaken for the expression of bipotential differentiation into neoplastic epithelium. The frequently observed stratification of RMT cells around engulfed tubules adds to the illusion of bipotential organoid formation. In addition, the condensations of mesenchymal cells sometimes seen within fibrosarcomatous sheets simulate the blastemal clusters in nephroblastoma. Despite these confusing aspects there are a number of differences which clearly discriminate RMT from nephroblastoma. In

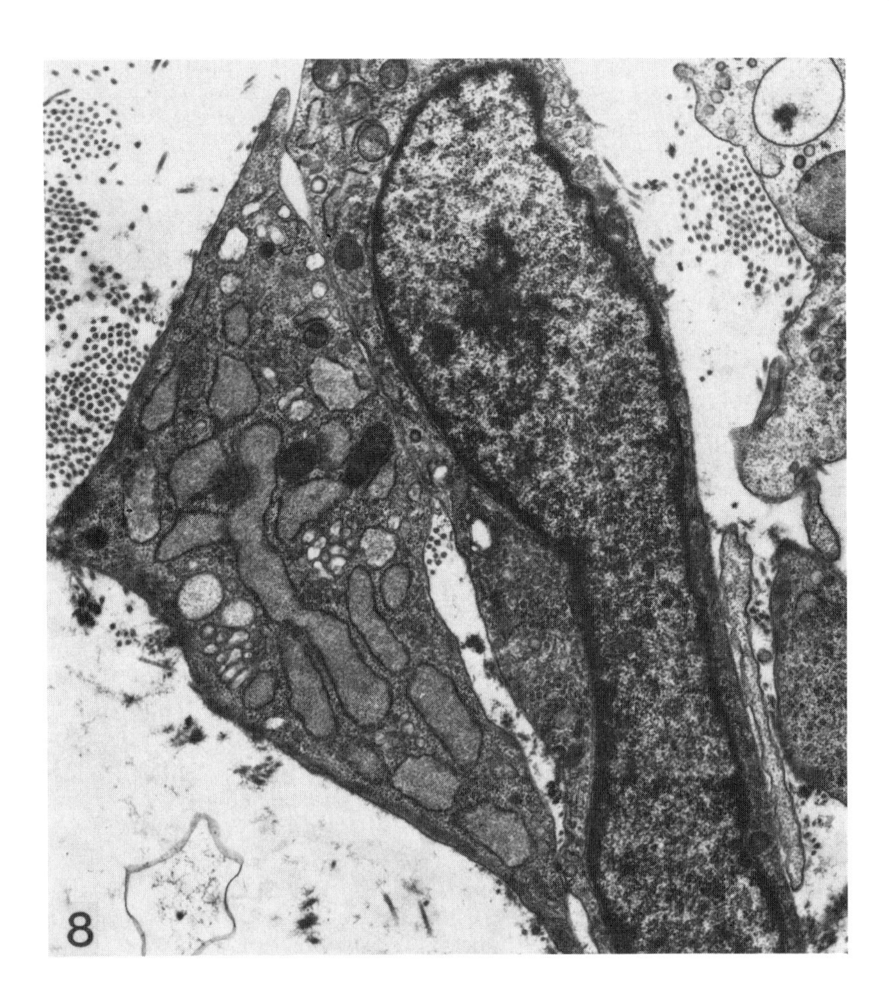

FIGURE 8. Mesenchymal tumor. Electronmicrograph of closely apposed tumor cells resembling active fibroblasts associated with collagen formation. Anastomosing channels of rough endoplasmic reticulum and bundles of actin-like microfilaments are characteristic. (Magnification × 12,250.) (From Hard, G. C. and Butler, W. H., *Cancer Res.*, 31, 348, 1971. With permission.)

the rat, nephroblastoma is a purely epithelial neoplasm presenting a rather uniform histological pattern in which neoplastic blastema is the hallmark, along with differentiation into tubules of varying stages of maturation and often, primitive avascular glomeruli.[11,18,19] RMT, on the other hand, is a purely mesenchymal neoplasm consisting of a heterogeneous range of secondary mesenchymal cells of neoplastic type which are never found in rat nephroblastoma.

The induction of nephroblastomas and mesenchymal tumors in parallel experiments by two separate carcinogens in the Nb rat[11] serves to emphasize the many distinctions between these two tumor types, not only from the histological standpoint but from the aspects of growth and behavior. The mode of local growth is different in nephroblastomas and mesenchymal tumors. The latter is characterized by an infiltrative form of growth with individual tumor cells penetrating between the elements of the kidney parenchyma and, in the process, causing their progressive sequestration within the tumor tissue. In nephroblastoma, growth is largely expansive as is typical for epithelial neoplasms, with only a limited tendency for local infiltration at the tumor periphery. In nephroblastomas, engulfed tubular or glomerular profiles do not survive beyond the very periphery of the tumor, whereas in RMT survival

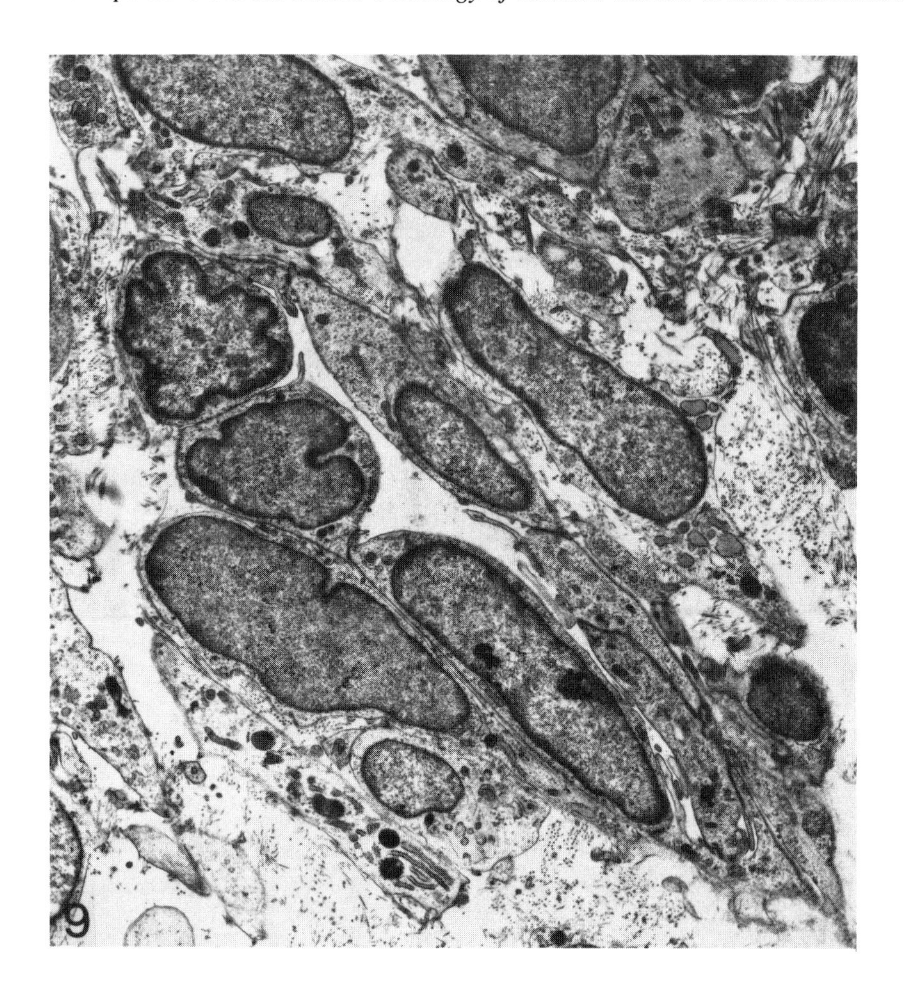

FIGURE 9. Mesenchymal tumor. Electronmicrograph of tumor cells aggregated to form a small vascular channel. The cells are joined by intercellular junctions and surrounded by basement membrane. (Magnification ×5000.) (From Hard, G. C. and Butler, W. H., *Cancer Res.*, 31, 348, 1971. With permission.)

of epithelial profiles representing preexisting parenchyma is a pathognomonic feature throughout the tumor tissue. Finally, the site of origin is clearly different for the two tumor types. In RMT many of the very early lesions appear to occupy zone 2, or the outer stripe of the outer medulla of the rat kidney,[15] i.e., the zone which contains the pars recta segments of proximal tubules, but no glomeruli. In contrast, the earliest nephroblastomas are located in zone 1, the cortex proper, and often in a subcapsular location.[11]

As frank vascular neoplasia is a frequent aspect of RMT, it could be argued that hemangiomas and other vascular neoplasms occasionally diagnosed in the kidney represent variants of RMT. However, RMT appears to be a separate entity from lipomatous tumors of the rat kidney. Although the latter have the same irregular and poorly demarcated form of RMT, their pathognomonic feature is the prominence of lipoblasts and lipocytes. In a combined personal experience of almost 2000 RMTs induced by renal carcinogens, we have never encountered fat cells as a differentiated component. It is on this basis that RMT is considered to be a distinct and unrelated entity from liposarcoma.

F. COMPARATIVE ASPECTS

RMT occurs spontaneously in the rodent but because of its confusion with nephroblastoma there is no information on the frequency rate for various strains of rat. On the other

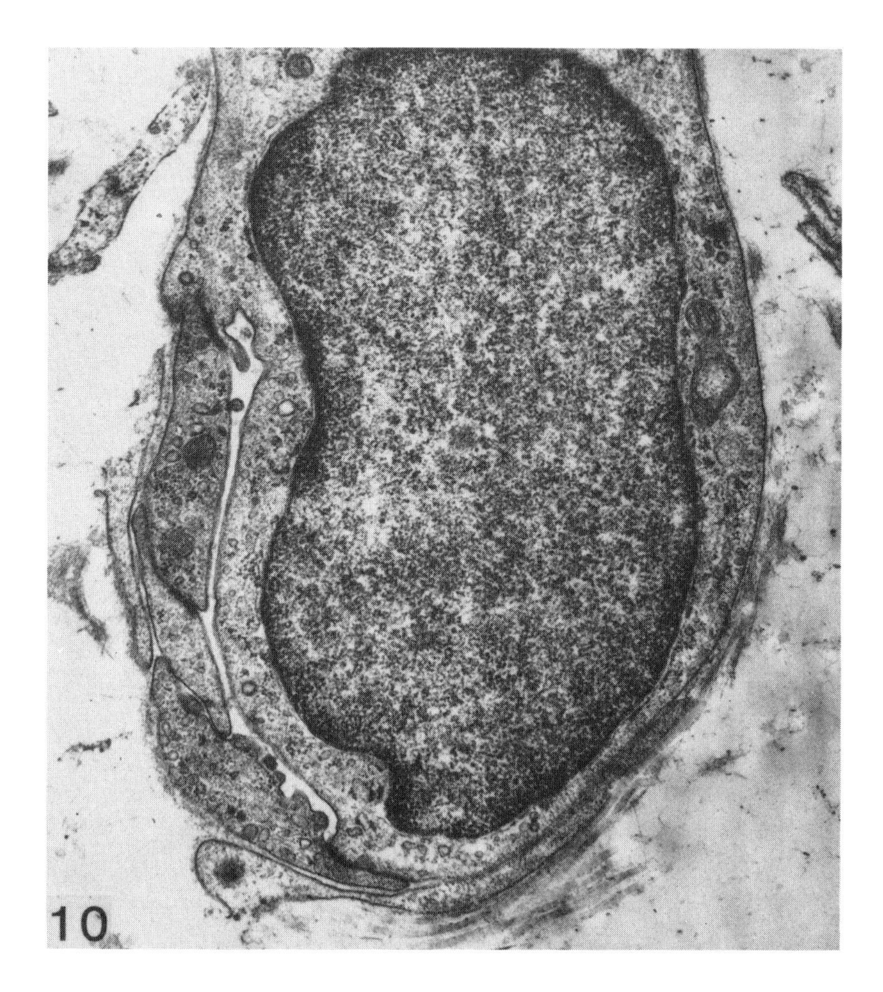

FIGURE 10. Mesenchymal tumor. Electronmicrograph of spindle cell processes joined by tight junctions and enclosing a small vessel-like cleft. (Magnification ×17,600.) (From Hard, G. C. and Butler, W. H., *Cancer Res.*, 31, 348, 1971. With permission.)

hand, next to epithelial renal cell tumors (adenoma/adenocarcinoma), RMT is the most common renal neoplasm of rats that is induced by chemical agents, particularly nitroso and related compounds. DMN has been the most extensively studied of the group and forms the basis of the model of chemical renal carcinogenesis described above. Other chemicals which induce RMT in relatively high incidence include cycasin and its derivative, methylazoxymethanol,[20,21] ethyl methanesulfonate,[22] 1,2-dimethylhydrazine,[23] streptozotocin,[24] *N*-methylnitrosourea,[25] and transplacentally administered *N*-ethylnitrosourea.[26] The latter compound is able to induce both RMT and nephroblastoma in rats, calling for particular care in diagnosis.[11]

Although there is no tumor of man which appears to reflect the entire histological complexity of the rat RMT, the human counterpart may be represented in several tumor types. First, in the small proportion of Wilms' tumors which display bipotential differentiation into neoplastic secondary mesenchyme as well as blastema/epithelium, there is a similarity between rat RMT and the human connective tissue component by virtue of the range of neoplastic cell types which can include spindle cells, smooth muscle fibers and striated muscle. Probably of closer relationship is congenital mesoblastic nephroma of infancy,[27] a mesenchymal tumor which displays a very similar heterogeneous profile to RMT including fibroblastic spindle cells, smooth muscle, and sequestered remnants of preexisting tubules.

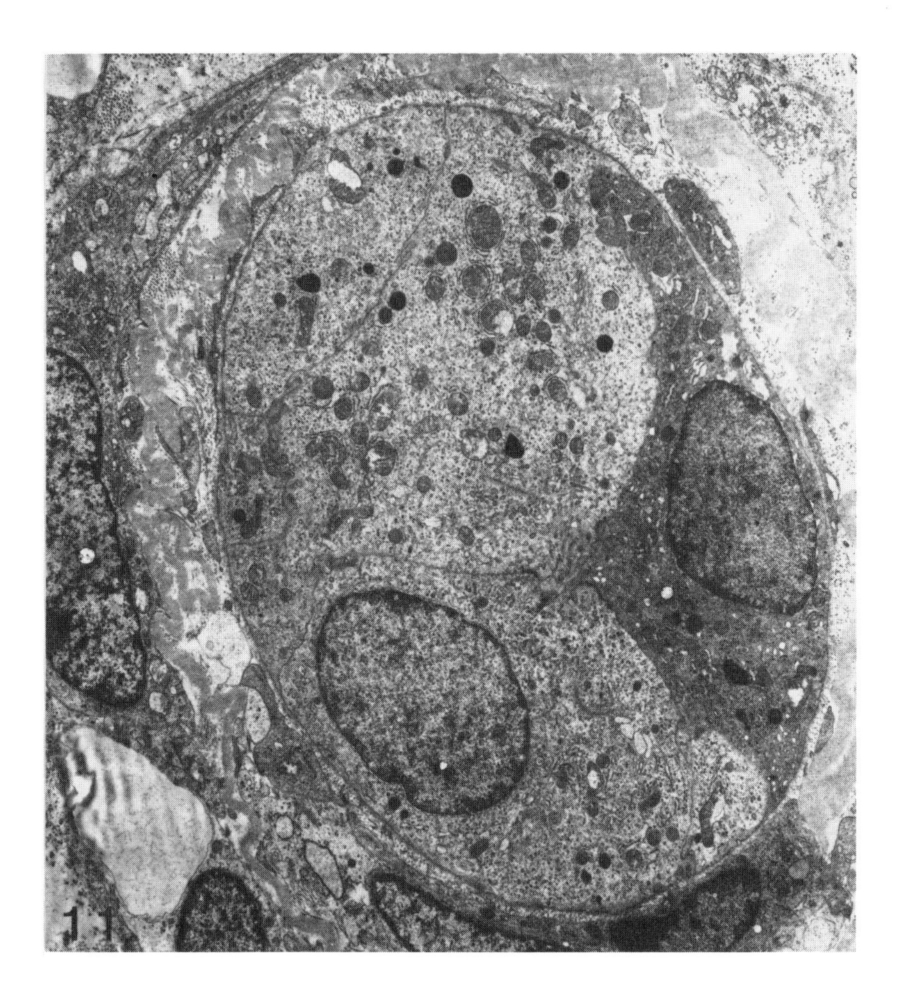

FIGURE 11. Mesenchymal tumor. Electronmicrograph showing a compressed residual tubule within a fibrosarcomatous area. (Magnification ×4500.) (From Hard, G. C. and Butler, W. H., *Carcer Res.*, 31, 348, 1971. With permission.)

This human counterpart does not usually behave as aggressively as in the rodent. Finally, the histological appearance of bone-metastasizing renal tumor of childhood[28] (or clear cell sarcoma[29]) is similar to aspects of rat RMT, in consisting of sheets of mesenchymal cells supported by a rich vascular network but also a tendency for heavy collagen deposition, liquefaction and inclusion of tubular and cystic profiles. This analogy has prompted radiographic examination of the skeletons of rats bearing large RMTs, but no evidence of an association with bone metastasis was found in the rodent tumor (Hard, unpublished observations). Nevertheless, the rat RMT induced by DMN appears to be a useful model representing a range of tumors of childhood at the mesenchymal end of what is still regarded by many as the Wilms' tumor spectrum.

Among the lower animals, renal tumors of mesenchymal type are rare and their contradistinction to nephroblastoma is not as certain as in the rat. Thus, adenoleiomyofibromatous hamartoma in the ring-tailed lemur, *Lemur catta,*[30] embryonal sarcoma in the cat,[31] and renal tumors of a polymorphic sarcomatous type in mice,[32] are possibly similar to the rat tumor. Likewise in sheep and dogs, but under the classification of nephroblastoma, are renal tumors consisting solely of sarcomatous tissue, smooth muscle, and/or cartilaginous metaplasia.

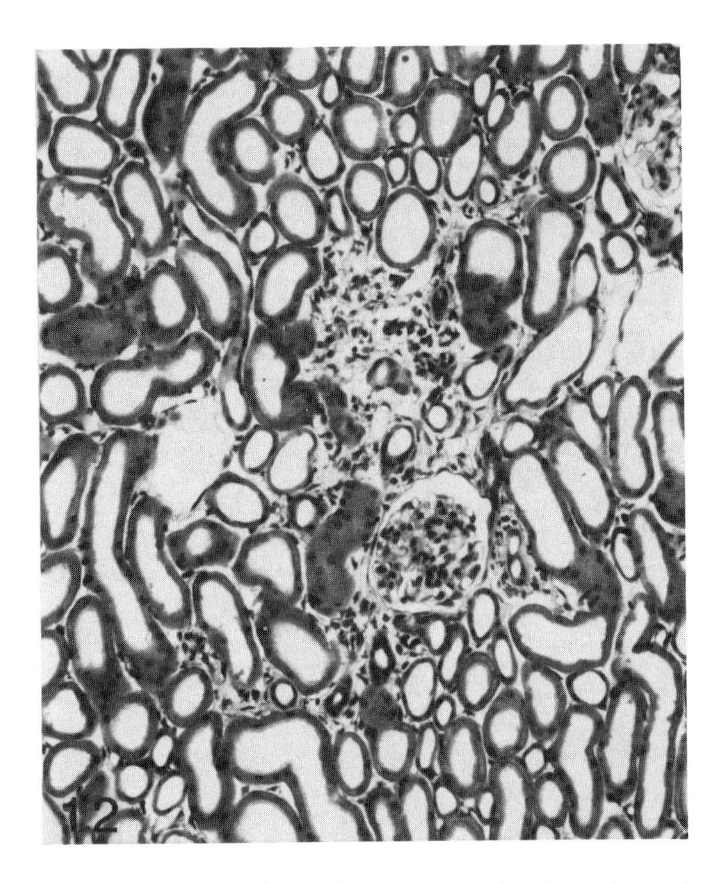

FIGURE 12. Hypercellular focus in the renal cortex 4 weeks after treatment with DMN. (H and E; magnification ×110.) (From Hard, G. C. and Butler, W. H., *Cancer Res.*, 30, 2806, 1970. With permission.)

IV. CORTICAL EPITHELIAL TUMOR

As determined by our investigation into the age dependency of renal tumors induced by DMN, cortical epithelial tumors of the renal parenchyma occur more frequently when the carcinogen is administered to the mature, rather than juvenile, rat.[9] Based on this age dependency, a high frequency model for renal cell tumors in the rat has been recently developed.[10] Using the outbred Crl:(W)BR rat at 9 to 10 weeks of age, 30 mg/kg of DMN administered intraperitoneally after a 5-d period of protein deprivation (high sugar-no protein diet) induces an epithelial renal tumor incidence in excess of 90%. Of the lesions, 70% were classifiable as adenocarcinomas or carcinomas thus providing a highly suitable model for subcellular investigations of renal carcinogenesis. Although preceding the refinement of this epithelial tumor system, our various studies have characterized the histopathology,[4,12] ultrastructure,[14] and aspects of the pathogenesis[34] of DMN-induced renal cell tumors.

A. HISTOPATHOLOGY

Epithelial tumors of the renal parenchyma induced by DMN are essentially identical to those induced by other renal carcinogens. When viewed grossly, they appear as rounded, well-circumscribed lesions (Figure 17) owing to their expansive mode of growth and, in the larger specimens, the formation of an encircling pseudocapsule. On section, these tumors are soft and fleshy. Such characteristics serve to distinguish them from RMT which are irregular and poorly delineated from the surrounding tissue.

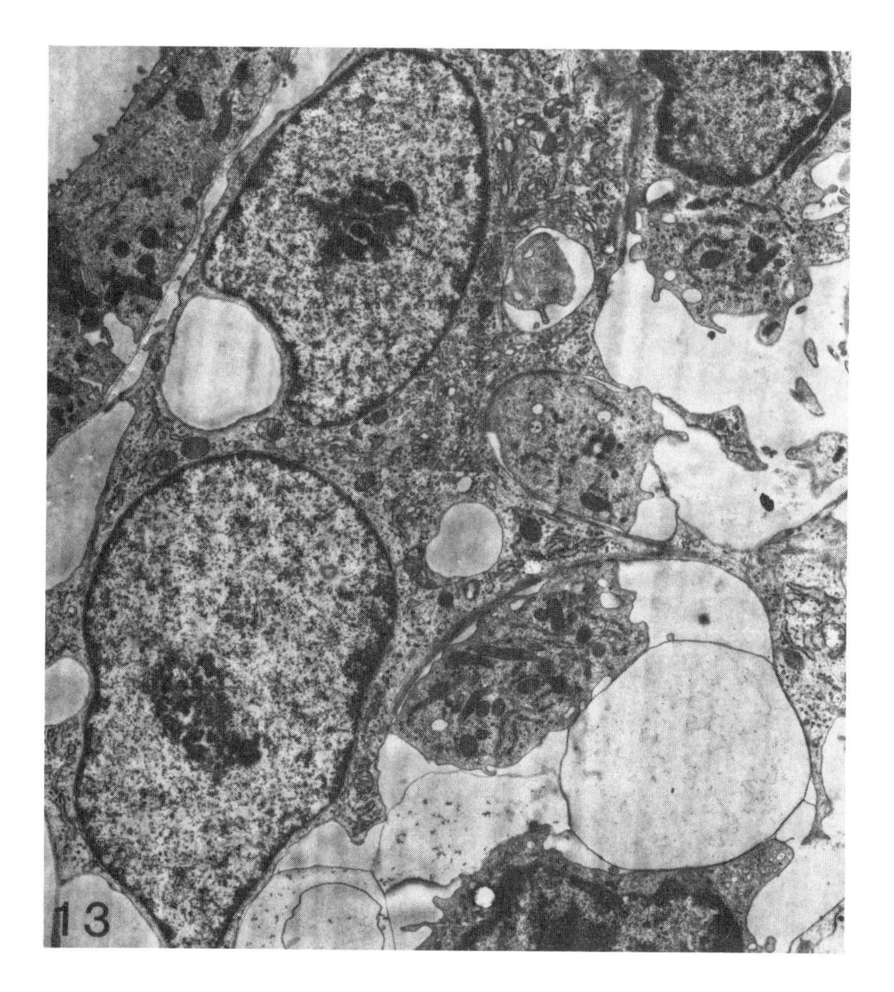

FIGURE 13. Large abnormal fibroblast-like cell within a hypercellular focus 3 weeks after treatment with DMN. The cytoplasm has abundant organelles, abnormal nucleus and hypertrophied nucleoli. (Magnification ×7750.) (From Hard, G. C. and Butler, W. H., *Cancer Res.*, 31, 337, 1971. With permission.)

Cortical epithelial tumors can be classified according to general type, according to the staining characteristics, texture, and shape of the tumor cells, and according to histological organization within the tumor. The cells may have granular cytoplasm staining in either acidophilic or basophilic fashion, or clear cytoplasm. Individual tumors may consist of an admixture of granular cells intermingled with islands of clear cells (Figure 18) either of which can display a uniformity in size or, conversely, pleomorphism.

The arrangement of neoplastic cells further characterizes the tumor morphologically. Tumors in which the cells form more or less distinct tubules are designated tubular adenoma or adenocarcinoma. The integrity of these structures varies from well-formed to ill-defined tubules; the latter arrangement can be referred to as acinar. Tumors in which the cells are arranged into solid aggregates without lumens each of which is separated by a scanty fibrous framework are described as lobular while those forming continuous sheets of cells without structural organization are generally designated as solid (Figure 18). Less well-differentiated tumors are highly pleomorphic and disorganized, and display such invasive properties as inclusion of normal renal elements within their substance. Frequently, tumor cells may form frond-like papillary structures (Figure 18). Such tumors are designated papillary adenoma or adenocarcinoma. If the tumor consists of a cystic space lined by adenomatous cells it

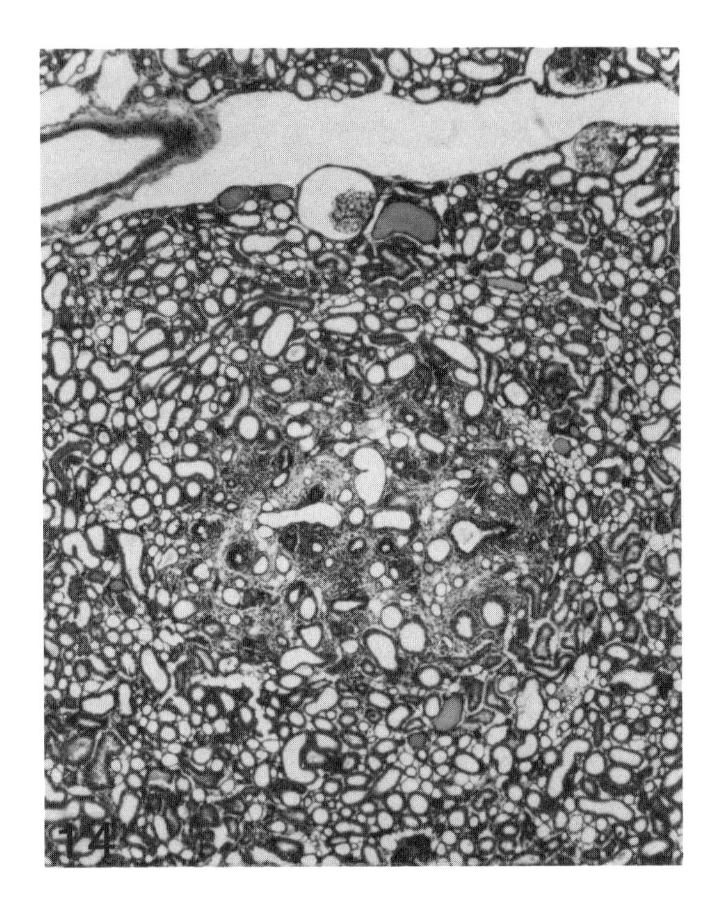

FIGURE 14. Developing focus of mesenchymal tumor within the zone 2 intertubular space at 12 wks after DMN treatment. (H and E; magnification ×32.)

may be called a cystadenoma or, if the lining is papillated, a papillary cystadenoma. These various morphological patterns may occur in different parts of the same tumor and composite terms such as tubulopapillary adenocarcinoma are frequently used. As in many animal epithelial tumors, the fibrous stroma is scanty and forms a ramifying network throughout the growth. Occasionally, however, the fibrous reaction is marked and of scirrhous form, widely separating the tubular or lobular structures of the neoplastic tissue.

B. ULTRASTRUCTURE

The cytological features of the epithelial tumors induced in the rat by DMN are similar to normal cells of the renal tubules, and in particular the proximal segments.[14] The granular tumor cells can vary from those in which there is a general paucity of organelles but for free ribosomal particles and vesicles, to darker cells with abundant mitochondria, stacks of rough endoplasmic reticulum and Golgi apparatus (Figure 19). The nucleoli are more prominent than in normal epithelium and lipid vacuoles, lysosomes, monoparticulate glycogen and cholesterol crystals can also be observed.

A further cytoplasmic feature which typifies the epithelial cells are cytoplasmic vesicles. These, often found in profusion, probably represent the apical vacuoles and vesicles of normal proximal tubule cells. In addition, peroxisomes, another organelle restricted to proximal tubule, are quite frequently observed. Cell membrane interdigitations of varying complexity are described less commonly than in the normal epithelium. Basement membrane is

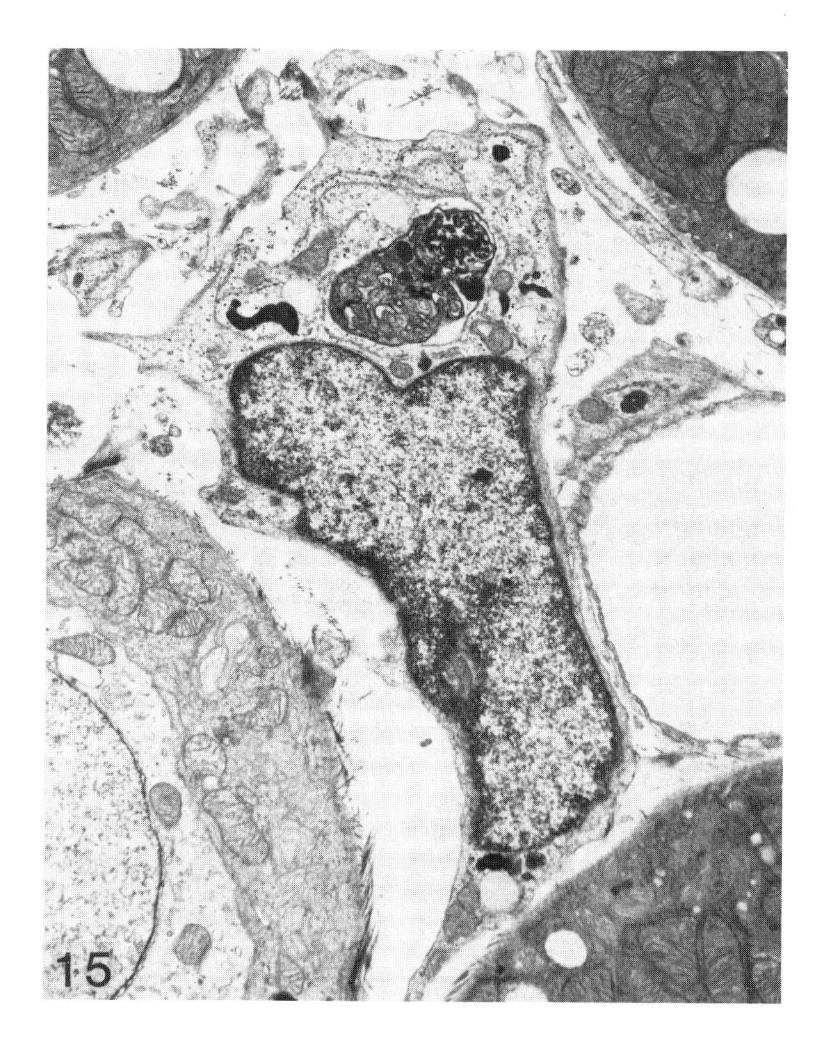

FIGURE 15. Electronmicrograph of resident cortical fibrocyte in zone 1 at 24 h after DMN treatment showing evidence of cytotoxicity with large autophagic vacuole and lipid droplets. (Magnification ×7000.) (From Hard, G. C., Mackay, R. L., and Kochhar, O. S., *Lab. Invest.*, 50, 659, 1984. With permission.)

always present, invariably as a surrounding investment to the acini and lobular aggregations of tumor cells, as well as between individual cells on occasion.

An ultrastructural hallmark in most tumors examined is the presence of microvilli organized into brush border. This feature, in particular, identifies the tumor with proximal tubule. In tumors where tubular or acinar differentiation is well developed, brush border can be oriented along the apical surfaces. However, in more solid or lobular tumors lacking a clear glandular pattern, brush border formation is equally evident but at inappropriate locations, such as the boundaries between adjacent cells or especially as focal intracellular profiles (Figure 20). Sometimes poorly organized brush border is seen as a haphazard tangle of transected microvilli within the cytoplasm just beneath the cell membrane (Figure 20). Another feature which has been frequently observed in the epithelial tumor is an intracytoplasmic lumen or canaliculus lined by sparse, nonorganized microvilli. These spaces may be solitary or multiple within individual tumor cells (Figure 21). Both the abnormal disposition of brush border and the intracytoplasmic lumina appear to be characteristic features of the neoplastic epithelium.

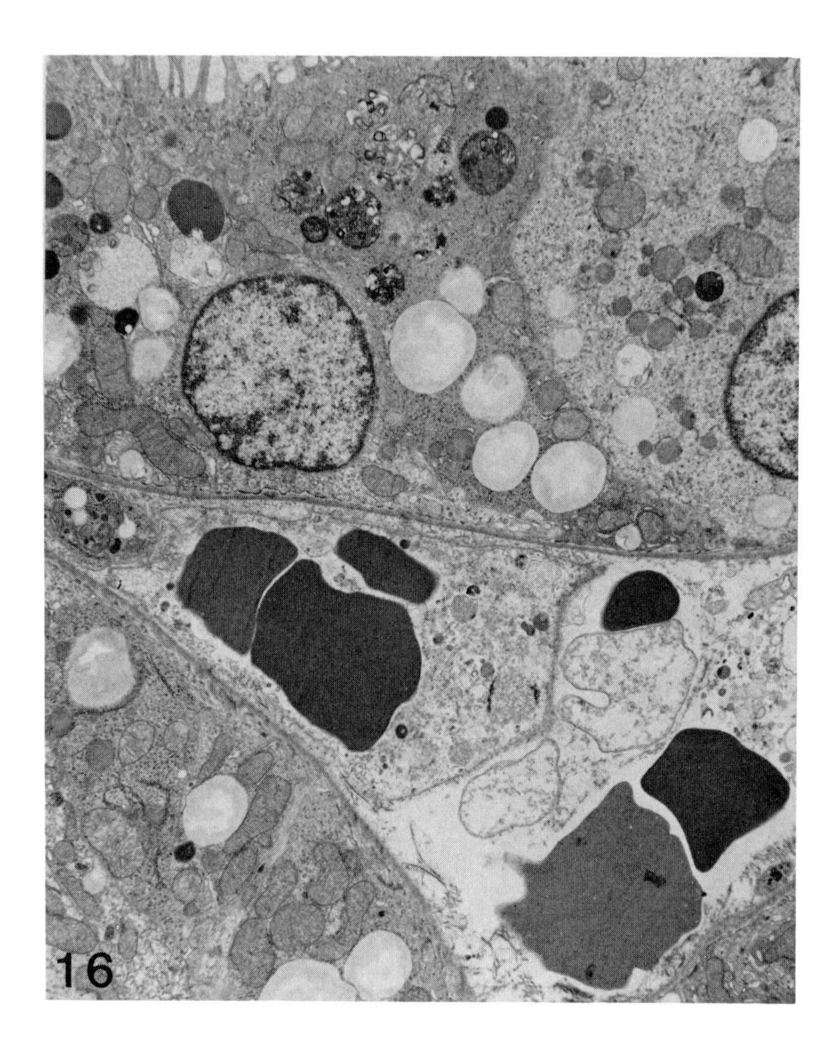

FIGURE 16. Electronmicrograph of peritubular capillary within zone 1 at 3 days after treatment with DMN showing necrosis of the endothelium and extravasation of erythrocytes and cellular debris. The adjacent P_2 segment of proximal tubule shows evidence of cytotoxicity. (Magnification ×7500.) (From Hard, G. C. and Mackay, R. L., in *Monographs in Applied Toxicology,* No. 2, John Wiley & Sons, New York, 1985. With permission.)

Thus, aspects of DMN-induced renal adenocarcinoma support the interpretation that these neoplasms arise from the proximal convoluted tubules. The ability of even poorly differentiated cells to synthesize brush border with associated vesicles that conform to apical structures of normal proximal tubules and the presence of microbodies in some well-differentiated cells represent persuasive evidence.

C. PATHOGENESIS

The latent period of induction to the stage of relatively large, palpable renal adenocarcinomas is long, at least requiring a term of 6 to 12 months.[34] Studies on the pathogenesis of renal neoplasms, using various chemical models, are in agreement that the proliferative sequence is hyperplastic tubule to microscopic proliferations usually designated as adenoma, to tumors with histological features of adenocarcinoma. Certain cellular changes may also precede this proliferative sequence.

In most models, acute toxicity is induced in the target cells of tubules prior to the appearance of proliferative foci. In the case of DMN, early ultrastructural changes of lipid

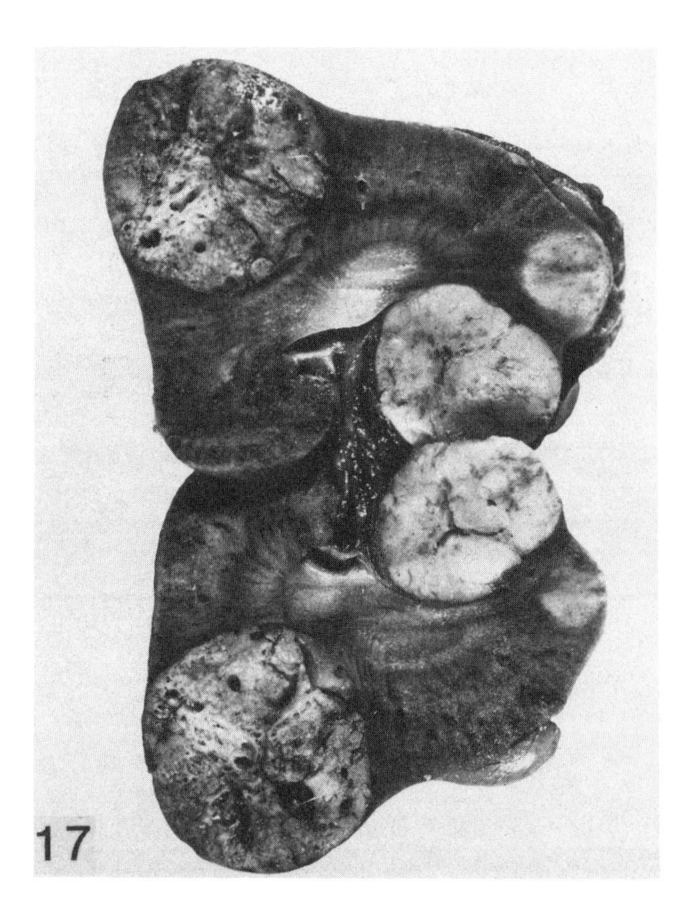

FIGURE 17. Rat kidney showing the rounded and well circumscribed appearance of several cortical epithelial tumors. (Magnification ×3.) (From Hard, G. C., in *Scientific Foundations of Urology,* Vol. 2, William Heinemann Medical Books, London, 1976. With permission.)

droplet accumulation and proliferation of smooth endoplasmic reticulum (SER) are seen within 24 h of the single carcinogenic dose in proximal convoluted tubules. This is followed by scattered single cell necrosis of the P2 tubule segment from 4 to 6 d, the same specific site where SER proliferation is observed.[17] In the ensuing period of repair, peaks of DNA synthesis and mitotic activity occur in the cortical (both proximal and distal) convoluted tubules at day 10,[35] while the acute phase of injury is fully resolved by 2 to 3 weeks. With continuously administered carcinogens such as *N*-(4′fluoro-4-biphenylyl)acetamide (FBPA)[36,37] and formic acid 2-[-4-(5-nitro-2-furyl)-2-thiazolyl]hydrazide (FNT),[38] cellular injury including lipid accumulation and scattered tubule cell necrosis is observed for an extended period from the initial weeks of exposure.

A morphological aberration preceding the development of proliferative foci with a number of renal carcinogens including DMN is conspicuous nuclear enlargement, termed karyomegaly or megalocytosis, affecting single cells within the renal tubules. DMN induces this change primarily in the proximal convoluted tubules of the cortex,[34] but the location is restricted to the P3 segment of proximal tubules with FBPA.[36] The nuclear anomaly appears to follow as a direct result of the carcinogens' early effects on cell replication and implies a block of G2 of the cell cycle resulting in progressive polyploidy. The development of karyomegaly takes several weeks in DMN carcinogenesis, becoming obvious by 6 weeks after the single treatment reflecting the slow cell turnover of tubular epithelium. Although

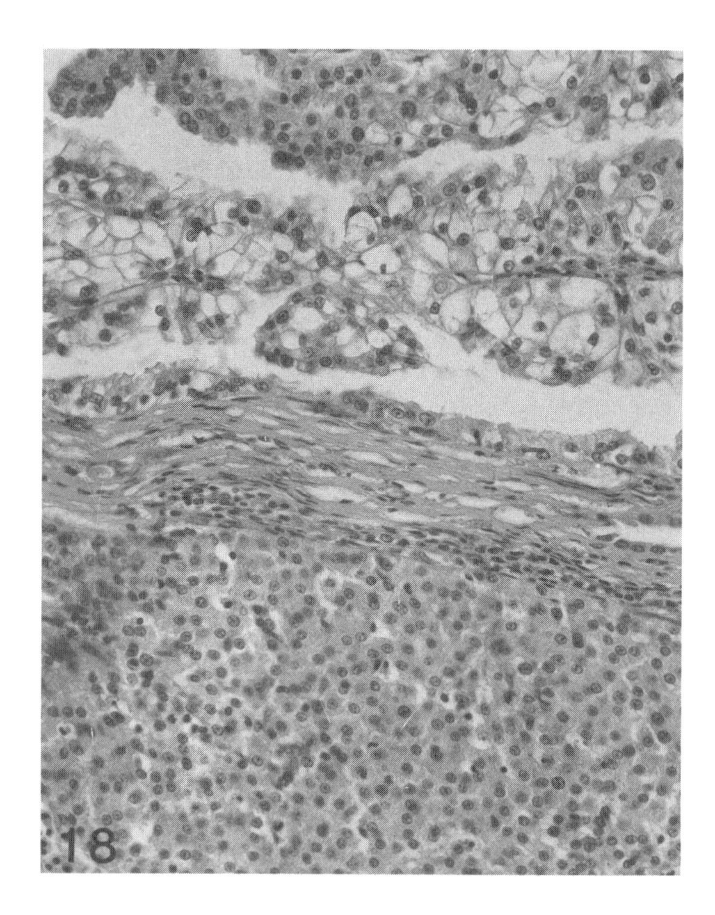

FIGURE 18. Adenocarcinoma. This tumor consists of a mixture of granular and clear cells. (H and E; magnification ×250.) (From Hard, G. C., *Carcinogenesis,* 5, 1047, 1984. With permission.)

the abnormality of cell division is not described as a constant kidney cell response to all renal carcinogens, it nevertheless might be indicative of the local action of a carcinogenic compound when encountered.[39] Despite this association, it seems unlikely that such cytokinetically arrested cells participate in the initial formation of proliferative foci.[37,39]

The time at which discrete hyperplastic foci within cortical tubules first appear depends on the chemical system. Where the carcinogen is administered in chronic dietary regimens tubular hyperplasia is not detected for approximately 6 months after the commencement of treatment. In the single dose DMN model, however, foci may be observed as early as 6 weeks but more frequently from 12 weeks after treatment[34] (Figure 22). The hyperplastic lesions occur within the proximal convoluted tubules of the cortex although their specific origin from P1 or P2 segments is not yet determined. The fact that the acute cellular changes associated with DMN toxicity in the kidney are restricted to the P2 segment of proximal tubule suggests that this might be their site of origin. In contrast, Dees and co-workers[36,37] describe the earliest tubule proliferations with FBPA as occurring in zone 2, the outer stripe of the outer medulla, as well as in the cortical medullary rays, indicating that the straight P3 segment of the proximal tubule is the site of origin for the resultant tumors. In *N*-nitrosomorpholine-induced renal carcinogenesis, Bannasch[40,41] considers that epithelial foci with glycogen accumulation in both proximal and distal tubules represent preneoplastic lesions for clear and eosinophilic renal cell tumors while basophilic tumors are preceded by proximal tubules with unusually large basophilic or chromophobic cells.

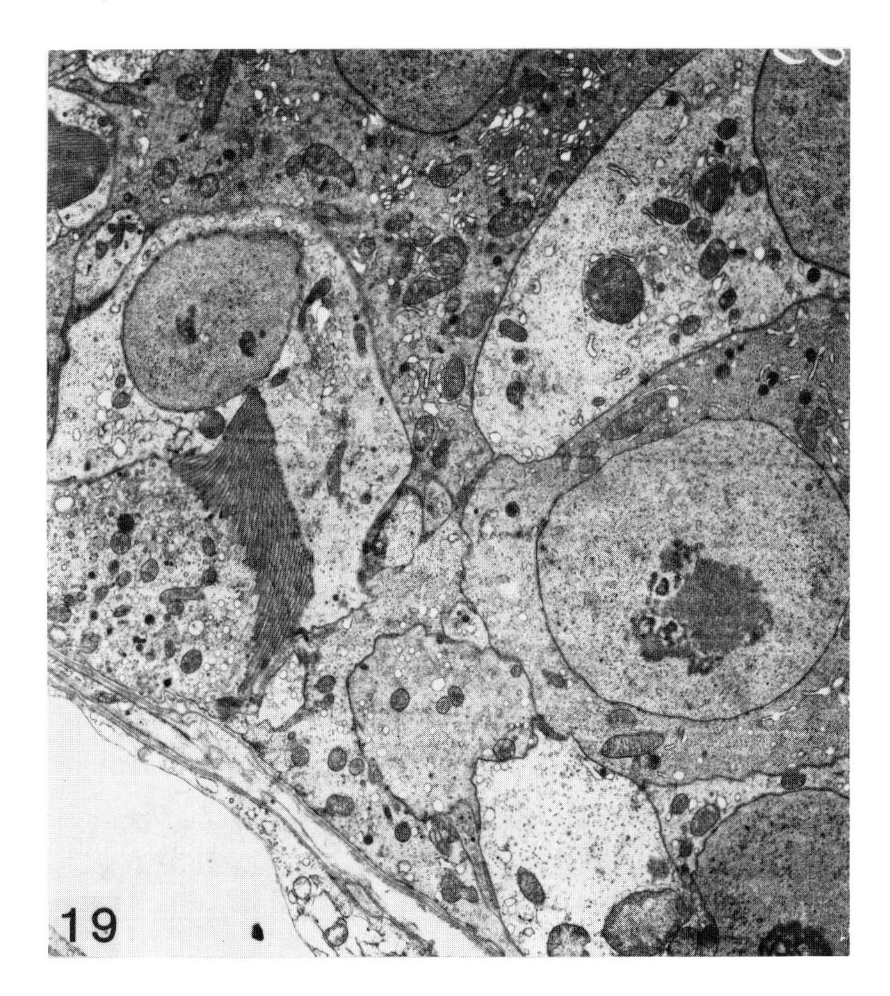

FIGURE 19. Adenocarcinoma. Electronmicrograph showing granular tumor cells with varying density of cytoplasmic organelles. Well developed profiles of brush border are present. (Magnification ×4700.) (From Hard, G. C. and Butler, W. H., *Cancer Res.*, 31, 366, 1971. With permission.)

Microscopic adenomas are presumed to develop as an increase in size of hyperplastic tubular foci. This invariably involves formation of a solid proliferating nodule of epithelial cells extending well beyond the dimensions of single tubules (Figure 23). An increase in nucleus to cytoplasm ratio is usually apparent in the nodules and mitotic activity is not infrequently observed. With FBPA this stage occurs from 36 weeks onwards[36] but may be present between 12 and 20 weeks after the single dose of DMN.[34] Electron microscopic examination of these foci does indeed reveal certain features that typify the later adenocarcinomas (Figure 24). Loss of cellular polarity, abnormal location, or internalization of microvilli and brush border, intercellular distribution of basement membrane within nodules, concentric arrangement of elongated mitochondrial profiles, and nucleolar hypertrophy and fragmentation have been described.[34,37]

Metastases from primary renal cell tumors are infrequent in most chemical models. Our recent work with DMN in the mature rat, however, indicates that epithelial tumors do metastasize providing the affected animal is able to survive for a sufficiently long period of time.[10] Statistical evaluation of tumor development has demonstrated that DMN-induced epithelial neoplasms of macroscopic dimensions increase in diameter by an average of 3 mm every 10 weeks and that the parameters of time and tumor size can be correlated with

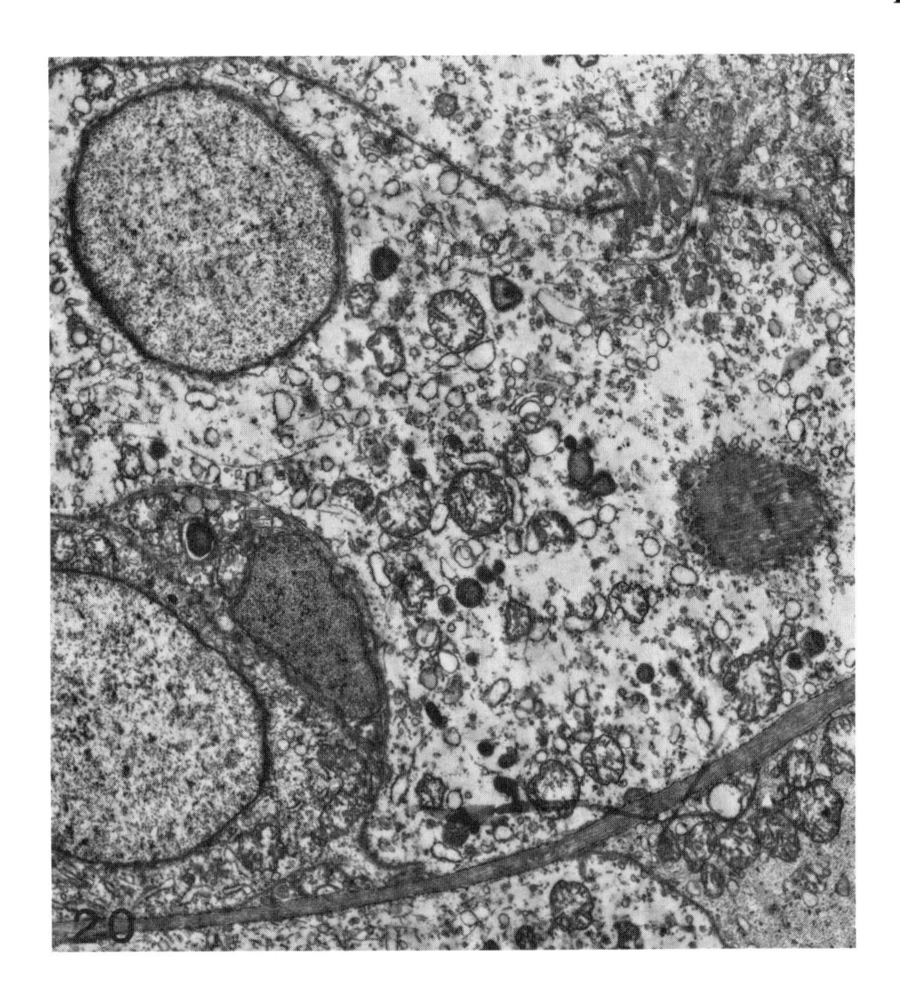

FIGURE 20. Adenocarcinoma. Electronmicrograph showing abnormal location of microvilli as an indentation within the cytoplasm. (Magnification ×7700.) (From Hard, G. C. and Butler, W. H., *Cancer Res.*, 31, 366, 1971. With permission.)

a capacity for metastasis, mainly to the lungs. Almost 50% of the tumors attaining a diameter of at least 2.5 cm metastasized. In the human disease, too, it is recognized that there is a linear relationship between the dimensions of renal carcinoma and the frequency of metastasis.[42,43]

With feeding regimes where it is necessary to administer the carcinogen to rats for extended periods, it is possible that there may not be sufficient time available from the initial induction of the neoplasm for progression to overt malignancy before the animal dies. This would explain the paucity of metastases recorded for chronic dose systems of renal carcinogenesis. The study with DMN provides, therefore, some reasonable justification for use of the terms adenocarcinoma and carcinoma, with the connotations of malignancy, to designate induced renal cell neoplasms of macroscopic size.

D. DIFFERENTIAL DIAGNOSIS

As in man, the distinction between adenomas and adenocarcinomas in the absence of invasion and metastasis is arbitrary. In the animal models studied, a sequence of lesions from small proliferative foci to large neoplasms suggest that small adenomas and adenocarcinomas are part of a continuum. Nevertheless, histological criteria are used for discriminating adenomas from adenocarcinomas. Nodular lesions, that is adenoma, less than 0.5

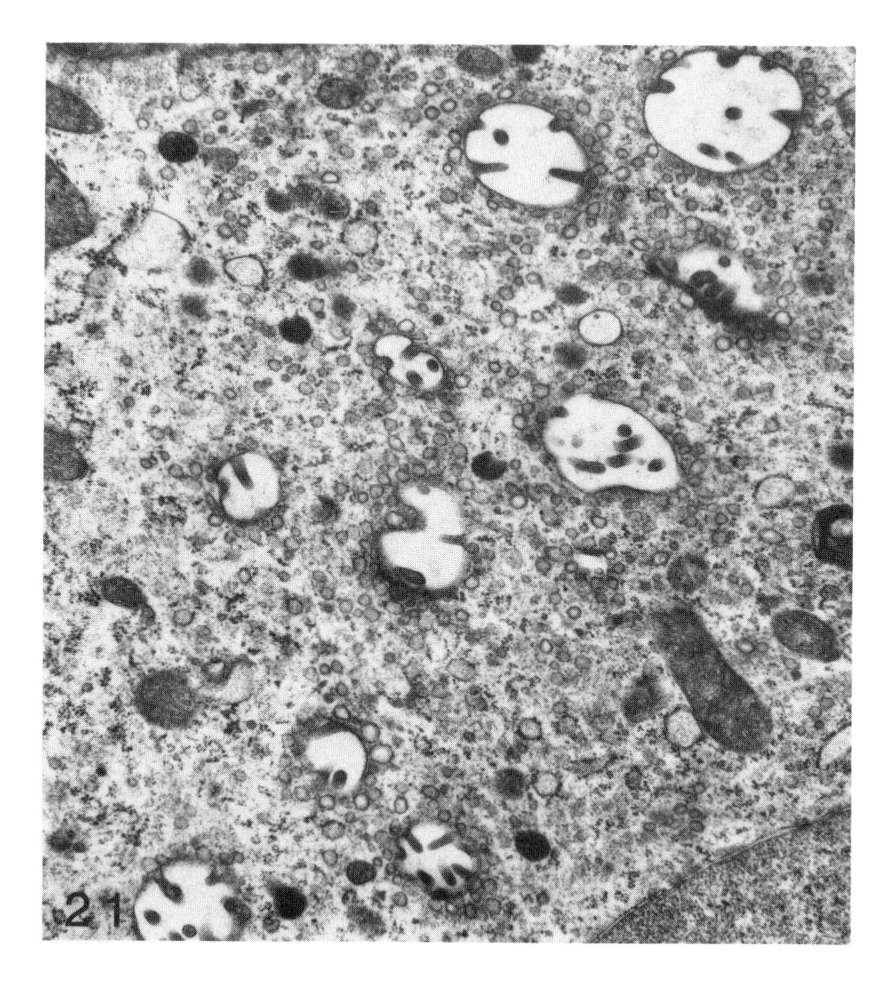

FIGURE 21. Adenocarcinoma. Electronmicrograph of cell containing multiple intracytoplasmic lumens lined by sparse microvilli. (Magnification ×22,000.) (From Hard, G. C., in *Nephron Toxicity in the Experimental and Clinical Situation,* Bach, P. H. and Lock, E. A., Eds., Martinus Nijhoff, The Hague, 1987. With permission.)

cm diameter invariably lack significant vascularization or evidence of hemorrhage or necrosis. Large neoplasms have a prominent vascular network and usually show mitotic activity, hemorrhage, necrosis, and local invasion of tumor cells beyond the pseudocapsule. As mentioned above, large adenocarcinomas metastasize to the lungs but the incidence of such metastasis is very low in most small laboratory animal systems.

It must be emphasized that the division of cortical epithelial tumors into adenoma or adenocarcinoma primarily upon size is an operational convenience and does not necessarily imply that the lesions represent separate pathways of development. However, in single dose models it is clear that not all small nodular lesions progress to invasive adenocarcinomas. Further study is necessary to characterize nonprogressive, adenoma-like lesions in order to recognize the features which distinguish them from developing adenocarcinomas.

A similar difficulty exists in the differentiation of adenoma from nodular hyperplasia. Proliferation of the epithelial lining restricted to an individual tubule represents tubular hyperplasia. There may be an increase in dimensions due mainly to lumen dilatation. Frequently the change does not appear focal, that is as a single tubule cross-section, but extends along much of the length of an affected nephron. Where the convoluted segments of a single proximal tubule are involved, several adjacent profiles of the same tubule may give the

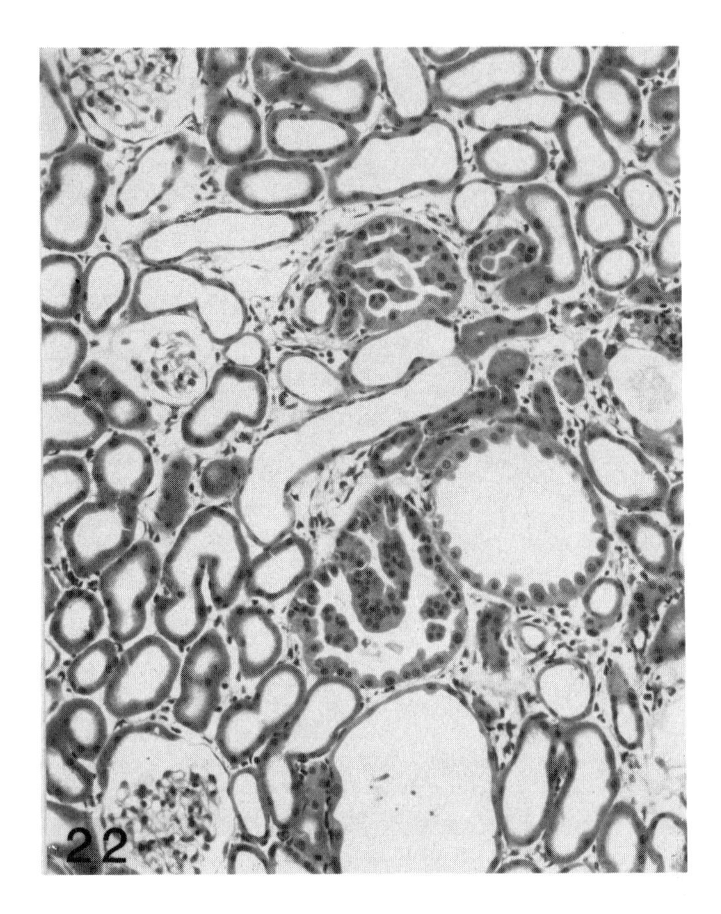

FIGURE 22. Atypical hyperplastic foci involving the proximal convoluted tubules at 12 weeks after treatment with DMN. (H and E; magnification × 190.) (From Hard, G. C. and Butler, W. H., *Cancer Res.*, 31, 1496, 1971. With permission.)

erroneous impression of a lobular adenoma. As epithelial hyperplasia extends in nodular fashion beyond the limits of a single tubule, the lesion can be viewed as an adenoma. The necessity for distinction between hyperplasia and adenoma formation is not merely of academic interest as studies with nitrilotriacetic acid indicate that tubule hyperplasias can be reversible.[44]

E. COMPARATIVE ASPECTS

Since Sempronj and Morelli[45] first showed in 1939 that kidney tumors could be produced "at a distance" in laboratory rats by a parenterally administered chemical, β-anthraquinoline, over 100 chemical compounds of diverse structure have been recorded as inducing kidney tumors in this species. On the other hand, the frequency of spontaneous adenoma/adenocarcinoma recorded for different strains of rats is low with Wistar,[46] Fischer 344,[47,48] and Osborne-Mendel,[49] all showing similar incidences of 0.24% for males and 0.06% for females. This is higher than the spontaneous frequency (0.01%) found for feral rats earlier in this century.[50,51] In contrast to rats, mice, and hamsters, spontaneous renal cell tumor is extremely rare in the laboratory rabbit, there being only one report in the literature.[52] With the exception of a single "well differentiated" kidney neoplasm encountered in a group of rabbits treated chronically with DMN,[53] we are not aware of any unequivocal reports of the chemical induction of this renal tumor type in the rabbit. Furthermore, the entity appears yet to be recorded in the guinea pig.

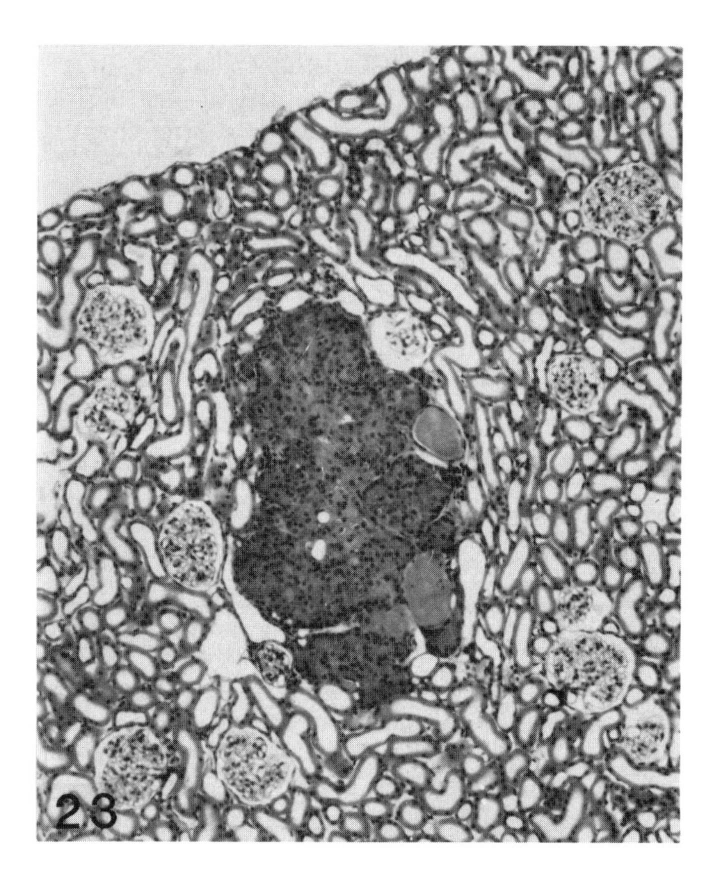

FIGURE 23. Developing cortical epithelial tumor within the renal cortex forming a solid nodule of proliferating epithelial cells at 14 weeks after treatment with DMN. (H and E; magnification ×40.) (From Hard, G. C., *Toxicol. Pathol.*, 14, 112, 1986. With permission.)

The various epithelial tumors of renal parenchyma closely resemble those of man in their morphology and progression. Ultrastructurally, human clear cell carcinomas show the same features as rat adenocarcinomas, including abnormal brush border formation, bizarre mitochondria, hypertrophied nucleoli, glycogen, and lipid.[54]

V. CELL CULTURE MODELS

In order to gain further insight into the nature of the target cell populations in DMN carcinogenesis, the technique of cell culture has been applied to the rat kidney tumor system. Because of the inexorable tumor response initiated by a single dose of DMN, which itself is undetectable by 20 h after administration,[2] the possibility of culturing renal cell populations that would express neoplastic transformation *in vitro* was investigated. In some of the first experiments to exploit the *in vivo-in vitro* system of cell culture for organ-specific carcinogenicity, kidney cortex was isolated for conventional monolayer culture from rats receiving the single tumor-inducing dose of DMN.[55,56]

Cell lines were cultured from DMN-treated rats at two stages in the carcinogenic sequence. In the first place, cells isolated from advanced renal mesenchymal tumors were established *in vitro* as continuously growing mesenchymal cell lines.[57,58] Second, the isolation of kidney cells into culture from rats within a few hours or days of the single dose of DMN resulted in the expression of morphological transformation in mesenchymal cells usually at

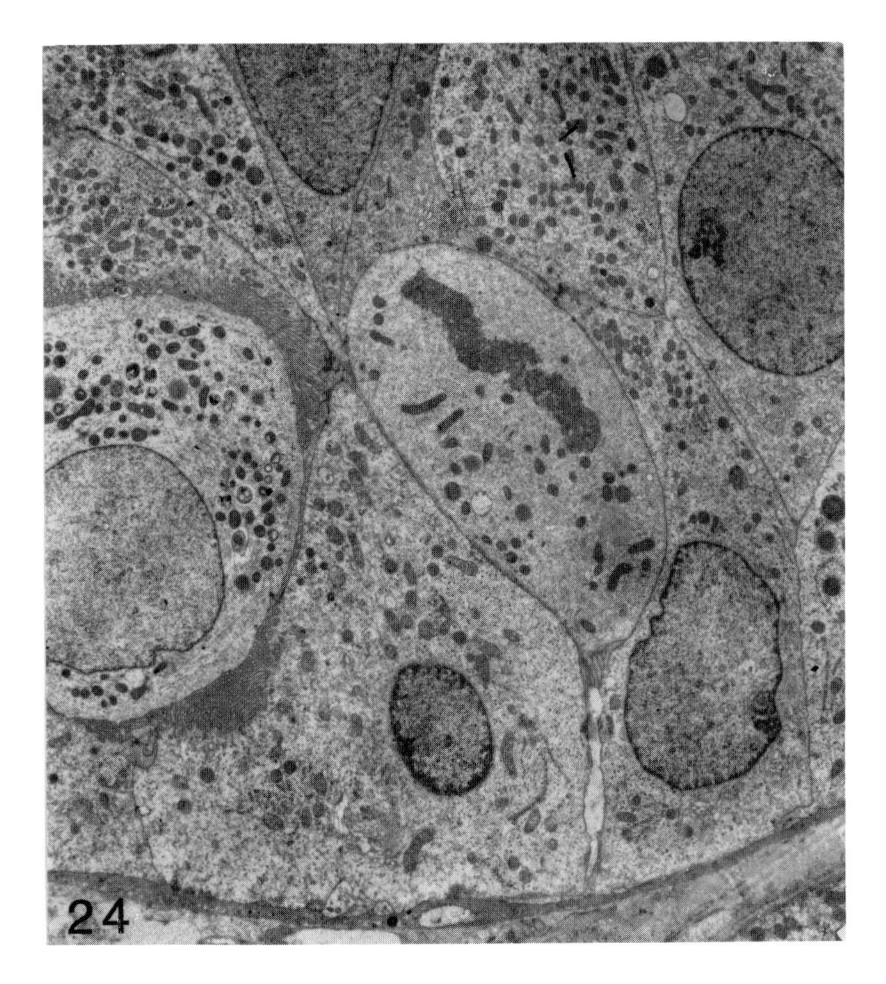

FIGURE 24. Electronmicrograph of an early cortical epithelial tumor showing features similar to Figure 19 and including abnormal brush border and a mitotic cell. (Magnification ×3800.) (From Hard, G. C. and Butler, W. H., *Cancer Res.*, 31, 1496, 1971. With permission.)

the fifth subculture.[56,59,60] These cell populations of differing temporal isolation possessed in common, fibroblast morphology (Figure 25), and altered growth properties consistent with neoplastic cells *in vitro,* including increased plating efficiencies, enhanced incorporation of [^{3}H]thymidine, colony formation in semisolid media, formation of multicellular tumor spheroids in suspension, and cytoagglutinability by concanavalin A. In addition, the malignant potential in each case was validated by injecting single cell suspensions intravenously into neonatal outbred rats within 24 h of birth.[61] The major sites for tumor growth with this method were lungs, heart, and eye, and the latency frequently was very short.

The transformed mesenchymal rat kidney cell lines (designated TRKM) derived very soon after the carcinogenic insult presumably represented the same neoplastic populations of cells as were obtained many months later from the DMN-induced renal mesenchymal tumors (RRMT lines). In support, the TRKM and RRMT cell lines were comparable in terms of morphology, chromosomal complement, and polypeptide profiles.[62,63] Transmission and scanning electron microscopy on both monolayer and suspension cultures showed ultrastructural conformity consistent with mesenchymal lineage of fibroblast type (Figure 26), reminiscent of the cortical fibrocyte resident in the interstitial space of the normal kidney. Cytogenetic analysis demonstrated the presence of a marker, a large metacentric chromosome present in the majority of cells in each neoplastic line but absent from mesenchyme derived

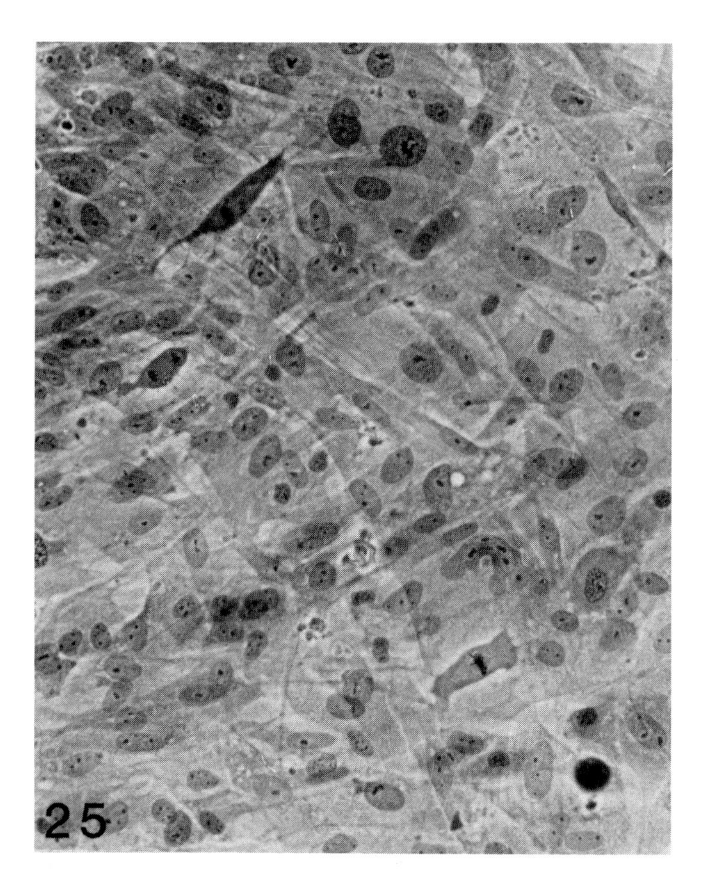

FIGURE 25. Transformed mesenchymal cell-line with fibroblastic morphology in monolayer culture originally derived from rat kidney 24 h after treatment *in vivo*. (H and E; magnification ×200.)

from normal, untreated rats.[62] Polyacrylamide gel electrophoresis and quantitative gel densitometry further demonstrated the common identity of the polypeptide profiles of transformed and tumor cell lines.[63] In contrast, normal rat kidney cells gave rise to polypeptide profiles which differed significantly in several respects from the common profiles of the tumor and transformed cell lines. The overall similarity between transformed and tumor cell lines which had been derived at temporally distinct periods from the kidneys of rats exposed to DMN *in situ* not only supported the notion that they were homologous populations, but that the *in vivo-in vitro* system of cell culture was selecting out clones of mesenchymal targets cells which had been programmed very early in the process by DMN to express malignancy *in vivo* at a later time.

In keeping with the ability of DMN to induce renal cortical epithelial tumors of tubule origin, as well as mesenchymal tumors, the *in vivo-in vitro* culture system has also produced a pure cell line of transformed epithelium, designated TRKE 1.[64] Derived 48 h after treatment with DMN using culture conditions which provided epithelium with a selective survival advantage over fibroblasts, the cell line is characterized by cohesive growth behavior typical for epithelium and the formation of hemicysts (domes) at post confluence as a manifestation of the differentiated function of transepithelial fluid transport (Figure 27). TRKE-1 cells are further characterized by structural features found in mammalian renal tubule epithelium *in vivo*, including microvilli, prominent junctional complexes (Figure 28), endocytic vesicles, microfilament tracts and even basolateral cellular interdigitations (Figure 29). The normal

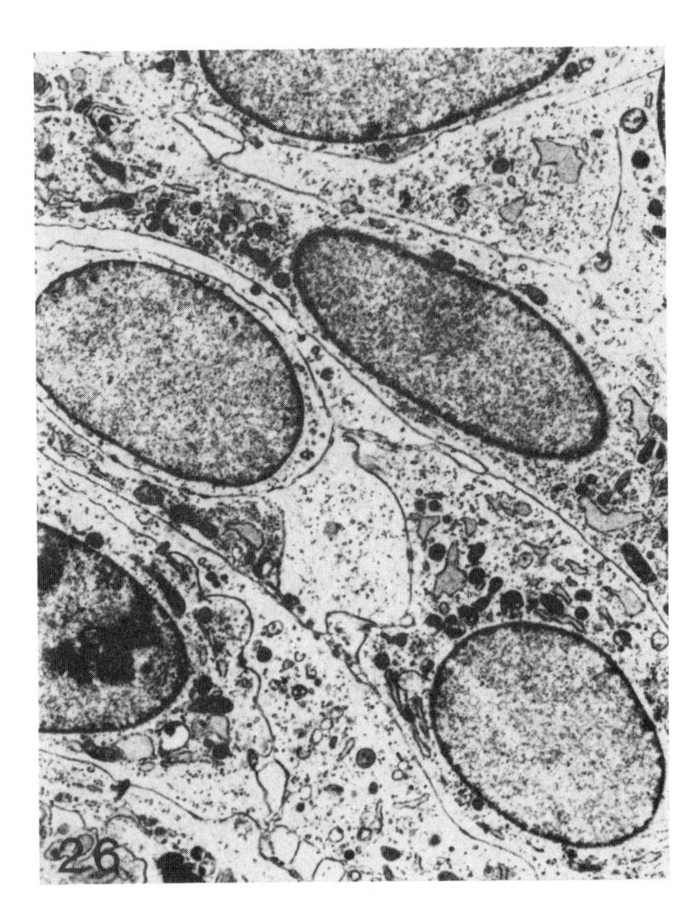

FIGURE 26. Electronmicrograph of transformed mesenchymal cells *in vitro* showing cytoplasmic features characteristic of fibroblasts (originally derived from rat kidney after treatment with DMN *in vivo*). (Magnification × 7350.)

polarity of these structures is preserved in monolayer culture, identifying the cell line in particular with proximal tubule epithelium.[65] When growth in suspension culture as multicellular tumor spheroids, TRKE-1 cells not only display microvillus-lined intracytoplasmic lumina recalling the features of renal carcinoma *in vivo,* but also a capacity for organoid differentiation into acinar structures. Representing as it does, *in vivo* chemically transformed renal epithelium, this cell line at the present time provides the only *in vitro* animal model which may be analogous to human renal cell carcinoma.

VI. CONCLUDING REMARKS

It is through the medium of potent animal systems that much basic information relating to complex disease processes is acquired. The carcinogen DMN is rather unique in this respect, providing high incidence systems for the induction of tumor types originating from two distinct cellular lineages within a single target organ. Cortical epithelial tumors induced by a single dose of DMN provide a most appropriate model for investigating the pathogenesis, and the underlying subcellular mechanisms of renal cell cancer, a malignancy which is of increasing importance in man. Although the human counterparts of RMT are uncommon renal tumor types of childhood, knowledge of this neoplasm in rats will contribute to our ultimate understanding of the enigmatic Wilms' tumor complex. The RMT model also has

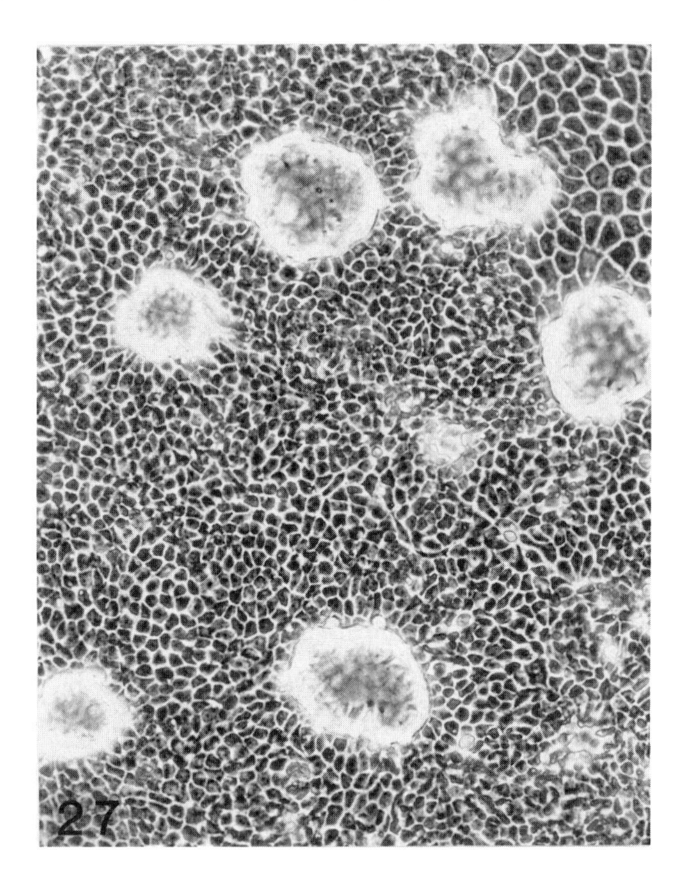

FIGURE 27. Epithelial cell-line in monolayer culture originally derived from rat kidney 48 hours after treatment with DMN. The culture is characterized by cohesive growth and the formation of hemicysts at post-confluence. (H and E; magnification ×160.) (From Hard, G. C. et al., *Cancer Lett.*, 10, 277, 1980. With permission.)

a wider application to the phenomenon of connective tissue carcinogenesis in more general terms because of the comprehensive mesenchymal spectrum which it represents. Furthermore, the capacity for deriving *in vitro* correlates comprising enriched populations which reflect the target cell types involved in DMN-induced renal carcinogenesis, provides a potentially powerful tool for elucidating cytogenetic and molecular mechanisms of cancer development relevant to the process occurring *in vivo*.

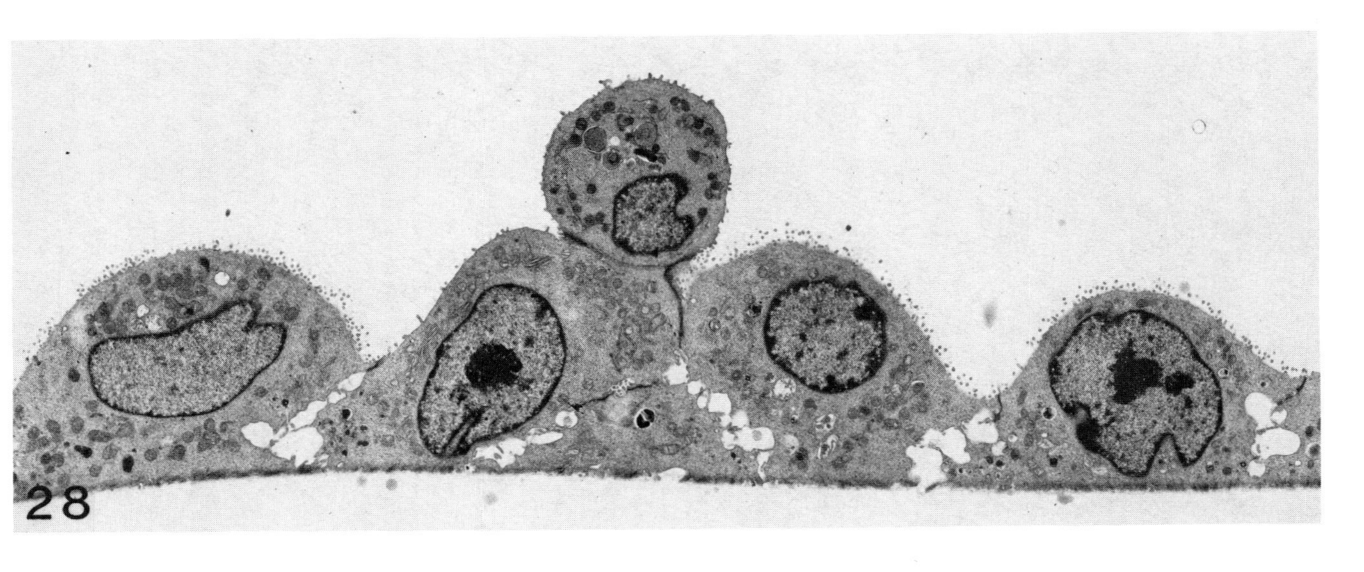

FIGURE 28. Electronmicrograph of the epithelial cell-line in Figure 27 showing cellular polarization with apical microvilli and junctional complexes. (Magnification ×3300.) (From Hard, G. C. et al., *Cancer Res.*, 43, 6045, 1983. With permission.)

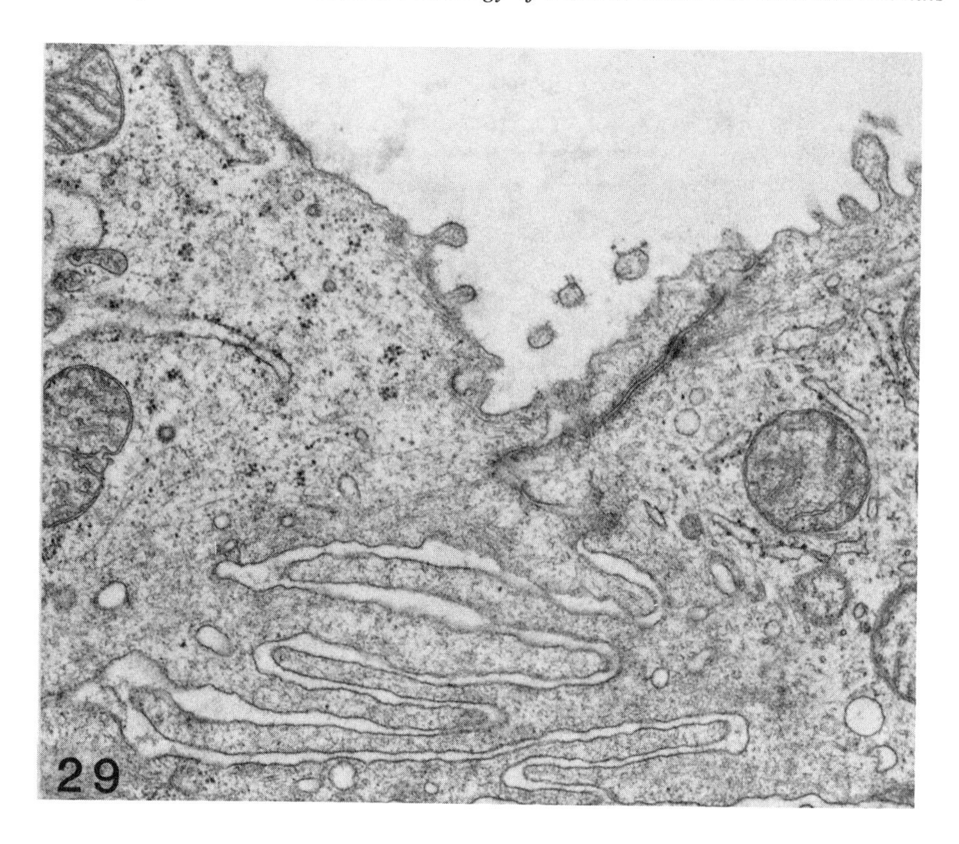

FIGURE 29. Electronmicrograph of the epithelial cell-line depicted in Figure 27 showing basolateral cellular interdigitations reminiscent of kidney parenchyma *in vivo*. (Magnification ×37,000.)

REFERENCES

1. **Magee, P. N. and Barnes, J. M.,** Induction of kidney tumours in the rat with dimethylnitrosamine (*N*-nitrosodimethylamine), *J. Pathol. Bacteriol.*, 84, 19, 1962.
2. **Swann, P. F. and McLean, A. E. M.,** Cellular injury and carcinogenesis. The effect of protein-free high carbohydrate diet on the metabolism of dimethylnitrosamine in the rat, *Biochem. J.*, 124, 283, 1971.
3. **McLean, A. E. M. and Magee, P. N.,** Increased renal carcinogenesis by dimethylnitrosamine in protein deficient rats, *Br. J. Exp. Pathol.*, 51, 587, 1970.
4. **Hard, G. C. and Butler, W. H.,** Cellular analysis of renal neoplasia. Induction of renal tumors in dietary-conditioned rats by dimethylnitrosamine with reappraisal of morphological characteristics, *Cancer Res.*, 30, 2796, 1970.
5. **Swann, P. F., Kaufman, D. G., Magee, P. N., and Mace, R.,** Induction of kidney tumours by single dose of dimethylnitrosamine: dose response and influence of diet and benzo(a)pyrene pretreatment, *Br. J. Cancer*, 41, 285, 1980.
6. **McLean, A. E. M. and Verschuuren, H. G.,** Effects of diet and microsomal enzyme induction on the toxicity of dimethylnitrosamine, *Br. J. Exp. Pathol.*, 50, 22, 1969.
7. **Terracini, B. and Magee, P. N.,** Renal tumours in rats following injection of dimethylnitrosamine at birth, *Nature (London)*, 202, 502, 1964.
8. **Murphy, G. P., Mirand, E. A., Johnston, G. S., Schmidt, J. D., and Scott, W. W.,** Renal tumors induced by a single dose of dimethylnitrosamine: morphologic, functional, enzymatic and hormonal characterizations, *Invest. Urol.*, 4, 39, 1966.
9. **Hard, G. C.,** Effect of age at treatment on incidence and type of renal neoplasm induced in the rat by a single dose of dimethylnitrosamine, *Cancer Res.*, 39, 4965, 1979.
10. **Hard, G. C.,** High frequency, single-dose model of renal adenoma/carcinoma induction using dimethylnitrosamine in Crl:(W)BR rats, *Carcinogenesis*, 5, 1047, 1984.

11. **Hard, G. C.,** Differential renal tumor response to *N*-ethylnitrosourea and dimethylnitrosamine in the Nb rat: basis for a new rodent model of nephroblastoma, *Carcinogenesis,* 6, 1551, 1985.
12. **Hard, G. C.,** Tumours of the kidney, renal pelvis and ureter, in *Pathology of Tumours in Laboratory Animals — Tumours of the Rat,* Vol. 1, Part 2, Turusov, V. S., Ed., International Agency for Research on Cancer, Lyon, IARC Scientific Publications No. 6, 1976, 73.
13. **Hard, G. C. and Butler, W. H.,** Ultrastructural analysis of renal mesenchymal tumor induced in the rat by dimethylnitrosamine, *Cancer Res.,* 31, 348, 1971.
14. **Hard, G. C. and Butler, W. H.,** Ultrastructural aspects of renal adenocarcinoma induced in the rat by dimethylnitrosamine, *Cancer Res.,* 31, 366, 1971.
15. **Hard, G. C. and Butler, W. H.,** Cellular analysis of renal neoplasia. Light microscope study of the development of interstitial lesions induced in the rat kidney by a single, carcinogenic dose of dimethylnitrosamine, *Cancer Res.,* 30, 2806, 1970.
16. **Hard, G. C. and Butler, W. H.,** Ultrastructural study of the development of interstitial lesions leading to mesenchymal neoplasia induced in the rat renal cortex by dimethylnitrosamine, *Cancer Res.,* 31, 337, 1971.
17. **Hard, G. C., Mackay, R. L., and Kochhar, O. S.,** Electron microscopic determination of the sequence of acute tubular and vascular injury induced in the rat kidney by a carcinogenic dose of dimethylnitrosamine, *Lab. Invest.,* 50, 659, 1984.
18. **Hard, G. C. and Grasso, P.,** Nephroblastoma in the rat; histology of a spontaneous tumor, identity with respect to renal mesenchymal neoplasms, and a review of previously recorded cases, *J. Natl. Cancer Inst.,* 57, 323, 1976.
19. **Hard, G. C. and Noble, R. L.,** Occurrence, transplantation, and histological characteristics of nephroblastoma in the Nb hooded rat, *Invest. Urol.,* 18, 371, 1981.
20. **Laqueur, G. L., Mickelson, O., Whiting, M. G., and Kurland, L. T.,** Carcinogenic properties of nuts from *Cycas circinalis,* L., indigenous to Guam, *J. Natl. Cancer Inst.,* 31, 919, 1963.
21. **Laqueur, G. L. and Matsumoto, H.,** Neoplasms in female Fischer rats following intraperitoneal injection of methylazoxymethanol, *J. Natl. Cancer Inst.,* 37, 217, 1966.
22. **Swann, P. F. and Magee, P. N.,** Induction of rat kidney tumours by ethyl methanesulphonate and nervous tissue tumours by methyl methanesulphonate and ethyl methanesulphonate, *Nature,* 223, 947, 1969.
23. **Sunter, J. P. and Senior, P. V.,** Induction of renal tumours in rats by the administration of 1,2 dimethylhydrazine, *J. Pathol.,* 140, 69, 1983.
24. **Horton, L., Fox, C., Corrin, B., and Sönksen, P. H.,** Streptozotocin-induced renal tumours in rats, *Br. J. Cancer,* 36, 692, 1977.
25. **Leaver, D. D., Swann, P. F., and Magee, P. N.,** The induction of tumours in the rat by a single oral dose of *N*-nitrosomethylurea, *Br. J. Cancer,* 23, 177, 1969.
26. **Turusov, V. A., Alexandrov, V. A., and Timoshenko, I. V.,** Nephroblastoma and renal mesenchymal tumor induced in rats by *N*-nitrosoethyl- and *N*-nitrosomethylurea, *Neoplasma,* 27, 229, 1980.
27. **Bolande, R. P., Brough, A. J., and Izant, R. J.,** Congenital mesoblastic nephroma of infancy. A report of 8 cases and the relationship to Wilms' tumor, *Pediatrics,* 40, 272, 1967.
28. **Marsden, H. B. and Lawler, W.,** Bone-metastasizing renal tumour of childhood. Histopathological and clinical review of 38 cases, *Virch. Arch. Pathol. Anat. Histol.,* 387, 341, 1980.
29. **Haas, J. E., Bonadio, J. F., and Beckwith, J. B.,** Clear cell sarcoma of the kidney with emphasis on ultrastructural studies, *Cancer,* 54, 2978, 1984.
30. **Jones, S. R. and Casey, H. W.,** Primary renal tumors in nonhuman primates, *Vet. Pathol.,* 18 (Suppl. 6), 89, 1981.
31. **Fitts, R. H.,** Bilateral feline embryonal sarcoma, *J. Am. Vet. Med. Assoc.,* 136, 616, 1960.
32. **Gúerin, M., Chouroulinkov, I., and Rivière, M. R.,** Experimental kidney tumours, in *The Kidney. Morphology, Biochemistry, Physiology,* Vol. 2, Rouiller, C. and Muller, A. F., Eds., Academic Press, New York, 1969, 199.
33. **Hard, G. C.,** Comparative oncology. II. Nephroblastoma in domesticated and wild animals, in *Wilms' Tumor. Clinical and Biological Manifestations,* Pochedly, C. and Baum, E. S., Eds., Elsevier, New York, 1984, 169.
34. **Hard, G. C. and Butler, W. H.,** Morphogenesis of epithelial neoplasms induced in the rat kidney by dimethylnitrosamine, *Cancer Res.,* 31, 1496, 1971.
35. **Hard, G. C.,** Autoradiographic analysis of proliferative activity in rat kidney epithelial and mesenchymal cell subpopulations following a carcinogenic dose of dimethylnitrosamine, *Cancer Res.,* 35, 3762, 1975.
36. **Dees, J. H., Heatfield, B. M., Reuber, M. D., and Trump, B. F.,** Adenocarcinoma of the kidney. III. Histogenesis of renal adenocarcinomas induced in rats by *N*-(4′-fluoro-4-biphenylyl)acetamide, *J. Natl. Cancer Inst.,* 64, 1537, 1980.
37. **Dees, J. H., Heatfield, B. M., and Trump, B. F.,** Adenocarcinoma of the kidney. IV. Electron microscopic study of the development of renal adenocarcinomas induced in rats by *N*-(4′fluoro-4-biphenylyl)acetamide, *J. Natl. Cancer Inst.,* 64, 1547, 1980.

38. **Ertürk, E., Cohen, S. M., and Bryan, G. T.,** Induction, histogenesis, and isotransplantability of renal tumors induced by formic acid 2-[4-(5-nitro-2-furyl)-2-thiazolyl]hydrazide in rats, *Cancer Res.*, 30, 2098, 1970.
39. **Hard, G. C.,** Morphological correlates of irreversible tumour formation, in *Mechanisms of Toxicity and Hazard Evaluation,* Holmstedt, B., Lauwerys, R., Mercier, M., and Roberfroid, M., Eds., Elsevier/North Holland, Amsterdam, 1980, 231.
40. **Bannasch, P., Krech, R., and Zerban, H.,** Morphogenese und Mikromorphologie epithelialer Nierentumoren bei Nitrosomorpholin-vergifteten Ratten. II. Tubuläre Glykogenose und die Genese von klar-oder acidophilzelligen Tumoren, *Z. Krebsforsch.*, 92, 63, 1978.
41. **Bannasch, P., Krech, R., and Zerban, H.,** Morphogenesis und Mikromorphologie epithelialer Nierentumoren bei Nitrosomorpholin-vergifteten Ratten. IV. Tubuläre Läsionen und basophile Tumoren, *J. Cancer Res. Clin. Oncol.*, 98, 243, 1980.
42. **Bennington, J. L.,** Cancer of the kidney—etiology, epidemiology, and pathology, *Cancer,* 32, 1017, 1973.
43. **Selli, C., Hinshaw, W. M., Woodard, B. H., and Paulson, D. F.,** Stratification of risk factors in renal cell carcinoma, *Cancer,* 52, 899, 1983.
44. **Alden, C. L. and Kanerva, R. L.,** The pathogenesis of renal cortical tumours in rats fed 2% trisodium nitrilotriacetate monohydrate, *Food Chem. Toxicol.*, 20, 441, 1982.
45. **Sempronj, A. and Morelli, E.,** Carcinoma of the kidney in rats treated with beta-anthraquinoline, *Am. J. Cancer,* 35, 534, 1939.
46. **Crain, R. C.,** Spontaneous tumors in the Rochester strain of the Wistar rat, *Am. J. Pathol.*, 34, 311, 1958.
47. **Goodman, D. G., Ward, J. M., Squire, R. A., Chu, K. C., and Linhart, M. S.,** Neoplastic and nonneoplastic lesions in aging F344 rats, *Toxicol. Appl. Pharmacol.*, 48, 237, 1979.
48. **Maekawa, A., Kurokawa, Y., Takahashi, M., Kokubo, T., Ogiu, T., Onodera, H., Tanigawa, H., Ohno, Y., Furukawa, F., and Hayashi, Y.,** Spontaneous tumors in F-344/DuCrj rats, *Gann,* 74, 365, 1983.
49. **Goodman, D. G., Ward, J. M., Squire, R. A., Paxton, M. B., Reichardt, W. D., Chu, K. C., and Linhart, M. S.,** Neoplastic and nonneoplastic lesions in aging Osborne-Mendel rats, *Toxicol. Appl. Pharmacol.*, 55, 433, 1980.
50. **McCoy, G. W.,** A preliminary report on tumors found in wild rats, *J. Med. Res.*, 21, 285, 1909.
51. **Woolley, P. G. and Wherry, W. B.,** Notes on twenty-two spontaneous tumors in wild rats, *(M. norvegicus), J. Med. Res.*, 25, 205, 1911.
52. **Kaufman, A. F. and Quist, K. D.,** Spontaneous renal carcinoma in a New Zealand white rabbit., *Lab. Anim. Care,* 20, 530, 1970.
53. **LePage, R. N. and Christie, G. S.,** Induction of liver tumors in the rabbit by feeding dimethylnitrosamine, *Br. J. Cancer,* 23, 125, 1969.
54. **Tannenbaum, M.,** Ultrastructural pathology of human renal cell tumours, in *Pathology Annual,* Sommers, S. C., Ed., Appleton-Century-Crofts, New York, 6, 249, 1971.
55. **Hard, G. C., Borland, R., and Butler, W. H.,** Altered morphology and behaviour of kidney fibroblasts *in vitro,* following *in vivo* treatment of rats with a carcinogenic dose of dimethylnitrosamine, *Experientia,* 27, 1208, 1971.
56. **Borland, R. and Hard, G. C.,** Early appearance of "transformed" cells from the kidneys of rats treated with a "single" carcinogenic dose of dimethylnitrosamine (DMN) detected by culture *in vitro, Eur. J. Cancer,* 10, 177, 1974.
57. **Hard, G. C. and Borland, R.,** *In vitro* culture of cells isolated from dimethylnitrosamine-induced renal mesenchymal tumors of the rat. I. Qualitative morphology, *J. Natl. Cancer Inst.*, 54, 1085, 1975.
58. **Hard, G. C. and Borland, R.,** In vitro culture of cells isolated from dimethylnitrosamine-induced renal mesenchymal tumors of the rat. II. Behavior and morphometry, *Oncology (Basel),* 30, 485, 1974.
59. **Hard, G. C. and Borland, R.,** Morphological character of transforming renal cell cultures derived from rats dosed with dimethylnitrosamine, *J. Natl. Cancer Inst.*, 58, 1377, 1977.
60. **Hard, G. C., King, H., Borland, R., Stewart, B. W., and Dobrostanski, B.,** Length of *in vivo* exposure to a carcinogenic dose of dimethylnitrosamine necessary for subsequent expression of morphological transformation by rat kidney cells, *in vitro, Oncology (Basel),* 34, 16, 1977.
61. **Hard, G. C.,** Demonstration of the tumorigenicity of transformed rat kidney cell-lines by intravenous allotransplantation in the neonate, *Int. J. Cancer,* 30, 197, 1982.
62. **Hard, G. C., Clarke, F. M., and Brown, C.,** Homology of neoplastic rat kidney cell-lines derived during the early and terminal phases of *in vivo* renal carcinogenesis, *Proc. Am. Assoc. Cancer Res.*, 21, 88, 1980.
63. **Hard, G. C., Clarke, F. M., and Toh, B. H.,** Comparison of polypeptide profiles in normal and transformed kidney cell lines derived from control, dimethylnitrosamine-treated, and renal tumor-bearing rats, with particular reference to contractile proteins, *Cancer Res.*, 40, 3728, 1980.
64. **Hard, G. C., Brown, C., and King, H.,** Isolation of a morphologically-transformed epithelial cell-line from rat kidney following an *in vivo* dose of dimethylnitrosamine, *Cancer Lett.*, 10, 277, 1980.

65. **Hard, G. C., Mackay, R. L., Martin, J. T., and Inoue, K.,** Differentiated features of a transformed epithelial cell line (TRKE-1) derived from dimethylnitrosamine-treated rat kidney, *Cancer Res.,* 43, 6045, 1983.

Chapter 5

TUMORS OF THE NEUROENDOCRINE (APUD) SYSTEM

Timothy J. Triche

TABLE OF CONTENTS

I. INTRODUCTION

The group of tumors generally referred to as tumors of the dispersed neuroendocrine system or so-called APUD tumors have several common features which warrant their inclusion as a single group of tumors.[1] The use of the acronym "APUD tumors" is a reflection of these common phenotypic characteristics. Pearse first proposed the use of the terminology APUD in a series of papers commencing in 1968,[2] perhaps most comprehensively reviewed in 1969.[3] The terminology specifically refers to the ability of the cells of each of these tumor types to apparently take up amine precursors (thus the APU) and decarboxylate same (therefore the D). In addition they share a number of ultrastructural and cytochemical characteristics which have been confirmed in subsequent years as more antibodies and assays have become available. These common properties include (1) a content of flurogenic amine such as catecholamines, dihydroxytryptamine, or others; (2) the ability to take up DOPA (dihydroxyphenylalanine); and (3) the expression of cholinesterase or nonspecific esterase. The catecholamine content is routinely demonstrable by the so-called catecholamine fluorescence technique (formaldehyde induced fluorescence (FIF) or glyoxylic acid associated fluorescence).

Originally, Pearse and others proposed that tumors and normal tissues of this series were derived from neural crest primordial cells via physical migration and differentiation during embryogenesis in multiple diffuse sites of the body.[3] This initial work was based predominantly on embryologic considerations involving the ultimobranchial body or C-cells of the thyroid.[4] Subsequently, numerous additional studies have clearly indicated that many other so-called neuroendocrine cells[4] or precursor cells are in fact not derived by physical migration of neural crest cells.[5-8] Rather, they appear to acquire a neural crest type phenotype by local induction phenomena.[9] This latter concept allows for the combined neuroendocrine and nonneuroendocrine phenotypic expression observed in numerous so-called APUD tumors which have been described and studied in detail subsequent to Pearse's original description. Thus, current concepts of the dispersed neuroendocrine system would suggest that one should use the term APUD in a generic, all-encompassing sense, but with the understanding that there is now well-documented embryologic and other evidence that this is instead a genetic phenotype which may be expressed in multiple cells throughout the body distal from and apparently nowhere related to the primordial neural crest cells of the body. Nonetheless, certain tumors of the so-called APUD system probably do physically originate therefrom, such as melanocytes and melanoma cells. The same can be said for childhood neuroblastoma. In other cases, such as islet cell tumors and neuroendocrine carcinomas of skin, the tumor cells are not demonstrably derived from neural crest cells per se.

For the purposes of the following discussion, discussed first will be those tumors with clear-cut neural crest phenotypic traits, then those with a neuroendocrine phenotype but no known or documented association with neural crest cells per se, and finally a controversial group of tumors will be discussed (Table 1). The first category includes all obviously neuronal cell types such as neuroblastoma and paraganglioma; Schwann cell tumors; and most important clinically, melanocytic tumors, especially melanoma. The second category includes pulmonary carcinoid and oat cell carcinoma, islet cell tumors, cutaneous neuroendocrine tumors such as Merkel cell or trabecular carcinoma of skin, and tumors such as mediastinal carcinoids and enteric carcinoids.

Inclusion in one or the other category is a "best guess" based on conflicting views in the literature and the author's prejudices, not on rigorous scientific study. The concept that some cells may be physically derived from neural crest cells and the others may not itself be a simplification that will not stand the test of time. However, it is difficult to imagine that neurons, melanocytes, and Schwann cells are not physical derivatives of the neural crest and therefore warrant inclusion in the first category with no equivocation. Beyond that,

TABLE 1
Neuroendocrine Tumors

Presumed neural crest derived
- Pheochromocytoma
- Paraganglioma
- Neuroblastoma
- Melanoma
- Schwannoma

Neuroendocrine phenotype
- Medullary carcinoma thyroid
- Parathyroid adenoma/carcinoma
- Pituitary adenoma
- Carcinoids, diverse locations
- ''Oat cell'' (small cell) carcinoma, diverse locations
- Islet cell tumors
  - Alpha (glucagon)
  - Beta (insulin)
  - Gastrin
  - Somatostatin
  - Others
- Neurocutaneous
  - Merkel cell (trabecular carcinoma)
  - Mixed types

Probable neuroendocrine
- PNET (bone and soft tissue)
- Peripheral neuroepithelioma
- ''Askin'' tumor
- Ewing's ''sarcoma''

however, distinctions become less clear. For example, everyone would agree that pheochromocytes of pheochromocytoma, an adrenal medulla-derived tumor, are clearly neural crest in origin due to embryologic considerations in the development of the autonomic nervous system in general and the adrenal medulla in particular. Neural tumors associated with the adrenal medulla such as neuroblastoma are likewise presumably true neural crest derivatives. However, other neural tumors such as primitive neuroectodermal tumors (PNETs) are less clearly true neural crest derivatives. They do not occur anatomically within sites of known neural crest migration, (e.g., the autonomic nervous system, carotid bodies, or other sites). Second, they have several characteristics which appear to distinguish them from true neural crest derived neuroblasts (no catecholamines or catecholamine enzymes). Thus, this category of neural tumor may be more related to the neuroendocrine tumors described above, i.e., tumor cells where a neural crest phenotype has been induced.

In this chapter, then, three separate (and somewhat arbitrary) groups of APUD tumors will be discussed. The first group includes those whose phenotype is purely neural crest. Examples include melanoma, neuroblastoma, and Schwannoma. The second group includes those which have commonly shown mixed neural crest and non-neural crest (i.e., neuroendocrine and exocrine) phenotypic traits. These tumors have a less clear-cut connection to neural crest cells in terms of embryologic development. These so-called neuroendocrine neoplasia include carcinoids of diverse type including bronchial, mediastinal, hepatic, pancreatic, and other sites. This same group will include small cell carcinoma and intermediate forms of neuroendocrine tumors. Likewise, islet cell tumors of specific type such as alpha or glucagon tumors, beta or insulinomas, and others such as somatostatinoma and gastrinoma will be discussed separately and in context with exocrine tumors of the pancreas. Cutaneous neuroendocrine tumors including the so-called Merkel cell tumor or trabecular carcinoma of the skin will be discussed as a group, and, finally, complex tumors with multiple tissue elements will be discussed as a separate group. In the latter case clear-cut association with true neural crest neoplasia as opposed to neuroendocrine tumors as discussed here is less clear. This grouping is summarized in Table 1.

Finally, a third group of highly controversial tumors often referred to as a neuroepithelioma or primitive neuroectodermal tumors will be discussed in some detail because of their recently recognized importance among childhood tumors and tumors of young adults as well as their implications for understanding neural crest associated neoplasia. In this case, the distinction between neural tube derivatives, i.e., true central nervous system tumors, and neural crest derived tumors, i.e., the APUD tumors, is far less clear.

II. APUD TUMORS IN NONHUMAN SPECIES

Although the focus of the illustrative material and discussion in the chapter to follow is overwhelmingly human oriented, the phenomenon of APUD tumors in multiple animal species is well documented.[10,11] Further, virtually all of the embryologic considerations discussed herein are based on animal model systems. However, no body of literature of comparable magnitude, size, or sophistication exists to document the interrelationships of APUD tumors in nonhuman species. Consequently the human material described herein can be used as a model system with general applicability to all vertebrate species in general. Certainly, diverse APUD tumors in multiple vertebrate species have been documented by comparative oncologists and there appears to be no particular incidence of these tumors in one species or another.[12] Thus, in general, the observations on humans can be generalized to other animal species as well with certain obvious distinctions and cautions. In particular, familial syndromes, such as familial medullary carcinoma of the thyroid, are difficult to document in animal series and specific clinical relationships between tumor product and tumor type are difficult, if not impossible, to document due to the lack of correlative laboratory studies and pathologic examination in most animal studies. Nonetheless, the principles herein described in detail should provide a guide for those who study nonhuman species as well.

III. APPARENT NEURAL CREST *DERIVED* APUD TUMORS

In this group three major categories of tumors will be discussed:

1. Pheochromocytoma and neural tumors of adrenal medullary origin
2. Pigmented tumors including melanoma
3. Schwannoma and related nerve sheath tumors

A. PHEOCHROMOCYTOMA AND RELATED ADRENAL TUMORS

Adrenal medullary cells and tumors derived therefrom have been considered true neural crest-derived APUD cells and tumors. Multiple adrenal medullary tumors exist. In children, the most common form is neuroblastoma and its differentiated forms including ganglioneuroblastoma and ganglioneuroma.[13] In adults, the most common adrenal medullary tumor is pheochromocytoma.[14] In fact, pheochromocytoma can occur in other, extraadrenal anatomic sites, but these are generally nonetheless part of the sympathetic autonomic nervous system, such as the Organ of Zuckerkandl and other sympathetic ganglia. In other sites tumors such as paraganglioma are identified; these are related in some respects to pheochromocytoma, but unlike them are generally nonchromaffin, i.e., they fail to secrete epinephrine or norepinephrine.

1. Pheochromocytoma

The classic tumor in this group then is the pheochromocytoma. This tumor distinguishes itself from all other APUD neoplasia by its overwhelming content of neurosecretory gran-

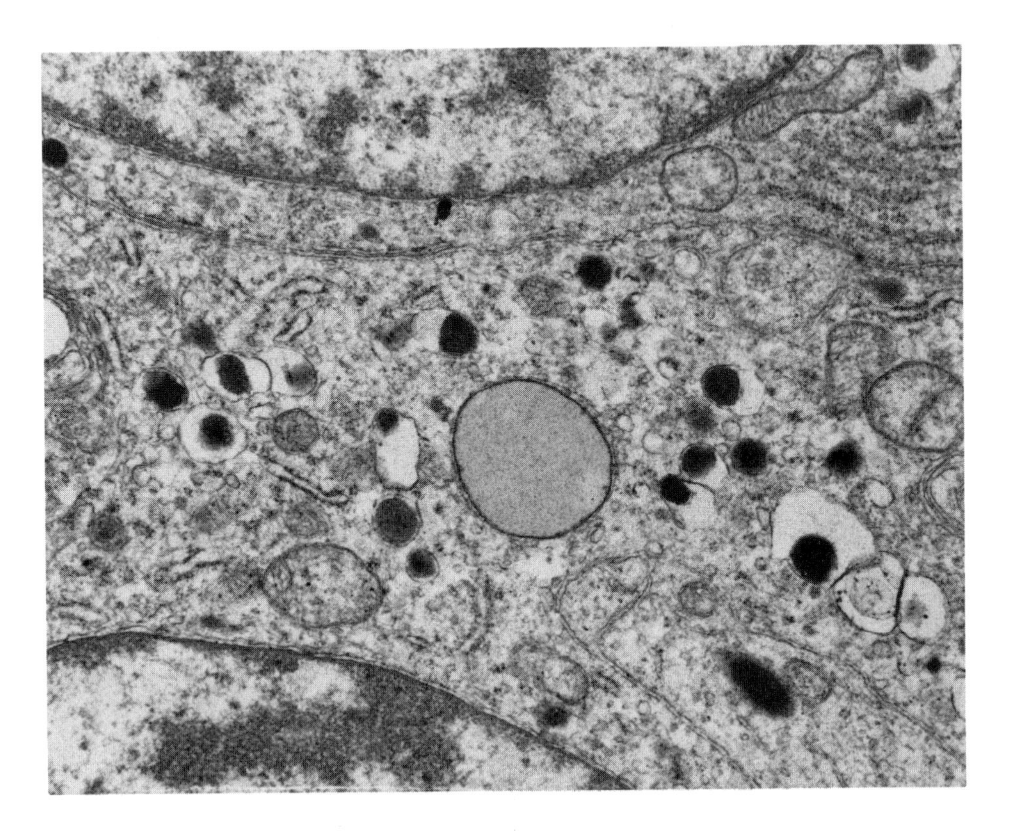

FIGURE 1. Pheochromocytoma, norepinephrine secreting. Note that the norepinephrine-containing granules, scattered throughout the cytoplasm in the center of the picture, have electron-dense cores which in favorable planes of section are noted to be eccentric. These loose-fitting granules with clear haloes are typical of norepinephrine secreting pheochromocytoma.

ules.[15] In this case they are not the typical unit membrane bound type with clear regular submembraneous halo of less than 200 nm in diameter, nor are they the larger, regular carcinoid type granules as seen in medullary carcinoma of the thyroid and carcinoid tumors of diverse location. Rather, these granules are distinctive for their content and appearance. Those which secrete norepinephrine typically are large and elongate to oval with eccentric, dense nucleoids, frequently adherent to the granule membrane (Figure 1). In contrast, those which secrete predominantly epinephrine contain granules with a central dense nucleoid and often loose-fitting membranous envelope (Figure 2). However, most pheochromocytoma have a mixture of granules (Figure 3) and complex pheochromocytomas secreting other hormonal products such as ACTH have also been described (Figure 4).[16,17] In the latter case, the clear-cut concordance between granule morphology and secretory product is less apparent, since no particular granule morphology associated with ACTH secretion has been identified.

2. Paraganglioma

Paragangliomas, though related to pheochromocytomas, are distinctly different for their content of two major conspicuous cell populations. First, two cell types occur within the tumor: a large, clear neural cell, or chief cell, and a supporting, nonneural cell, the sustentacular cell (Figure 5).[18] In some cases, dark chief cells are also seen. In this case, the structure of the tumor exactly recapitulates the normal structure of the associated specialized organ, such as carotid body.[18] The granules, as indicated in Figure 6, are typically neurosecretory in character, with a dense nucleoid, unit membrane envelope, and occasionally a clear submembranous halo.

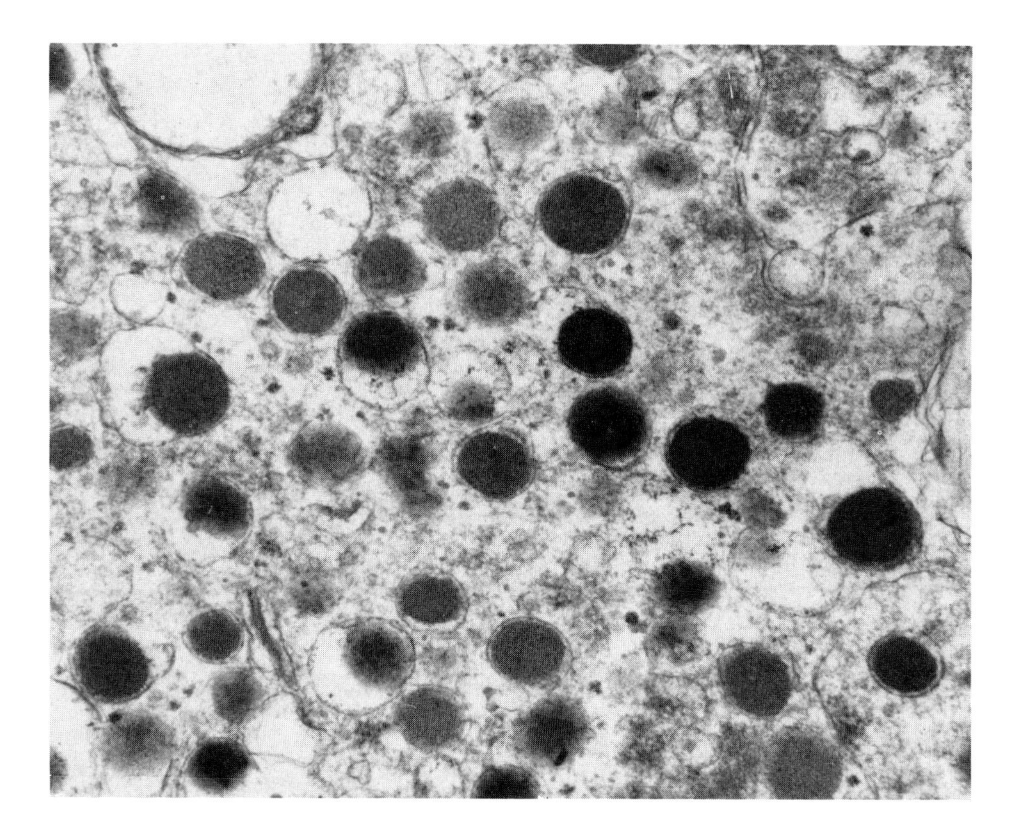

FIGURE 2. Pheochromocytoma, epinephrine secreting. In this case the epinephrine-containing granules are more tightly fitting with very little lucent halo surrounding the electron-dense core; however, the overwhelming majority appear to be concentric with the membranous envelope.

3. Neuroblastoma

In general, pheochromocytoma and paraganglioma represent differentiated neoplastic forms of sympathetic autonomic nervous system tissues. In contrast, the blastic or embryonal forms of sympathetic autonomic nervous system tumors occur almost solely in children,[19] generally less than 5 years of age, and are conspicuous for their similarity to the normal developing adrenal medulla.[20] As a group these are referred to as childhood neuroblastoma, though, in fact, the gamut of morphology in these tumors varies from those which are indistinguishable from developing adrenal medulla to those which are exact analogs of the normal tissue derivatives.[21] In the former case (primitive childhood neuroblastoma, Figure 7), round cells with numerous dense core neurosecretory granules less than 200 nm in diameter are generally observed. Also, occasional neuritic extensions are seen under favorable circumstances (Figure 8). With increasing differentiation, many of these tumors develop distinctive cell populations including cells with obvious ganglion cell (Figure 9) and Schwann cell characteristics (Figure 10). Nonetheless, these tumor cells are typically surrounded by other less-differentiated tumor cells, confirming the persistent malignant character of the tumor. In some cases, the tumor cells appear to undergo spontaneous maturation to a benign counterpart, a so-called ganlioneuroma (Figures 11 and 12).[22] In this case large numbers of ganglion cells and Schwann cells are readily identified, but no persistent blastic elements are seen.

4. Melanoma and Other Pigmented Tumors

Perhaps the most dramatic APUD tumor or family of tumors are the pigmented tumors including, most conspicuously, melanoma.[23] Generally acknowledged to be true neural crest

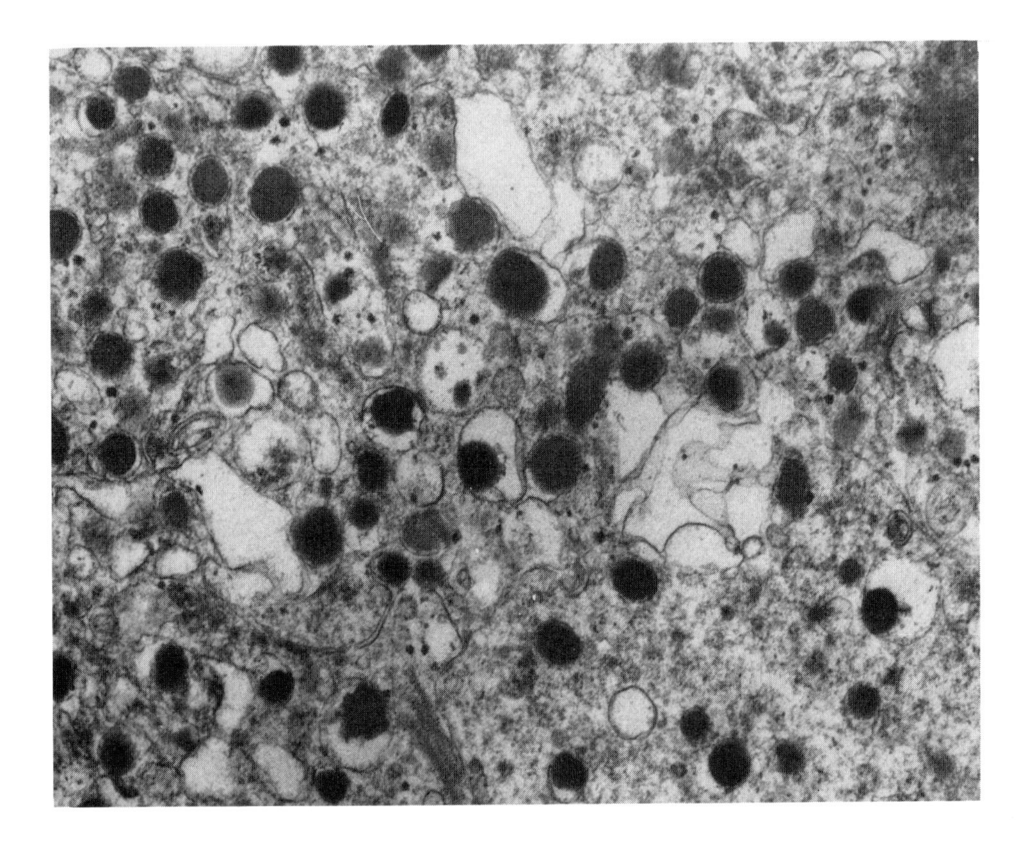

FIGURE 3. Pheochromocytoma, mixed epinephrine and norepinephrine secreting. Note that in this case about half of the population shows dense cored concentric granules with very little clear halo and the other half are predominantly more electron dense, looser fitting, and eccentric. Biochemical assay of this tumor showed a mixed epinephrine/norepinephrine biochemical phenotype.

cell derivatives, these tumors as a group all contain pigment granules known as melanosomes which may have a highly variable ultrastructure. Immature forms referred to as premelanosomes typically contain readily identifiable striated substructure (Figure 13) but more mature forms are generally uniformly or irregularly electron dense with little if any identifiable substructure (Figure 14). In addition to obvious classic forms of melanosomes, many of the malignant pigmented tumors (i.e., melanoma) contain atypical forms of melanosomes not seen in normal tissue. These are evident in Figure 15. One of the remarkable characteristics of these tumors is their origin from diverse sites throughout the body and a remarkable propensity for recurrence anytime, anywhere in the body, even as much as a quarter of a century after initial removal. Clearly, these cells appear to be migratory throughout the body, as expected of a true neural crest cell.

Unfortunately, many of the ''pigmented tumors'', especially melanoma, are not in fact pigmented. Many cases of amelanotic melanoma have been described; in such cases, premelanosomes are also often absent.[24] Figure 16 illustrates a typical amelanotic melanoma cell lacking any identifiable melanosomes. In this case, only subsequent recurrence of the tumor with obvious melanosomes and pigmentation at the time of death allow identification of the tumor cells as melanoma. For this reason, a number of alternative techniques have been employed to identify all forms of pigmented tumors. One common technique is the catecholamine fluorescence technique[25,26] which identifies all APUD tumors on the basis on their content of catecholamine precursor uptake, as noted by several authors.[27,28] Figure 17 illustrates a typical result with a melanoma touch preparation. In this case the tumor cells

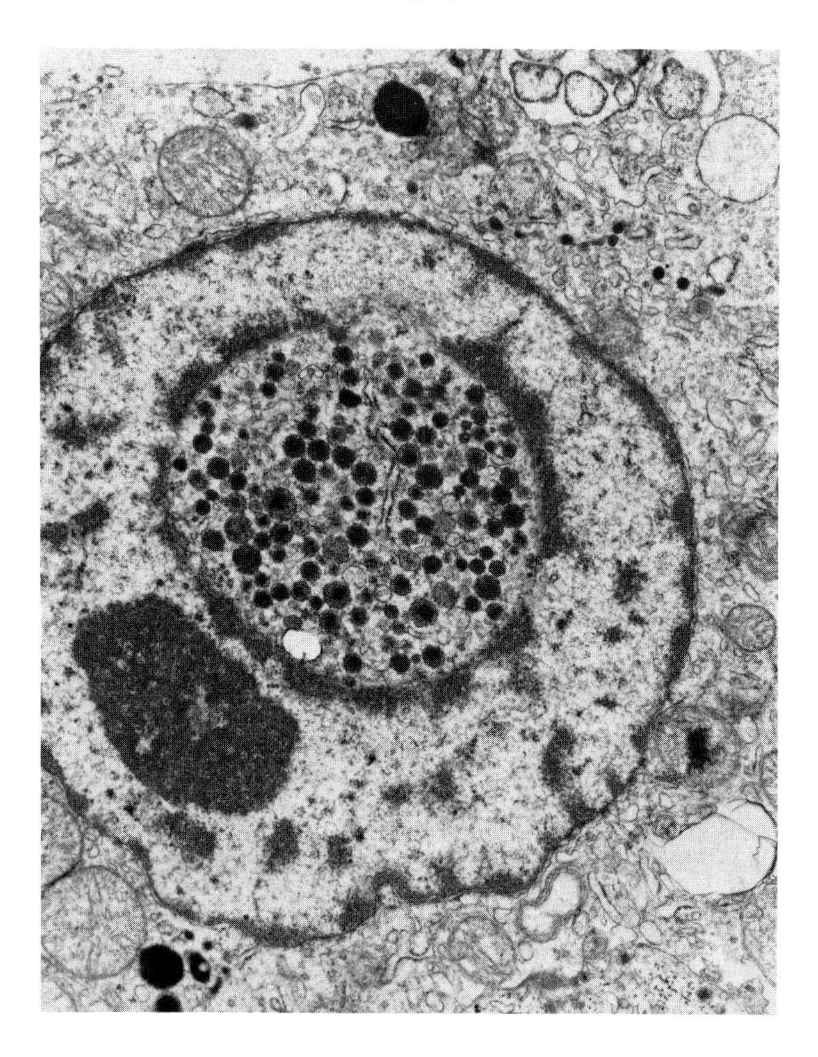

FIGURE 4. Pheochromocytoma, ACTH secreting. Note that in this particular tumor the granules are dissimilar from Figures 1 to 3, above. The secretory product, ACTH, is likewise different from those noted above. However, the specific secretory product cannot be discerned from the morphology of the granules. Note that the majority of the granules in this case happen to occur within a pseudonuclear inclusion; other granules are also noted in the surrounding cytoplasm.

were exposed to formaldehyde vapor and examined under a fluorescence microscope. Specific catecholamine fluorescence is readily appreciated in the melanoma (A) but not the lymphoma control (B).

More recently, other techniques requiring less specialized equipment have been introduced. Immunocytochemistry with antibodies against S100 protein or neuron specific enolase (NSE) has become the technique of choice. Melanoma is reportedly positive in virtually every case for S100 protein and in the majority of cases for NSE.[29,30,31] Figure 18 illustrates a typical result employing anti-S100 antibodies on paraffin embedded sections of tumor material. In this case, the patient had no antecedent history of any malignancy, yet developed a cutaneous nodule, which proved to be a lymph node infiltrated by tumor cells. As apparent here, the cells are decidedly S100 positive, yet electron microscopy of the patient's tumor cells from the same biopsy (Figure 19) revealed no evidence of melanosomes. The introduction of methodology such as this has vastly improved diagnostic reliability in cases of

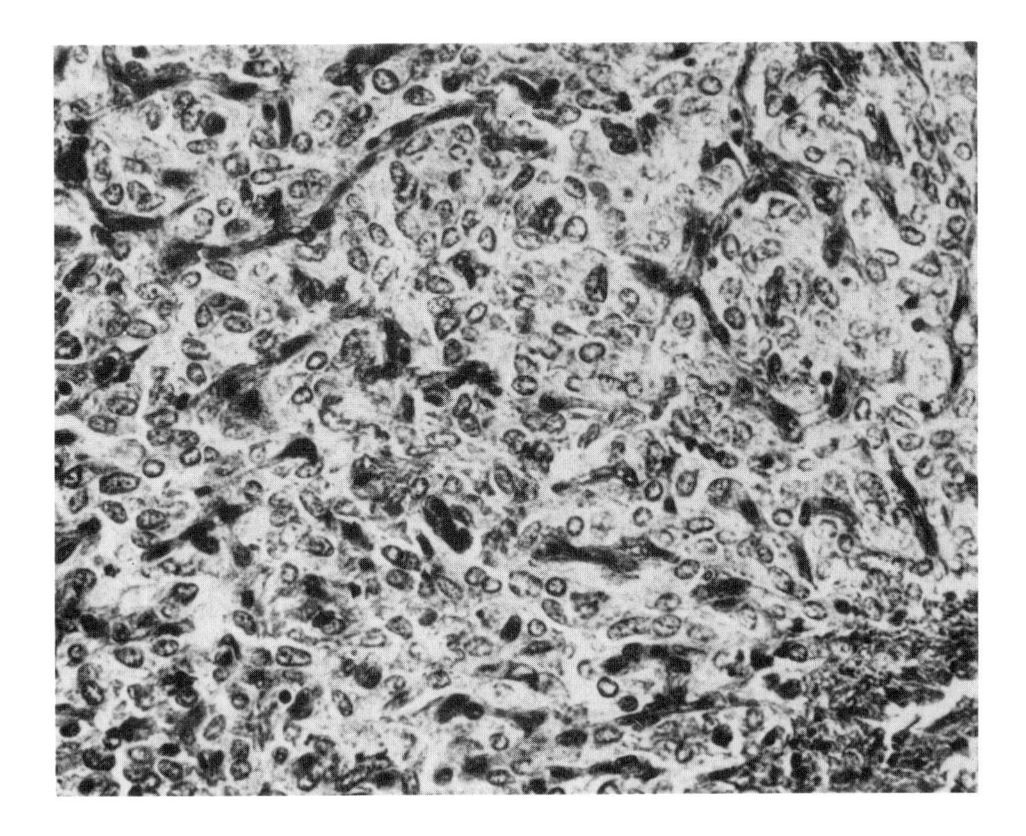

A

FIGURE 5. Paraganglioma. (A) Light microscopy. Note that a two-cell population consisting of a preponderant population of larger, clearer cells (chief cells) and a less numerous, smaller, more dense population (sustentacular cells) is noted. (B) Electron microscopy. The ultrastructure appearance of the same tumor shows a similar population of more numerous, larger, clearer cells with round nuclei and clear cytoplasm, and less frequent cells with more irregular nuclei and less abundant, more electron-dense cytoplasm. These appear to correspond to the chief and sustentacular respectively cells noted above.

amelanotic melanoma and has simplified the diagnosis of all forms of melanoma and pigmented tumors in general.

5. Clear Cell Sarcoma of Tendon Sheath and Aponeuroses (Malignant Melanoma of Soft Parts)

An intriguing intermediate tumor between the melanomas described above and the Schwann cell tumors to be described below is the so-called melanoma of soft parts. Originally described as clear cell sarcoma of tendon sheath and aponeuroses,[32] subsequent studies have clearly identified this tumor as a form of (generally) amelanotic melanoma.[33-35] However, other authors have noted a distinct propensity for Schwann cell differentiation.[36] In this respect, the tumor is no different than other reported classic melanomas which with recurrence have apparently undergone progressive ''neuritization'' or differentiation toward neural or nerve sheath elements.[37] Thus, this tumor differs from usual melanoma for its initial presentation as an apparent soft tissue sarcoma, with in some cases clear-cut Schwann cell or nerve sheath differentiation. Like melanoma, however, both the amelanotic and ''pigmented'' or melanotic forms of this tumor contain clear-cut premelanosomes (Figure 20). This tumor, then, is more appropriately considered a true neural crest-derived tumor with either intermediate or multiple histogenetic lineage. Unlike cutaneous melanoma, there is no racial

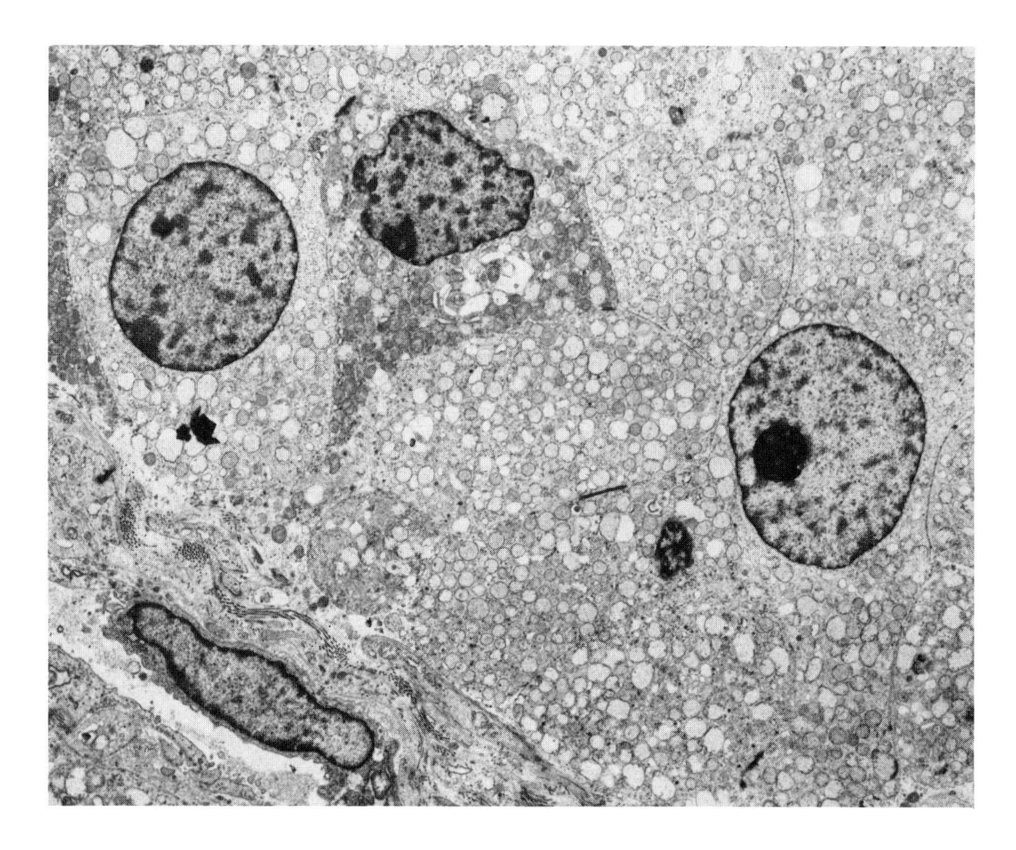

FIGURE 5B

predilection (melanoma occurring almost exclusively in fair-skinned individuals) but instead appears to occur in all racial groups with relatively equal frequency.

6. Schwannoma, Malignant Schwannoma, and Nerve Sheath Tumors

In view of the preceding, it is perhaps not surprising to encounter the third major component of known neural crest derivatives, that is the nerve sheath elements, as malignant tumors possessing Schwann cell or nerve sheath (perineural)[38] as well as intermediate forms with pigmentation as well.[39,40] In the latter category are so-called melanotic malignant Schwannoma where one encounters both clear-cut and conspicuous melanosomes (Figure 21) as well as very obvious features of Schwann cell differentiation including stellate cells surrounded by matrix with elongate slender processes typical of Schwann cells (Figure 22). Like normal Schwann cells, a continuous or (in malignant cells) discontinuous basal lamina (Figure 23) is usually present. In some cases, the cells even appear to possess unit membrane-bound neurosecretory-type granules in addition to the melanosomes (Figure 24).

IV. NEUROENDOCRINE TUMORS

This group of tumors apparently represents neuroendocrine-programmed tumors of equivocal or nonneural crest origin.[41] Alternatively, the cells may be partially induced with a neural crest phenotype or genetic program, but may persist with other ectodermal genetic expression.[42] Thus, a combined endocrine and exocrine expression and differentiation is not uncommon in this entire group of tumors, in contrast to those described above. These tumors include:

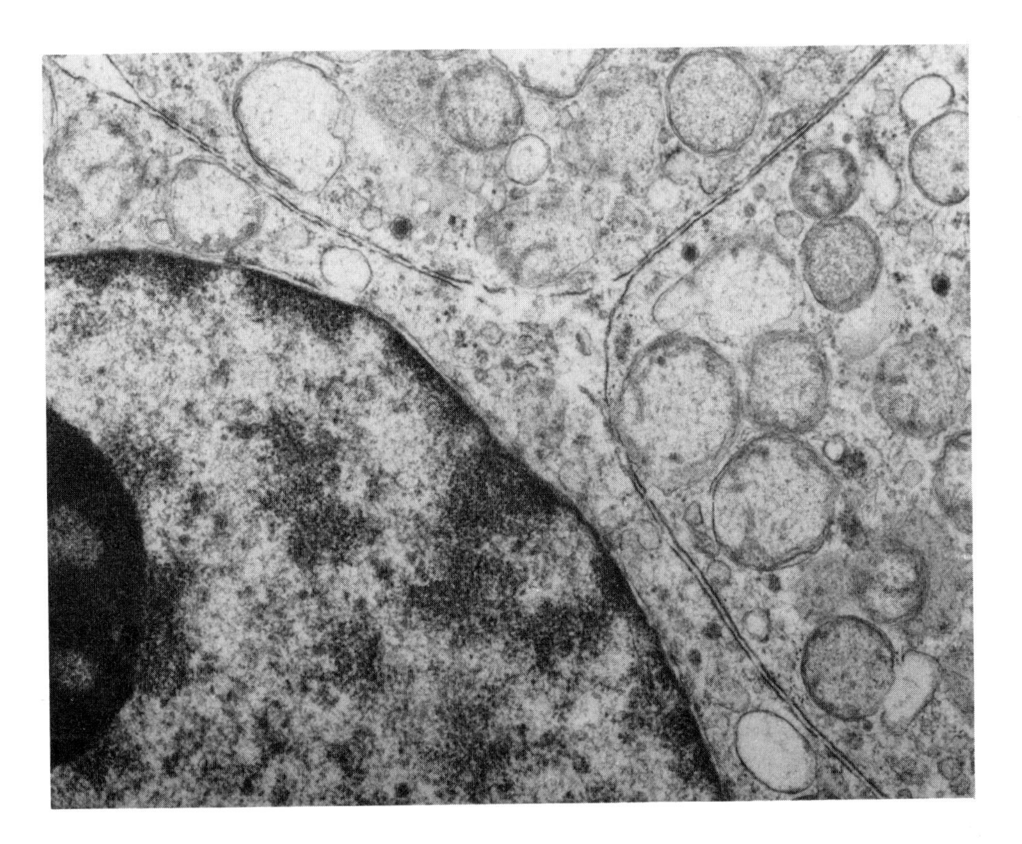

FIGURE 6. Paraganglioma. The electron microscopic appearance of chief cells will generally reveal variable numbers of atypical, small, neurosecretory granules (encircled), typical of true neural cells.

1. Medullary carcinoma of the thyroid
2. Carcinoids and "oat cell" (small cell) carcinomas occurring in diverse anatomic locations
3. Islet cell tumors
4. Cutaneous tumors with neuroendocrine features (Merkel cell or trabecular carcinoma of skin).

Very rare tumors also occur that manifest unambiguous neural crest characteristics and multiple tissue types, such as *malignant ectomesenchymoma* (typical neural cells and muscle) and *nerve sheath tumors with complex elements* (Triton tumors). Although of nosologic and histogenetic interest, they will not be considered here.

A. MEDULLARY CARCINOMA OF THE THYROID

Medullary carcinoma of the thyroid is perhaps the classic APUD tumor. Intensively studied by Williams[43] and multiple other authors,[44] this tumor is clearly neuroendocrine and even neural crest in character, but perhaps not in origin. Numerous embryologic studies have documented the migration of neural crest cells to the thyroid gland and their subsequent development into the neuroendocrine (calcitonin-secreting) cells of the thyroid.[45] In the case of human material these cells are embedded within and virtually indistinguishable from the exocrine thyroid follicular cells. However, early electron microscopic studies identified cells with classic APUD-type unit membrane-bound dense core granules in scattered cells within the parafollicular area of human thyroid.[46] Eventually it was appreciated that a particular type of thyroid carcinoma, so-called medullary carcinoma of the thyroid, was, in fact, the

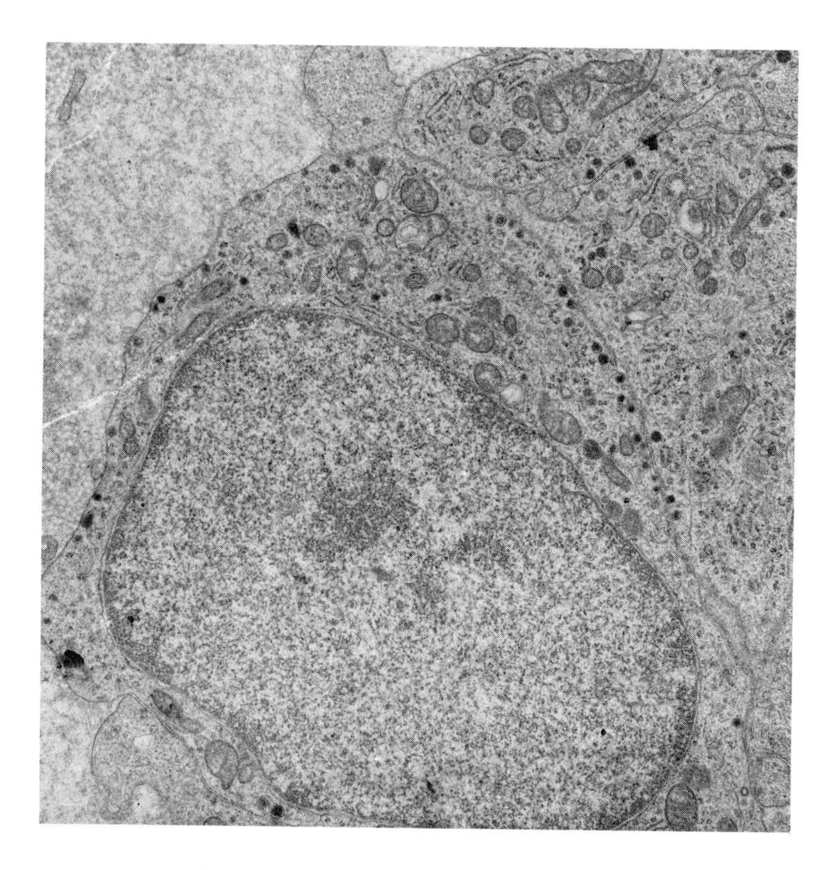

FIGURE 7. Primitive neuroblastoma with dense core granules. This bone marrow metastasis from a teen-age boy with neuroblastoma shows undifferentiated neuroblasts containing typical small electron dense neurosecretory granules scattered throughout the cell cytoplasm.

neoplastic counterpart of these parafollicular cells.[43] Such tumors, seen in Figure 25, are typically composed of small basophilic cells separated by large masses of extracellular material subsequently identified as tumor amyloid.[47] By electron microscopy both the characteristic neurosecretory granules and extracellular fibrillar amyloid material are readily demonstrable (Figure 26).

It should be noted that despite the strong evidence implicating physical migration of neural crest cells to the thyroid and their subsequent neoplastic transformation into medullary carcinoma,[4,45] complex thyroid neoplasia with both medullary (i.e., neuroendocrine) and exocrine (follicular epithelial cell) characteristics have been identified,[48,49] In these tumors, both calcitonin (the normal parafollicular cell product) and thyroglobulin (the normal follicular cell secretory product) have been identified. Thus, the actual neural crest derivation of even this apparent classic APUD tumor is in some question. It is precisely examples such as this that have caused Pearse[9] and others[1] to suggest that the APUD genotype is *programmed* into uncommitted stem cells, as opposed to actual migration of neural crest cells to sites such as thyroid, despite experimental evidence from quail-chick chimeras, as published by Le Dourain and colleagues.[4]

B. CARCINOIDS

The most classic member of this family of neuroendocrine neoplasia is certainly the bronchial carcinoid.[50] However, these tumors occur in numerous other anatomic sites as well.[51] Gut carcinoids are a second well-characterized group.[52] Classically, three groups are

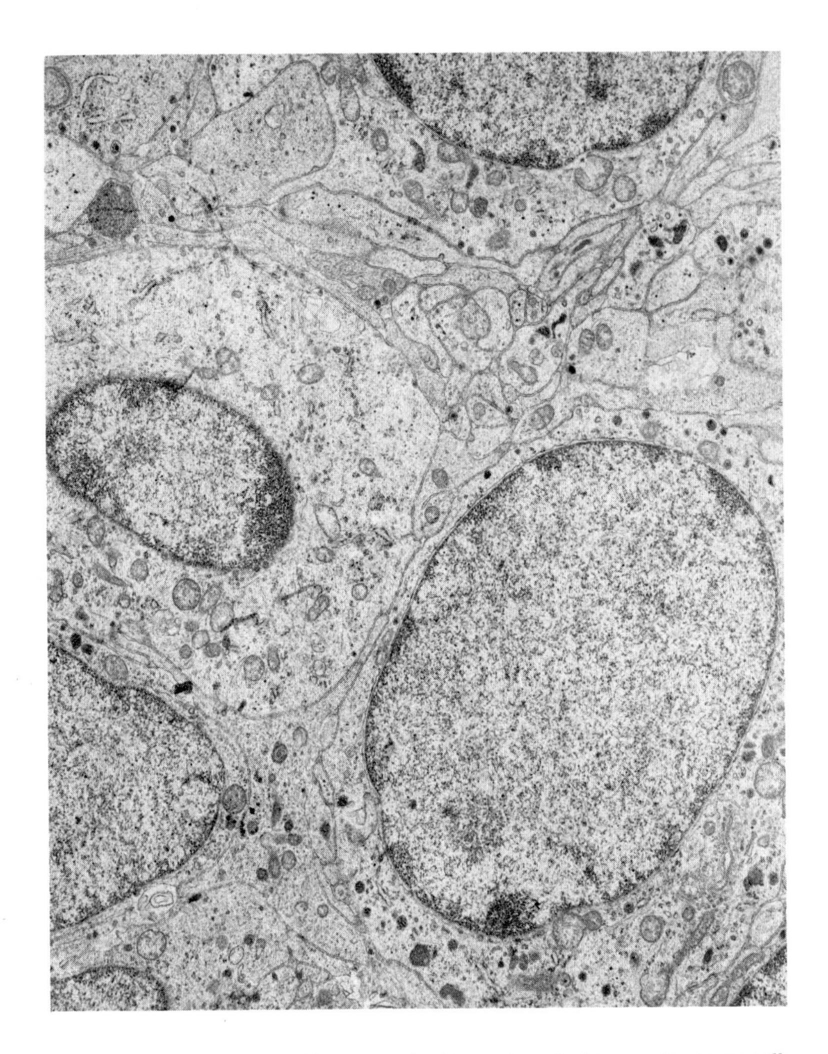

FIGURE 8. Primitive neuroblastoma cell with neurites. In this case the tumor cells show more definitive neural differentiation. In addition to dense core granules, tangled masses of cell processes or neurites are seen.

recognized: (1) foregut, (2) midgut, and (3) hindgut. These differ from one another in terms of granule morphology (with foregut containing small, neurosecretory type granules; midgut, larger, pleomorphic granules; and hindgut, larger neuroendocrine type granules), as well as clinical symptomatology secondary to granule content (typically serotonin or 5-HIAA). Many are asymptomatic, with no biologically active secretory product, but, nonetheless, contain the distinctive APUD granules by EM, a useful diagnostic feature in cases of metastatic tumor of liver with no known primary. Often, these large granules are demonstrable by light microscopy alone, using the Grimelius stain (Figure 27), though EM is clearly more reliable, definitive, and able to detect small numbers of granules, or smaller granules, too small to be resolved by light microscopy (Figure 28).

Beyond these two well-defined groups of carcinoids are innumerable reports of anatomically disparate carcinoids, such as those of mediastinum,[53,54] pancreas (where they blend imperceptibly with the classic islet cell tumors),[55] and rarely, liver and biliary tree[56,57] to name only a few. Morphologically, they are no different than the spectrum of tumors seen in gut and lung, for example, and will therefore not be discussed separately, other than mediastinum and liver.

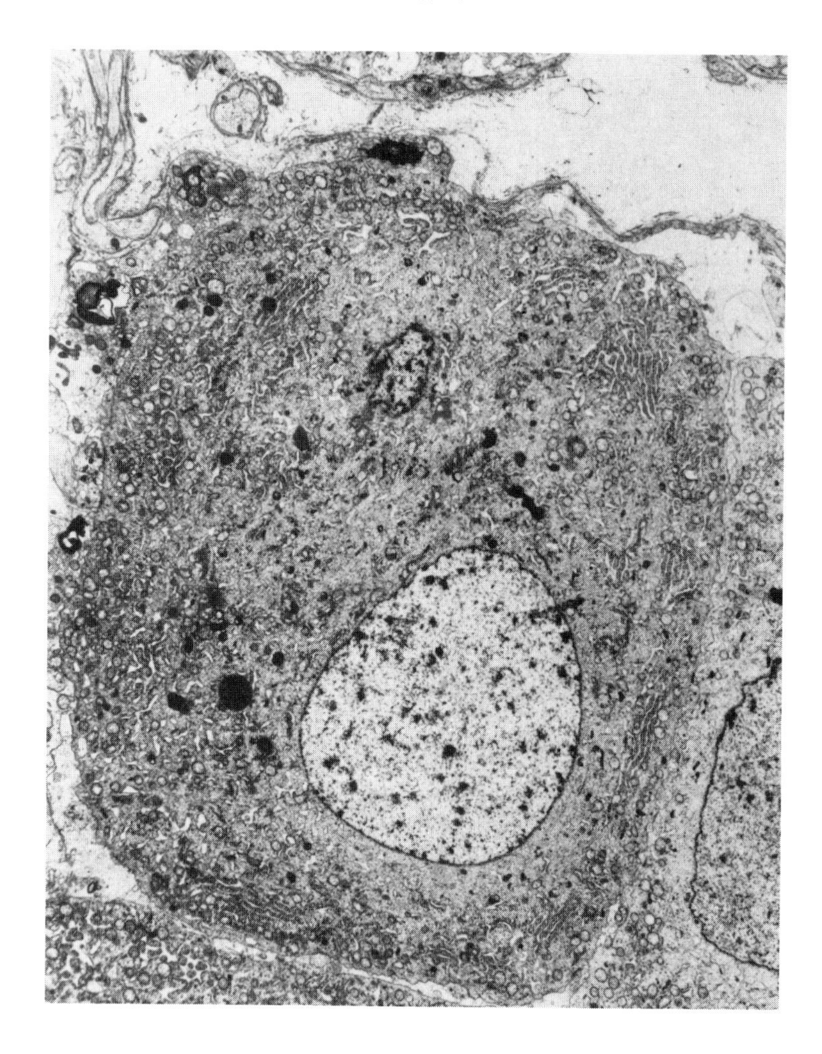

FIGURE 9. Ganglion cell, ganglioneuroblastoma. With extreme maturation, terminally differentiated neuroblastoma cells become ganglion cells, as illustrated here. Neurosecretory granules in the perinuclear cytoplasm are far less numerous and are instead found more abundantly arrayed in the distal neuritic processes, not seen here.

C. BRONCHIAL CARCINOIDS

Perhaps the clinically relevant and biologically interesting among this group of tumors is the bronchial carcinoid. This tumor is most likely closely related to small cell carcinoma of lung, and an intermediate "hybrid" tumor exists, aggressive carcinoid, which appears to link these two seemingly distinct and different entities.[58] Universally accepted terminology does not exist, and some authors prefer "well differentiated neuroendocrine carcinoma."[59] This group is clinically far more common and important than other carcinoids, and is almost certainly related to smoking, with the possible exception of classic intrabronchial carcinoid, which may occasionally be familial. The classic EM feature of bronchial carcinoid is the presence of innumerable, very large (>400 nm), very electron-dense unit membrane-bound cytoplasmic granules (Figure 29). Generally, these tumors also show marked evidence of glandular differentiation (Figure 30), an observation which has been interpreted as evidence of origin from peribronchial glands. This is in contrast to oat cell carcinoma (*vide infra*), thought to arise from uncommitted bronchial epithelial cells and with no known association with peribronchial glands.

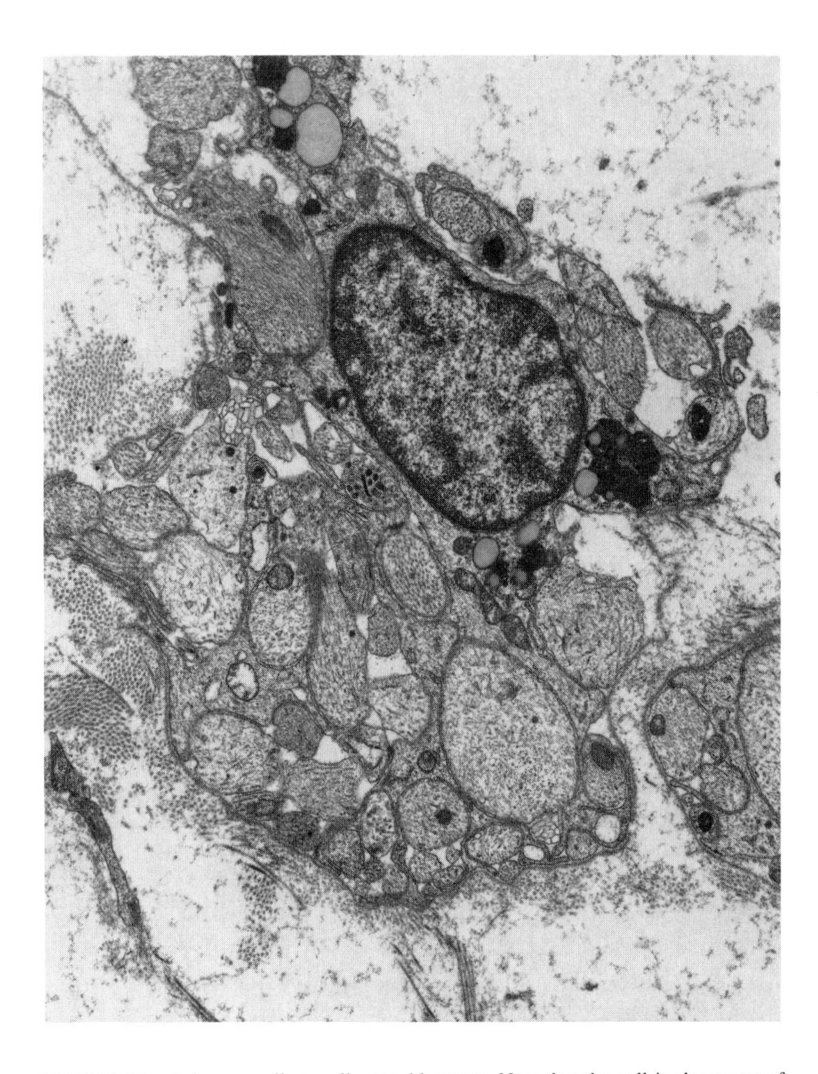

FIGURE 10. Schwann cell, ganglioneuroblastoma. Note that the cell in the center of the field with conspicuous nucleus envelopes large numbers of cross-sectioned cell processes which correspond to the neurites originating from the ganglion cells illustrated in Figure 9. Scattered clusters of dense core granules are noted in some of the neuritic processes. A basal lamina envelopes these bundles of unmyelinated neurites. This tumor closely recapitulates normal unmyelinated nerve development.

D. OAT CELL (SMALL CELL) CARCINOMA

Bronchial carcinoids as illustrated above are typically conspicuous for their large growth within the bronchus and indolent clinical behavior. However, far more aggressive tumors, often presenting clinically as mediastinal metastasis, are the far more common and clinically problematic group of related pulmonary tumors. The most common form, small cell carcinoma of lung, shows little, if any, glandular differentiation,[60] but does indeed contain smaller, scattered, less numerous dense core neurosecretory granules (Figure 31). In this tumor the cells are often neural in character, with slender cell processes containing these neurosecretory granules (Figure 32).

Once again, similar to the situation with medullary carcinoma of the thyroid with mixed endocrine-neuroendocrine characteristics, oat cell is frequently not simply an APUD tumor. Most cases can be demonstrated to express morphologic and biologic (i.e., intermediate filament) characteristics of squamous cell carcinoma of the lung as well. The latter is distinctly

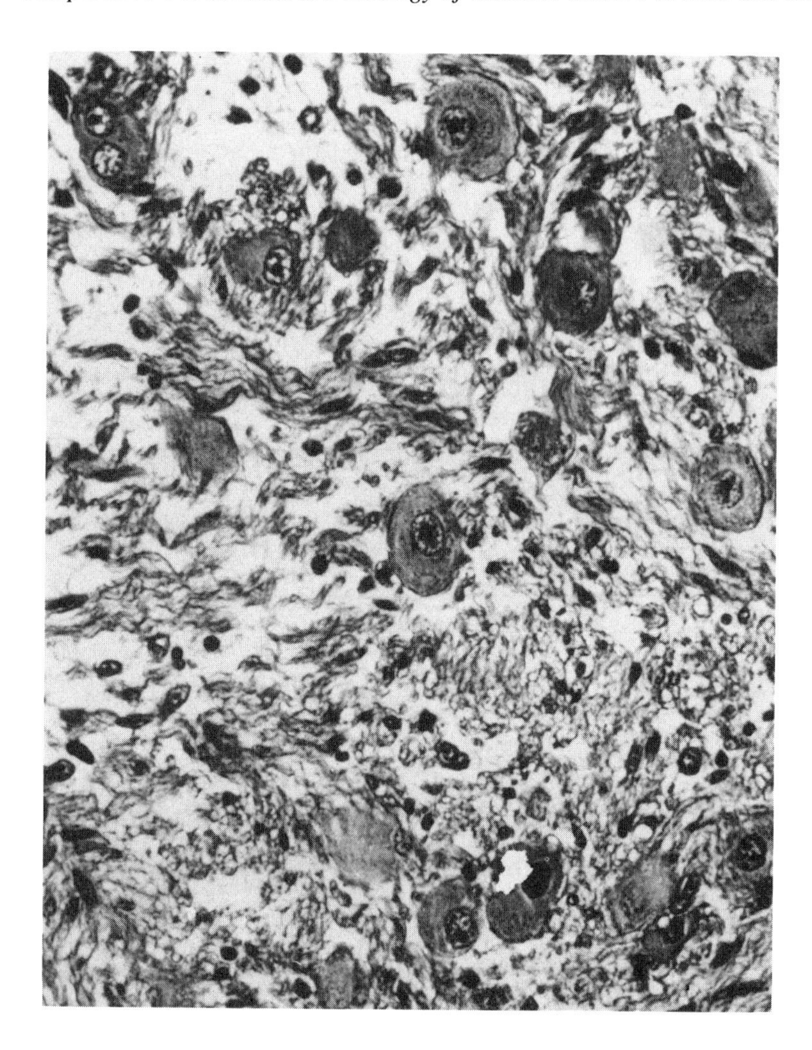

FIGURE 11. Ganglioneuroma. With terminal differentiation, neuroblastoma occasionally becomes ganglioneuroma, a benign tumorous counterpart of the malignant neuroblastoma. Here terminally differentiated ganglion cells are scattered throughout the field and are conspicuous as large cells with abundant cytoplasm. Also noted are large numbers of less conspicuous and smaller cells, which correspond to the Schwann cells containing or enveloping unmyelinated neuritic processes, illustrated in Figure 10. This, however, is benign, unlike the structures illustrated in Figures 9 and 10.

nonneuroendocrine as classically understood, yet the common mixture of neuroendocrine and entodermal features in oat cell carcinoma suggests a common histogenesis for both, as noted by several authors.[61-68] Thus, the theme of mixed APUD and non-APUD features of this second group of tumors, now generally termed neuroendocrine tumors, is that of an incompletely assumed neural crest phenotype, apparently superimposed on or co-existent with other, nonneuroendocrine phenotypic traits. Clearly, existing concepts and diagnostic criteria are in need of revision.[69]

E. AGGRESSIVE CARCINOID OF LUNG

Numerous studies of small cell carcinoma of lung have also revealed a third important category of tumors, i.e., those with some characteristics of carcinoid but with aggressive clinical behavior.[58,59] These so-called aggressive carcinoids frequently present with medias-

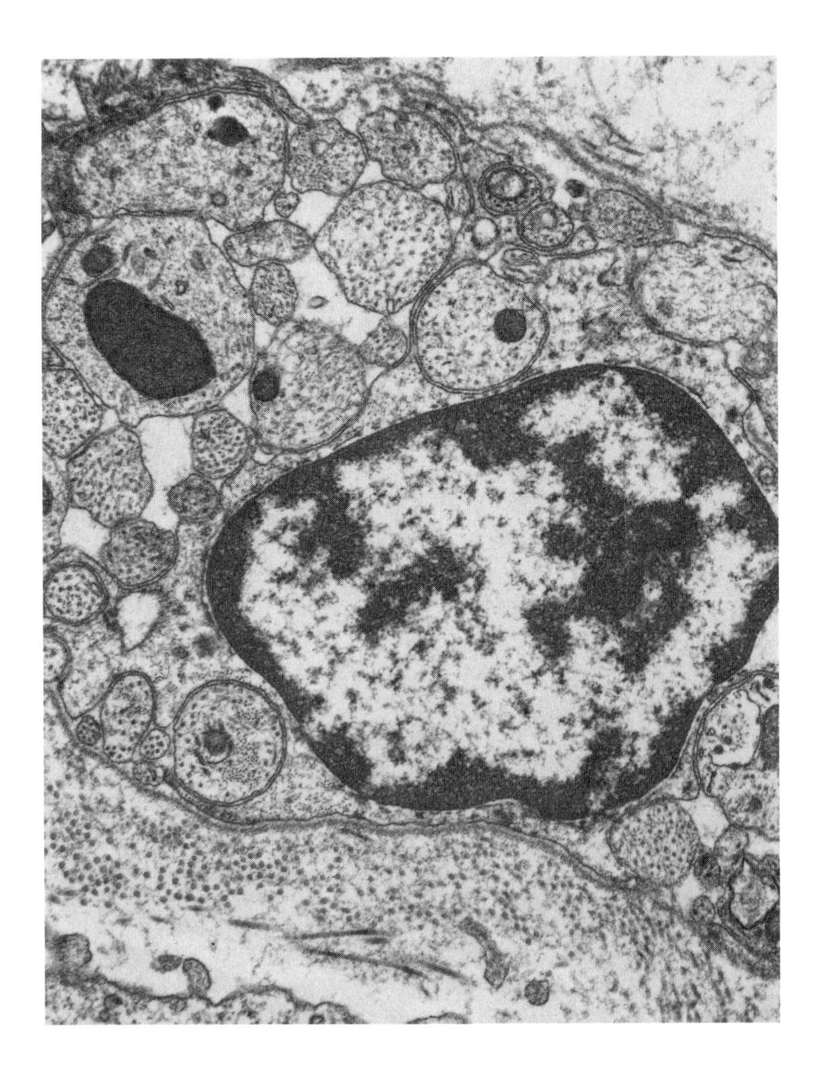

FIGURE 12. Ganglioneuroma, electron microscopy. Once the tumor has become terminally differentiated and benign, the structures virtually recapitulate normal unmyelinated nerve. Note here that the Schwann cell cytoplasm forms perfectly developed mesaxons surrounding the cross-sectioned neurites. A continuous basal lamina surrounds the entire mass, precisely as noted in normal unmyelinated nerves. The entire structure is enveloped by collagenous stroma noted outside the basal lamina.

tinal metastasis similar to that of oat cell carcinoma of lung (Figure 33), but upon electron microscopic examination show clear cut evidence of, for example, glandular formation including microvilli and lumen formation (Figure 34). Such tumors appear to represent an immediate form between classic oat cell and typical indolent bronchial carcinoid.

F. MEDIASTINAL CARCINOID

Mediastinal carcinoid was not reported until the 1970s when it was first popularized by Levine and Rosai in the AFIP fascicle on mediastinum (see Reference 69 for review). In these patients, a mediastinal tumor distinguishable from thymoma, lymphoma, germ cell tumor, or teratoma is identified predominantly on the basis of ultrastructural characteristics including conspicuous dense core granules. Like aggressive carcinoid of pulmonary origin, lumen formation may be noted and the granule content is typically small (<400 nm). The spectrum of appearance varies from that of an oat cell type lesion (Figure 35) to that of a

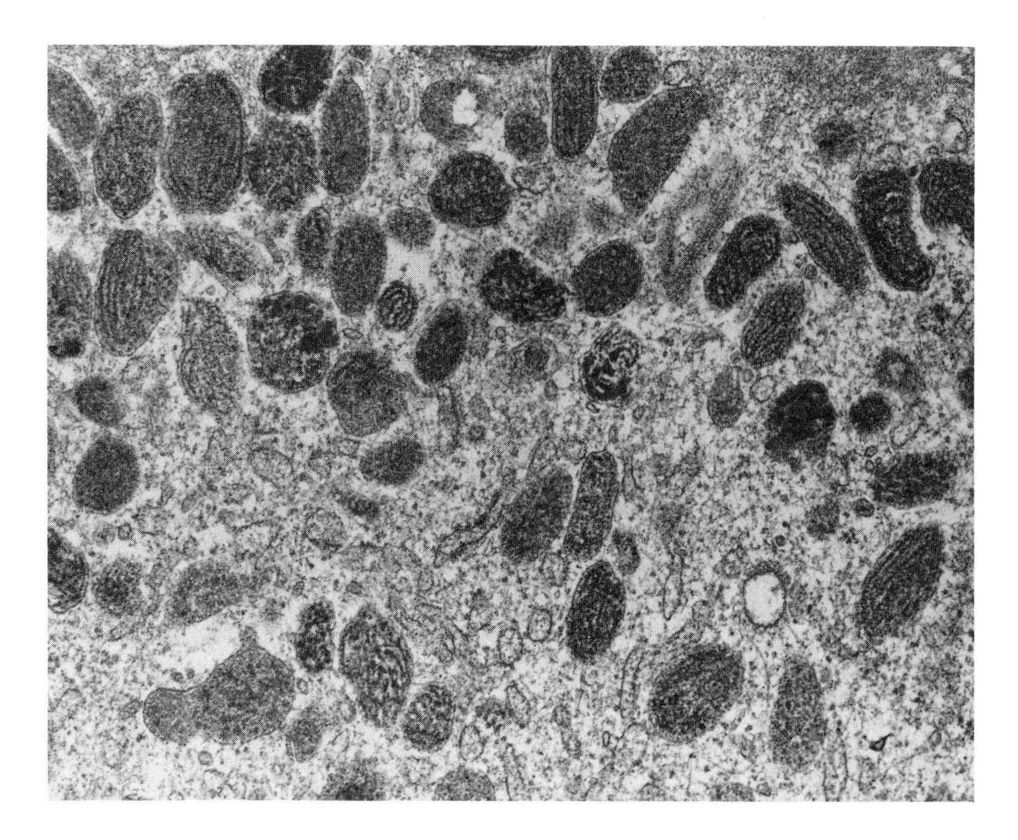

FIGURE 13. Melanoma, electron microscopy, premelanosomes. This detail photograph of cytoplasm from a melanoma cell reveals large numbers of melanosomes in various stages of melanogenesis. However, virtually all are in the premelanosome stage, and a fibrillar or ''herringbone'' substructure is noted in almost all cases. Only the few showing striking electron dense material within the premelanosome are undergoing actual melanogenesis. By light microscopy or gross examination such tumors are generally unpigmented.

true carcinoid-like entity (Figure 36). In the latter case, a mixture of neurosecretory and carcinoid granules can be found. An association with Hassell's corpuscles or pharyngeal mucosal-type epithelial cells has also been noted, suggested an origin from mucosal epithelial cells, like thymoma. With sufficient numbers of cases, one suspects mixed epithelial/neuroendocrine tumors will be described.

G. HEPATIC CARCINOID

Very rare examples of primary hepatic carcinoids have also been reported.[57,58,70,71] Often, these are difficult to distinguish from metastatic carcinoids in liver, but unequivocal cases of tumors arising in the liver have been documented. Often such tumors are suspected to arise from biliary tree ducts, especially terminal ductules.[72] A possible association with such ductal elements is rarely found (Figure 37).

The other histogenetic consideration is that hepatic carcinoids may be more closely related to hepatic epithelial cells. Tumors with combined hepatocellular and carcinoid features, including dense core granules, have indeed been described[70] (Figure 38). Nonetheless, hepatocytes themselves appear to regenerate in some cases from terminal ductal elements, and a clear cut distinction between the two possibilities may be entirely artificial.

4. ISLET CELL TUMORS

Certainly the most clearly documented examples of neuroendocrine tumors are those occurring within the pancreas and specifically within the islets of Langerhans.[73,74] This group

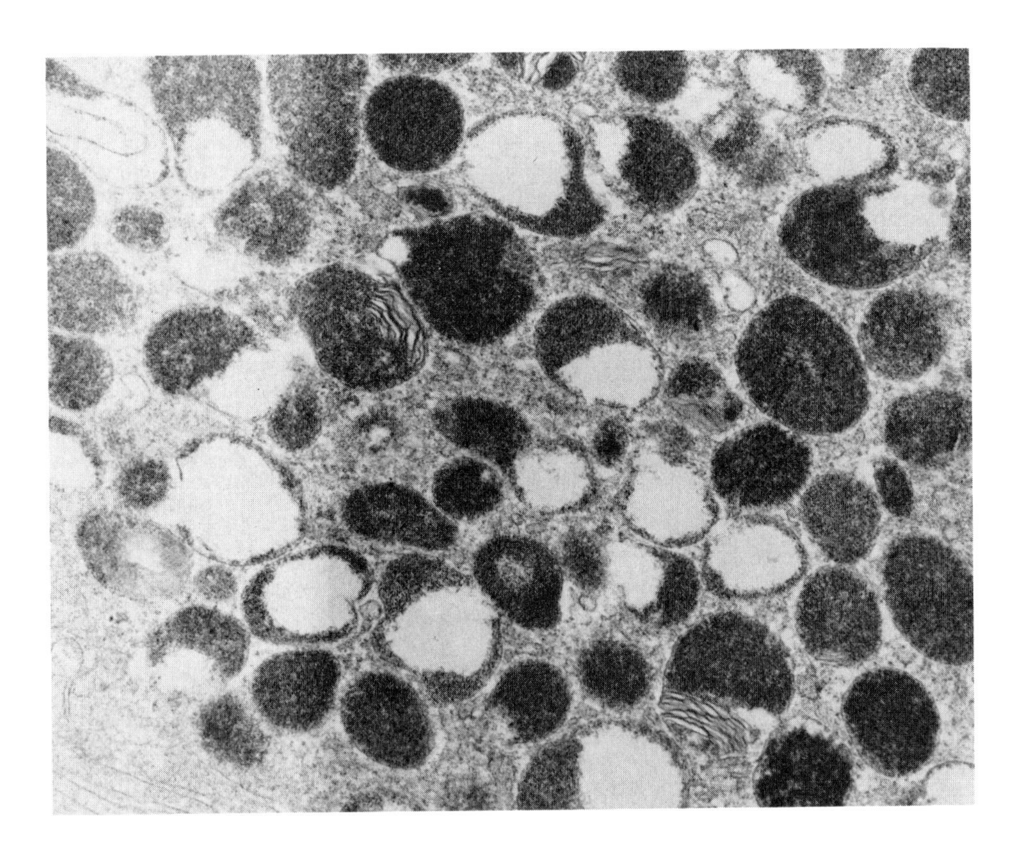

FIGURE 14. Melanoma, maturing premelanosomes. At higher magnification in this tumor cell, none of the melanosomes show the fibrillar or herringbone substructure noted in premelanosomes. Two, in fact, are virtually completely electron dense. This is typical of maturing and mature melanosomes in normal and malignant melanocytes.

of tumors, referred to as islet cell tumors, includes those with clear-cut endocrine function and morphology, such as alpha and beta cell tumors (producing glucagon and insulin, respectively), as well as those with a specific secretory product, including somatostatinoma and gastrinoma, and other tumors with no known secretory product but comparable morphology, referred to as islet cell carcinomas, islet cell tumors, or pancreatic carcinoids.

I. INSULINOMA

Originally, identification of these tumors was based solely on their content of classic islet cell type granules.[74] In the case of insulinomas or β-cell tumors, the relationship between granule morphology and secretory product has been well documented. Most insulinomas contain β-type, insulin-containing granules indistinguishable from or at least similar to those identified in normal β cells. Figure 39 documents the appearance of alpha and beta granules from normal islets; Figure 40, comparable granules from an insulinoma. Note that in the latter case the classic β granules seen are disordered and disarrayed and in many cases atypical. Some tumors fail to contain *any* typical beta cell granules, such as that illustrated in Figure 41.

An observation in connection with islets cell tumors in general, and insulinomas in particular, was one of the original reasons to suspect that not all APUD tumors were derived from neural crest cells. In particular, the childhood phenomenon of islet cells scattered in acinar areas,[75] with co-existence of zymogen and β granules within the these cells (so-called nesidioblastosis),[76] and the development of beta cells from pancreatic ductal elements[77] as

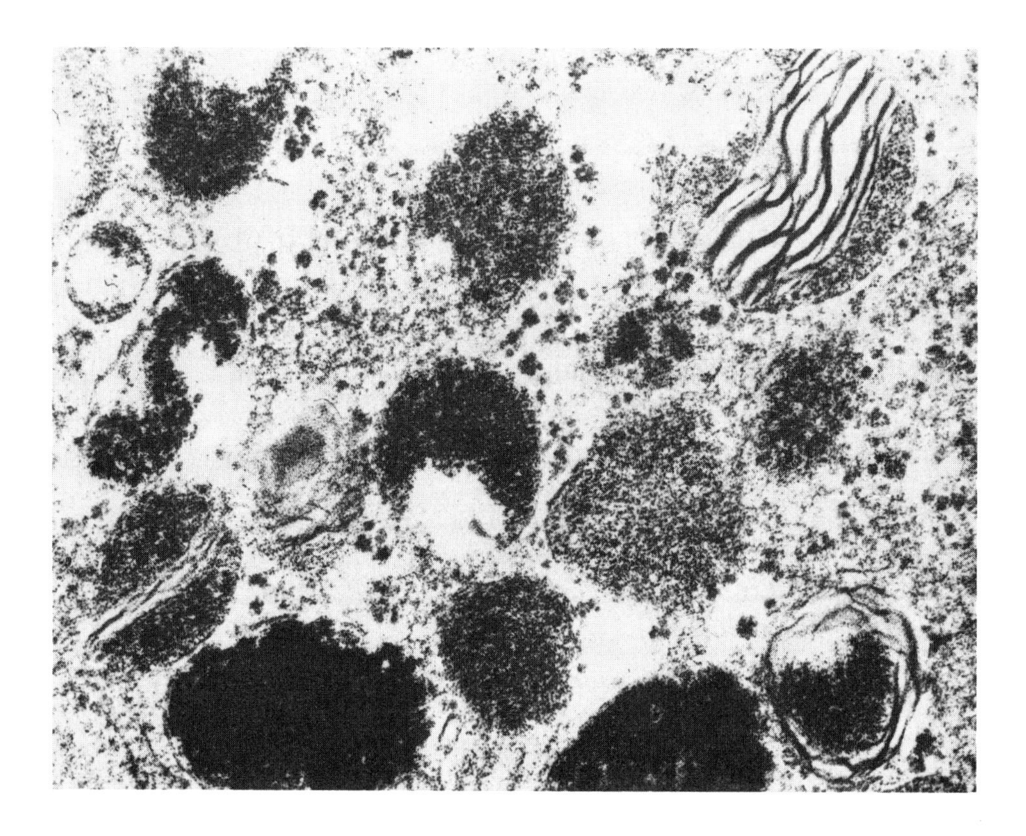

FIGURE 15. Melanoma, atypical melanosomes. In this detail of tumor cell cytoplasm, numerous electron-dense structures are noted. However, neither the typical fibrillar or herringbone substructure of premelanosomes, nor the coarsely electron dense or opaque structure of mature melanosomes, as illustrated, above is seen. Instead, a lamellar substructure in some of the granules exactly mimics that noted in the previous figure among mature melanosomes. This is presumptive evidence of melanogenesis.

well as the observation that certain insulinomas or β cell tumors appear to contain both beta and zymogen granules[78] (Figure 42) have strongly suggested that these tumors have a mixed potential phenotype composed of exocrine pancreas with acinar or ductal elements, as well as endocrine pancreas with β granules. Clearly, it is difficult to reconcile these observations with a purely neural crest cell origin of these tumors.

J. GLUCAGONOMA

A closely related islet cell tumor which illustrates a second principle is that of the glucagonoma or α cell tumor. In this case the tumor cell product is biologically active glucagon, but the ultrastructural appearance fails to identify typical anti-type granules within tumor cells,[76] as seen in Figure 43. Thus, the simple association between granule morphology and secretory product observed in insulinomas is not observed in other islet cell tumors.

K. GASTRINOMA

Comparable observations have been in connection with other islet cell tumors such as the so-called ulcerogenic tumor of the pancreas (or gastric antrum) of Zollinger-Ellison syndrome. These tumors are also termed gastrinoma (i.e., gastrin-producing islet cell tumors).[80] Again, as in the example illustrated here, although an obvious glandular morphology is identified (Figure 44), the granules contained within these glandular tumor cells are usually nondescript carcinoid or even neurosecretory type (Figure 45); only occasionally do they resemble normal gastrin-containing islet cell granules.

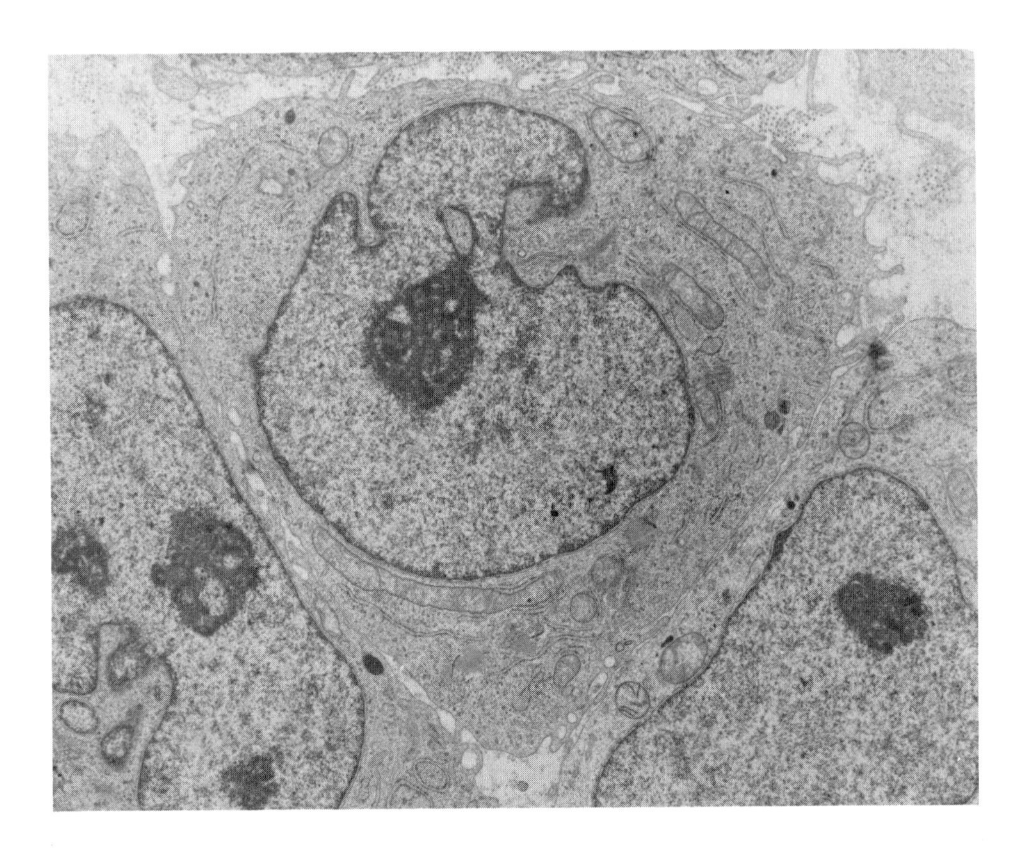

FIGURE 16. Amelanotic melanoma. Note that this tumor cell contains no structures identifiable as any form of premelanosomes or mature melanosomes. In this case there is no ultrastructural feature to precisely identify such tumor cells as melanocytic.

L. SOMATOSTATINOMA

Finally, other islet cell tumors possess distinctive granule morphology, which may or may not be associated with a specific secretory product. In the example depicted here, a somatostatinoma, the tumor cells have both conspicuous glandular differentiation (Figure 46) as well as another distinctive granule type resembling D-cell (somatostatin-containing) islet cells (Figure 47).[81]

M. CUTANEOUS NEUROENDOCRINE TUMORS

For some years the only cutaneous APUD tumors described were the melanomas discussed in detail above. In 1972, Toker described a so-called trabecular carcinoma of skin[82] which subsequently was reported independently by Tang and Toker[83] and by several other authors, under terms such as Merkel cell tumor and, more recently, neuroendocrine carcinoma of skin.[84-87] Significant controversy has surrounded both the terminology and the proper nosology of this distinctive neoplasm. Recent work has clarified the issue; it is now apparent that this is a legitimate neuroendocrine tumor, but one with distinctly mixed phenotype, usually expressing both neuroendocrine and squamous cell differentiation,[86,88,89] a situation similar to that described in connection with oat cell carcinoma of lung and mediastinal carcinoid. However, clearly distinctive forms of this tumor exist, such that a given case may express only neuroendocrine traits (i.e., neuroendocrine granules), while another will clearly display features of both epithelial (keratin filaments) and neuroendocrine (APUD-type granules) histogenesis. The classic neuroendocrine form of this tumor is illustrated in Figure 48, where classic neurosecretory type APUD granules are abundant. Other cases,

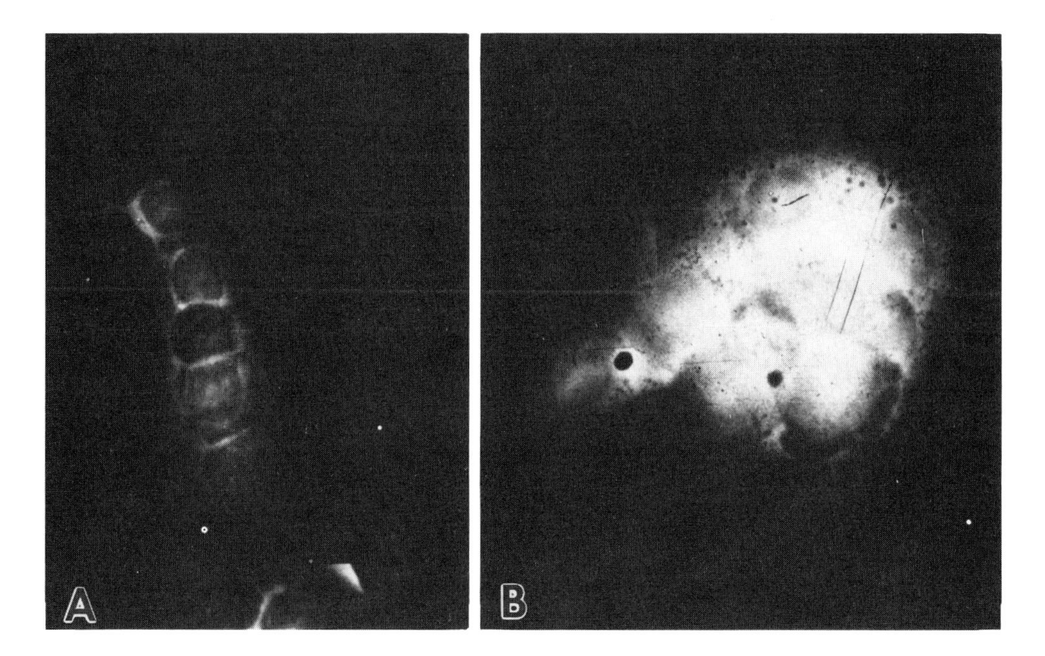

FIGURE 17. Melanoma, catecholamine fluorescence. (A) Negative control, lymphoma. Note that tumor cell cytoplasm is negative and only some extracellular material trapped in the crevices between and around tumor cells in this imprint show any fluorescence. (B) Tumor specimen, suspected melanoma. Note that the tumor cell cytoplasm is brightly fluorescent. Furthermore the fluorescence bleaches with time, unlike that illustrated in Figure 17 A. The bright but unstable fluorescence noted in B is diagnostic of a neural crest tumor. In the appropriate clinical setting, as here, it then becomes diagnostic of melanoma.

however, are less conspicuously neuroendocrine and may possess both cellular processes and pleomorphic dense core granules (Figure 49). In addition, these same tumors may simultaneously express conspicuous keratin bundles in their cytoplasm (Figure 50). Even multiple peptide secretion and melanosomes may be found together.[90]

N. PRIMITIVE NEUROECTODERMAL TUMORS (PNET)

The final group of tumors under consideration are a distinctive group of tumors which occur almost exclusively in children and young adults. Although they have been described both within the central nervous system and without,[91] the present discussion focuses exclusively on the extracranial PNETs. There is a substantial body of evidence to suggest that these are also legitimate neuroendocrine tumors. Unlike all preceding examples discussed above, these tumors as a rule are exceedingly primitive and the neuroendocrine character of the tumor is frequently difficult to ascertain. For that reason the terminology has varied greatly from one author to another and terms such as PNET (primitive neuroectodermal tumor),[92,93] peripheral neuroblastoma,[94] peripheral neuroectodermal tumor,[95,96] peripheral neuroepithelioma,[97,98] and more recently the so-called Askin tumor of chest wall[99] have all been independently described, yet appear to be members of the same family.

In all of these tumors, the degree of neural or neuroendocrine differentiation has been variably manifest. Sometimes there is no readily identifiable neural crest differentiation and in such cases morphologic evidence of a neural histogenesis is not evident. Here, recently developed laboratory techniques and data have allowed recategorization of such tumors within this group. The most conspicuous example of the latter is so-called Ewing sarcoma of bone which now clearly appears to be a primitive neuroectodermal tumor of childhood,[100,101,102] closely related to PNET, peripheral neuroepithelioma, and Askin tumor.[103,104]

Numerous common traits such as cholinesterase activity (common in all APUD tumors

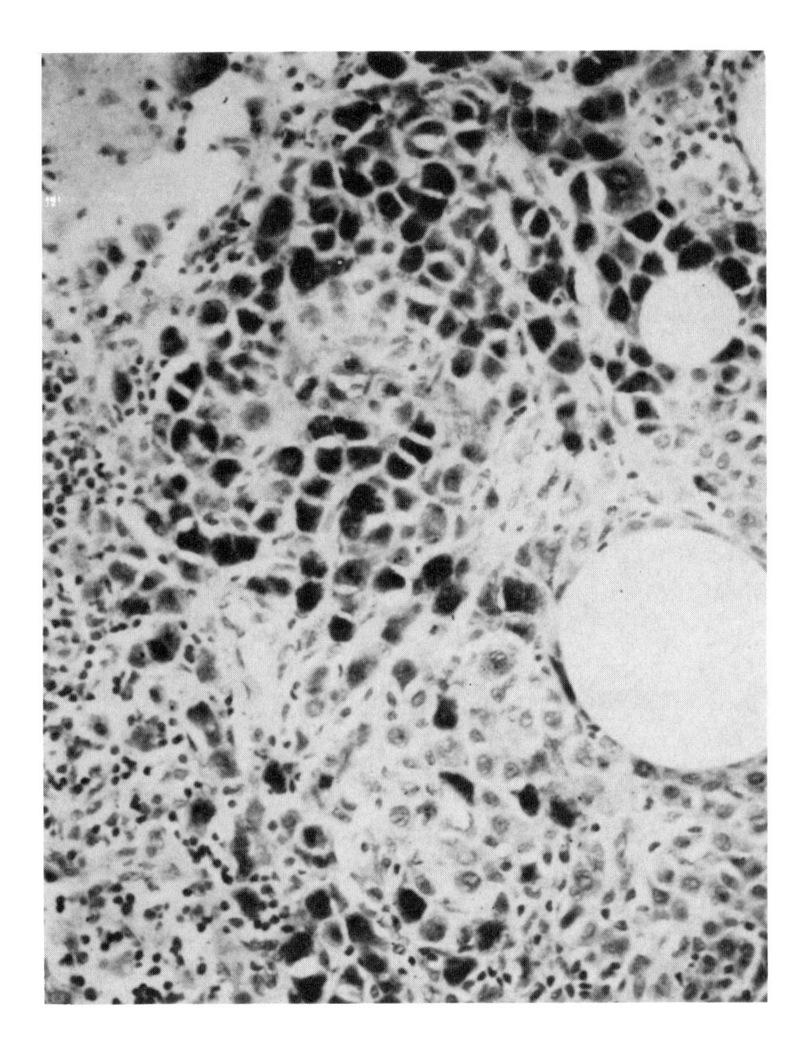

FIGURE 18. Melanoma, S-100 immunochemistry. The anaplastic tumor metastatic to lymph node illustrated here was of unknown origin. Clinical history was uninformative. However, the suspicion of melanoma was confirmed when immunocytochemistry with anti-S100 antibody was performed. As seen here, roughly half of the tumor cells show dense deposits of chromogen which identifies those tumor cells expressing S100 protein. This S100 reactivity is typical of virtually all cases of melanoma.

so examined), a propensity for neural differentiation when cultured *in vitro,* and especially a common cytogenetic abnormality, a reciprocal chromosome 11 and 22 translocation, so far reported for all members of this group (PNET, peripheral neuroepithelioma, Askin tumor, and Ewing sarcoma) appear to uniquely identify this particular group of tumors of childhood and adolescence,[104] but not the broad spectrum of APUD or neuroendocrine tumors[105,106] carcinoma of lung, for example, lacks the cytogenetic abnormality, as do all other APUD tumors so far examined. Thus, this finding alone serves as a unique tumor marker for this subset of APUD tumors.

The ultrastructural characteristics of these tumors are frequently subtle and difficult to appreciate. Nonetheless, an awareness of the neural crest characteristics described above allows ready recognition of certain of these features. Most conspicuously, the presence of dense core type granules (Figure 51) is highly suggestive though not in and of itself specific, since in certain circumstances lysosomes may have a similar appearance. However, the co-occurrence of slender cell processes resembling neuritic extensions (Figure 52) provide

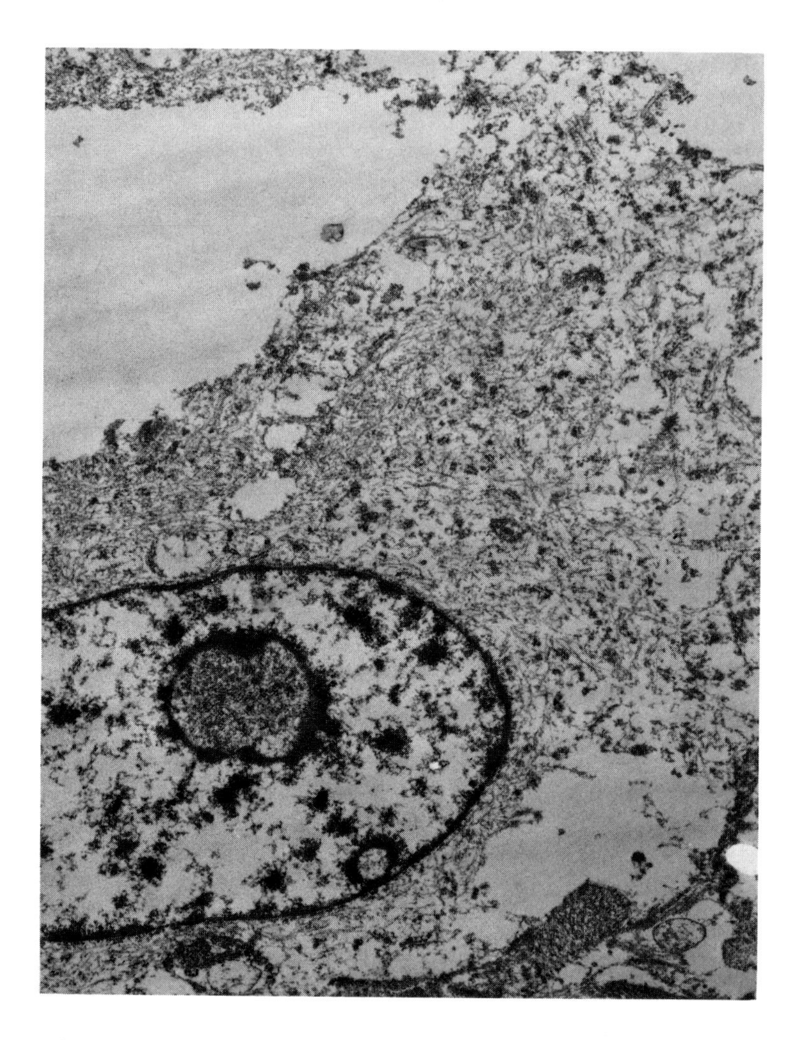

FIGURE 19. Melanoma, electron microscopy, same case as Figure 18. Note that the electron microscopic appearance of this tumor retrieved from formalin fixed, paraffin embedded tissue is not diagnostic by light microscopy. In this case, the ancillary studies, specifically immunocytochemistry, become the diagnostic procedure of choice.

substantial evidence for a neural character of these tumors. In other cases no such definitive ultrastructural evidence exists and only suggestive features common to all APUD tumors, such as hypertrophy of the Golgi apparatus and increased amounts of rough endoplasmic reticulum, even suggest such a histogenesis (Figure 53).

In extremely undifferentiated tumors, as exemplified by Ewing sarcoma, other nonultrastructural techniques are required to definitively identify these tumors as neuroendocrine in origin or character. In some cases expression of neuron specific enolase (NSE) readily identify these tumors as APUD in origin (Figure 54). In other cases only *potential* neural crest characteristics are present. In such tumors, neural differentiation must be induced spontaneously, most readily by short term tissue culture. Generally, nascent neural crest characteristics will be manifest *in vitro,* whereupon clear cut examples of neurites containing dense core granules are often manifest (Figure 55).

V. SUMMARY AND OVERVIEW

It is apparent from the foregoing description that the spectrum of so-called APUD tumors is enormously diverse, extending from clear cut analogs of normal neural crest derived

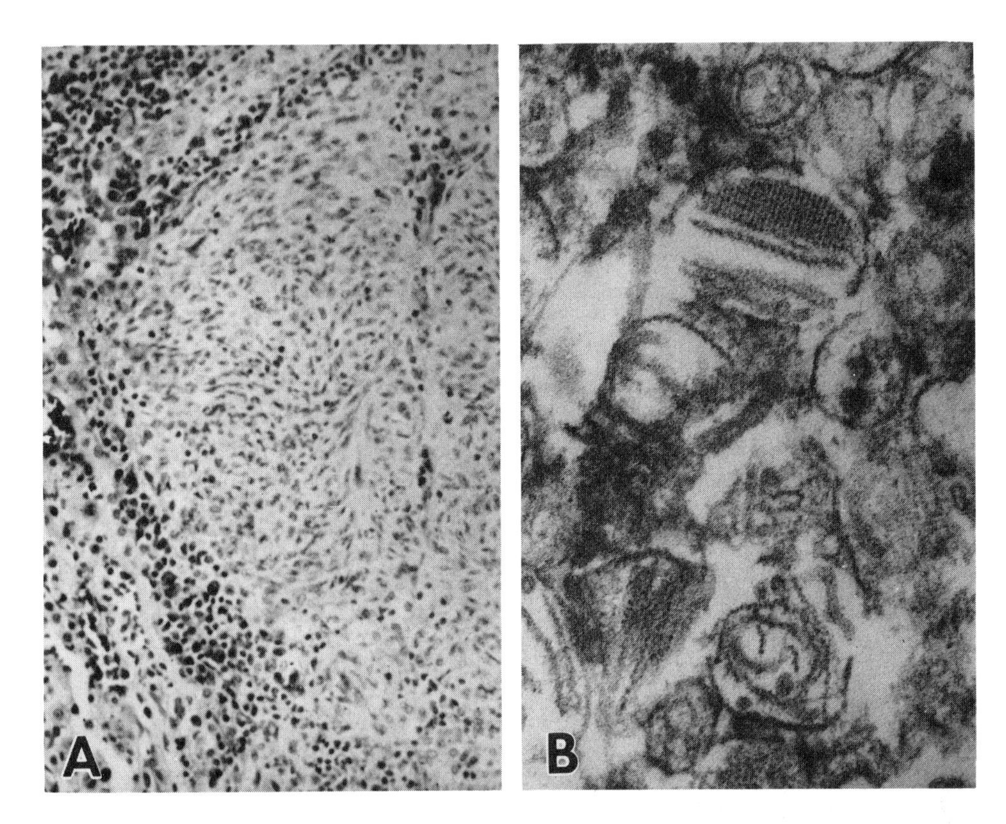

FIGURE 20. Melanoma of soft parts. (A) This spindle cell tumor occurred in the extremity of an adolescent male. The light microscopic appearance was not diagnostic of any specific soft tissue sarcoma although a clear cell sarcoma of tendon sheath and aponeuroses was suspected. (B) Electron microscopy, detail of cytoplasm. Note the unequivocal structure of premelanosomes noted in at least two areas of this micrograph. This is a typical finding in so-called clear cell sarcoma of tendon sheath and aponeuroses, and has provoked the revised terminology *melanoma of soft parts* for virtually all cases of this tumor.

tissues at one extreme (Islet cell tumors, for example) to those tumors having absolutely no definitive neural crest morphologic traits (Ewing sarcoma and primitive neuroectodermal tumors in general). In the latter case only nonmorphologic neural crest phenotypic traits are sufficient to identify these tumors within the broad family of APUD neoplasia.

The other observation of great import for the diagnosis and understanding of these tumors is the fact that the original concept of a distinct group of tumors which could be set apart from all other neoplasia is unfortunately invalid. APUD tumors in some cases may truly possess and express only neural crest phenotypic traits, but more commonly a mixed phenotypic expression of neuroendocrine or neural crest and nonneural crest traits is the rule. This is clear from consideration of oat cell carcinoma of lung (with squamous differentiation), cutaneous neuroendocrine tumors (with APUD granules and keratin bundles coresident within the same tumor cells), and even presumed medullary carcinoma of thyroid (with both calcitonin and thyroglobulin secretion). Clearly, our understanding of this group of tumors, though vastly improved over the situation at the time of its original description, is still far from complete. Nonetheless, this body of tumors remains perhaps the most biologically intriguing yet amenable to ultrastructural study.

Future studies will necessitate focus on multiple nonmorphologic characteristics of these tumors in conjunction with their ultrastructural manifestations. The biology of the tumors as a group remains enigmatic, especially when one considers the frequently bizarre and unanticipated regression or recurrence of tumors such as melanoma, and our complete inability to predict biologic aggressiveness (i.e., malignancy) among islet cell tumors. Clearly, we still have a great deal to learn about this fascinating group of tumors.

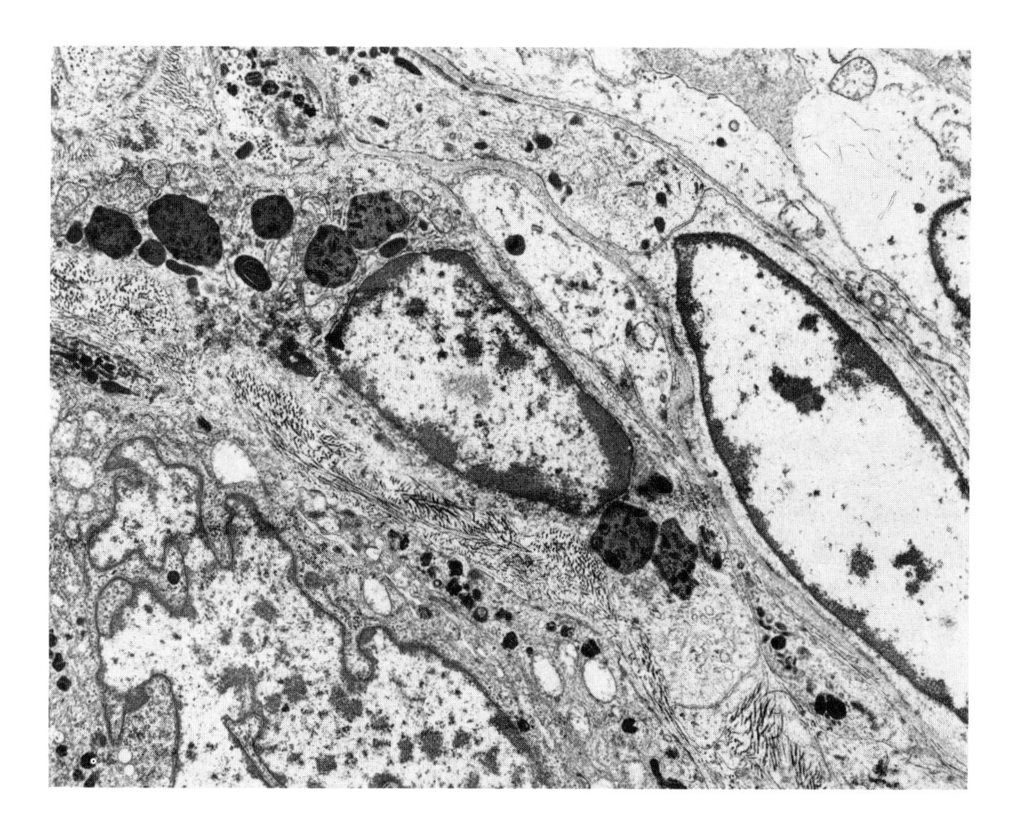

FIGURE 21. Malignant melanotic schwannoma. This spindle cell tumor arose in the rough vicinity of a major nerve but was not physically associated therewith. Upon light microscopy a high grade spindle cell sarcoma was noted; electron microscopy as illustrated here shows numbers of tumor cells invested by basal lamina and containing large numbers of membrane-bound vacuoles containing heterogeneous debris. These structures are diagnostic of melanophagosomes and are routinely found in many forms of malignant melanoma as well as related tumors such as malignant melanotic schwannoma, as illustrated here.

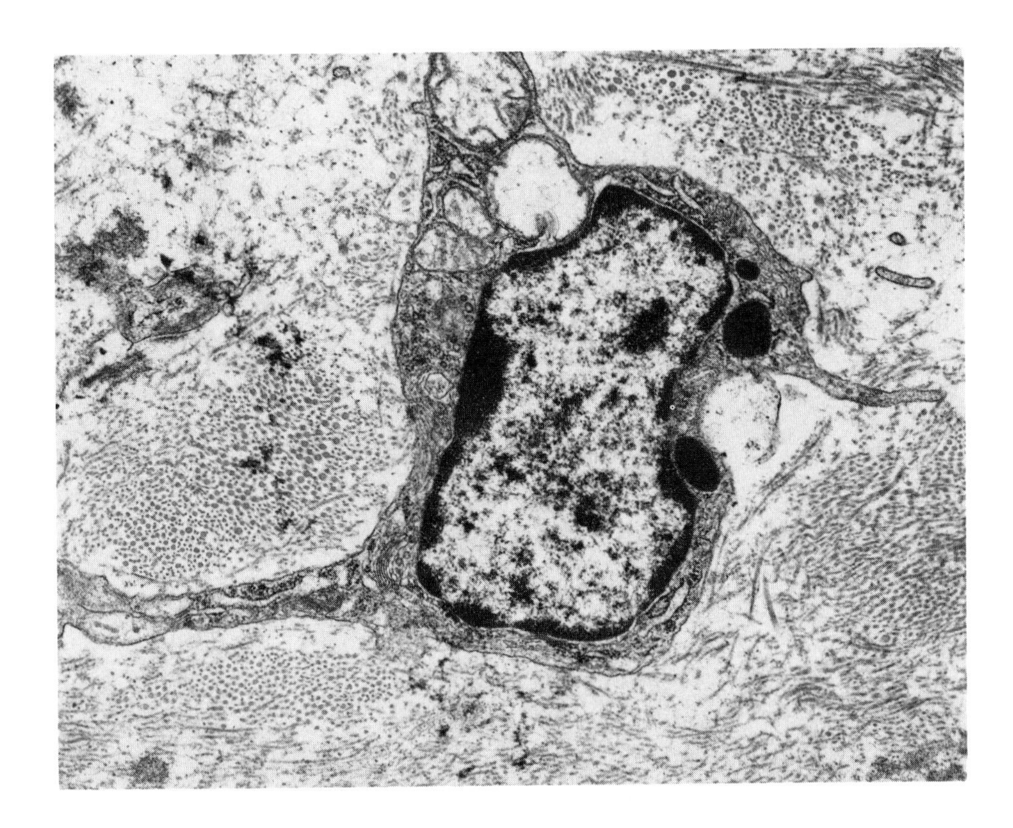

FIGURE 22. Malignant melanotic schwannoma, Schwann cell, no melanosomes. Other areas of the tumor illustrated in Figure 21 show pure Schwann cell differentiation with no evidence of melanogenesis. This heterogeneity is typical of this entity.

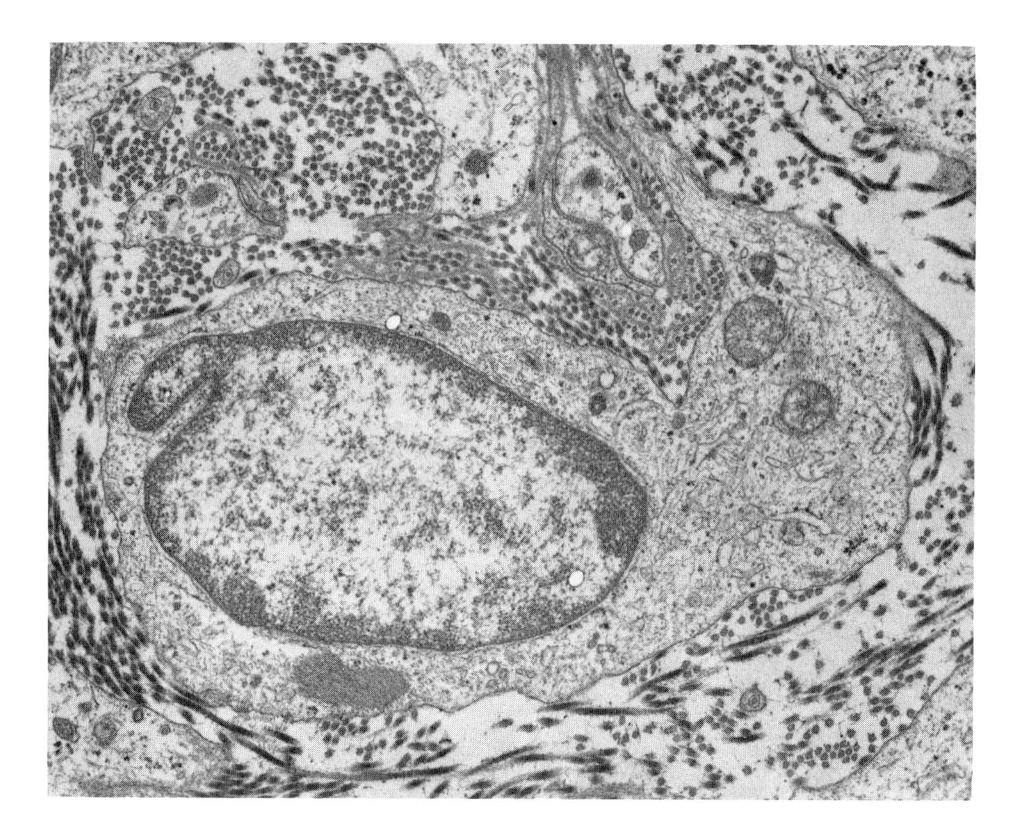

FIGURE 23. Malignant schwannoma, discontinuous basal lamina. In this particular case the tumor cell shows features often found in Schwann cells, such as elongate slender cell processes lacking any neural (neuronal) differentiation. However, a basal lamina is noted to be deposited around portions of the tumor cell, whereas most of the tumor cell surface is devoid of basal lamina. These also note that no neurites are found in association with the tumor cell and no mesaxons containing neurites are present anywhere in the tumor cell cytoplasm. This is typical of malignant nerve sheath tumors where the parallel to normal Schwann cell neuronal interaction is imperfect at best.

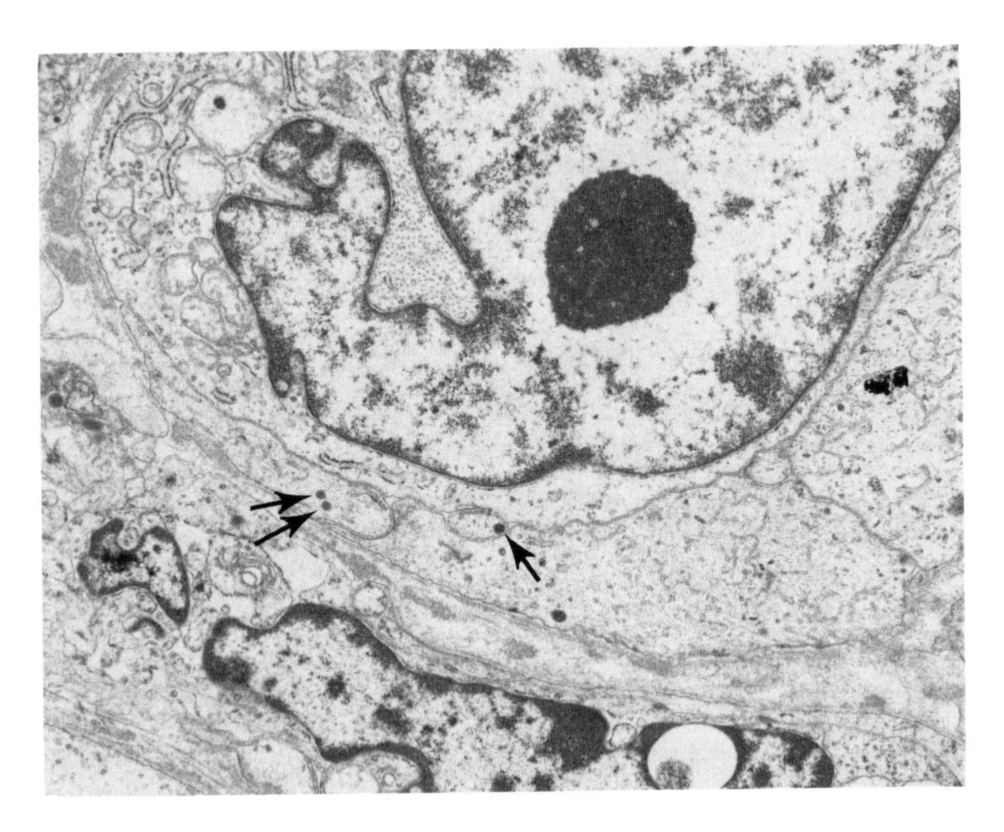

FIGURE 24. Malignant schwannoma, neurosecretory granules. In another tumor felt to be a malignant melanotic schwannoma, in addition to the Schwann cell appearance and areas of melanogenesis, areas such as illustrated here containing dense core granules were also found.

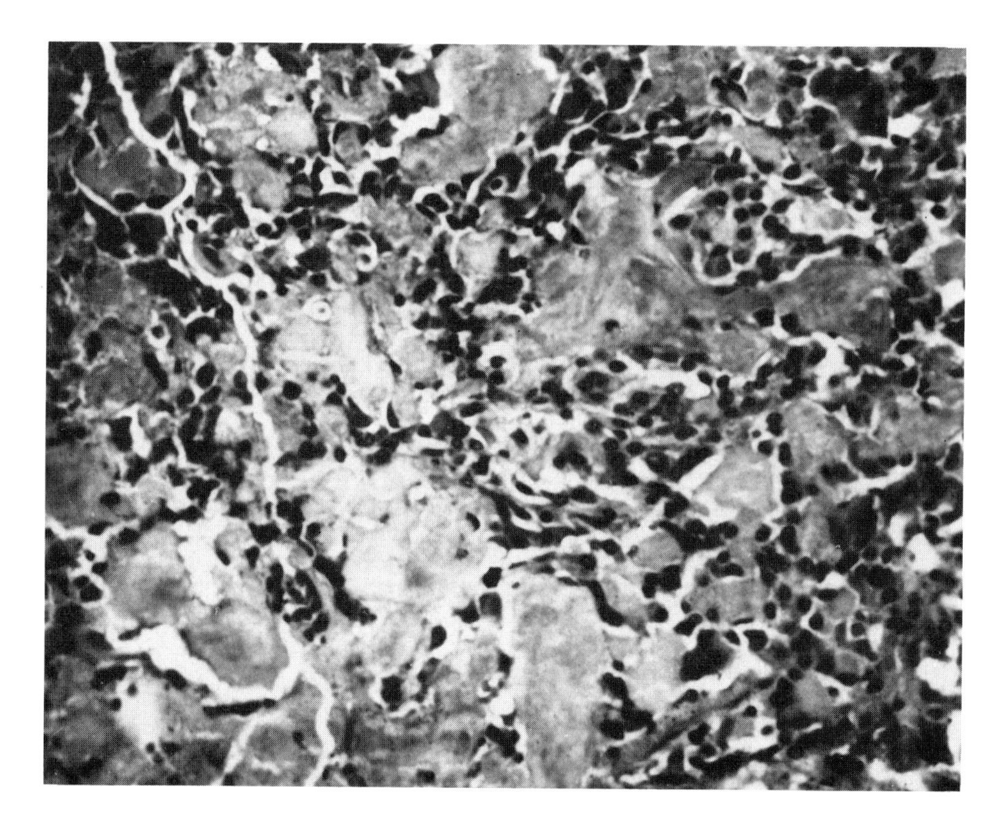

FIGURE 25. Medullary carcinoma of thyroid, light microscopy. Note that clusters of tumor cells are separated by islands of homogeneous eosinophilic extracellular material. By electron microscopy this proves to be amyloid fibrils.

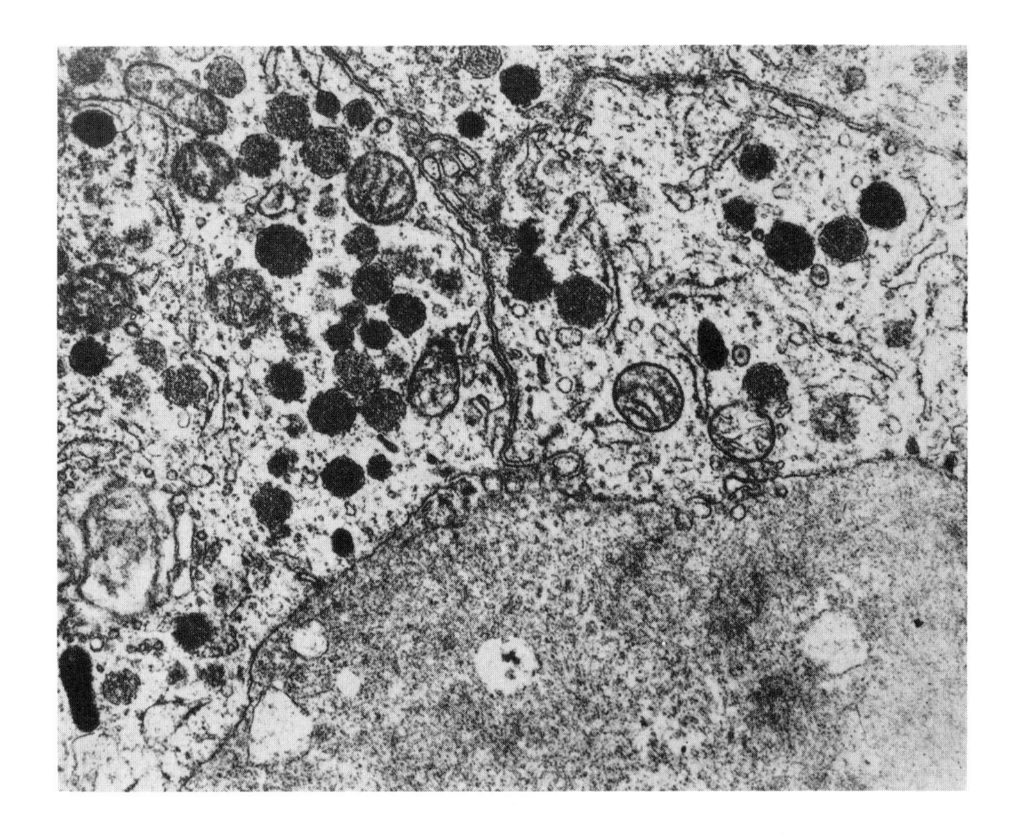

FIGURE 26. Same case as Figure 25, electron microscopy. Note here that the basal cell cytoplasm seen in the upper portion of the micrograph is replete with innumerable, very large, and generally quite electron-dense neurosecretory granules, in this case typical of medullary carcinoma of thyroid cells. Also note in the bottom right hand of the photomicrograph the finely fibrillar extracellular material which has been characterized as amyloid.

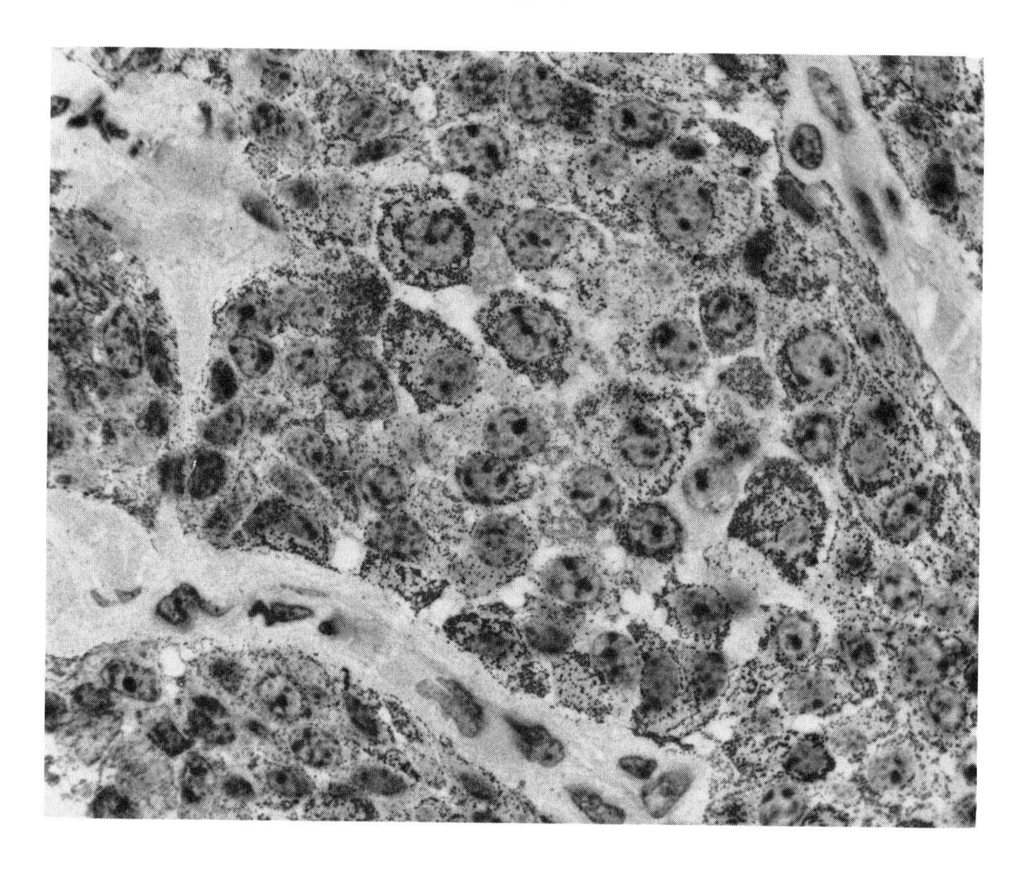

FIGURE 27. Carcinoid, Grimelius stain. A simple silver **impregnation** technique such as the Grimelius stain illustrated here will generally clearly identify the larger neurosecretory granules such as found in carcinoid, also illustrated here.

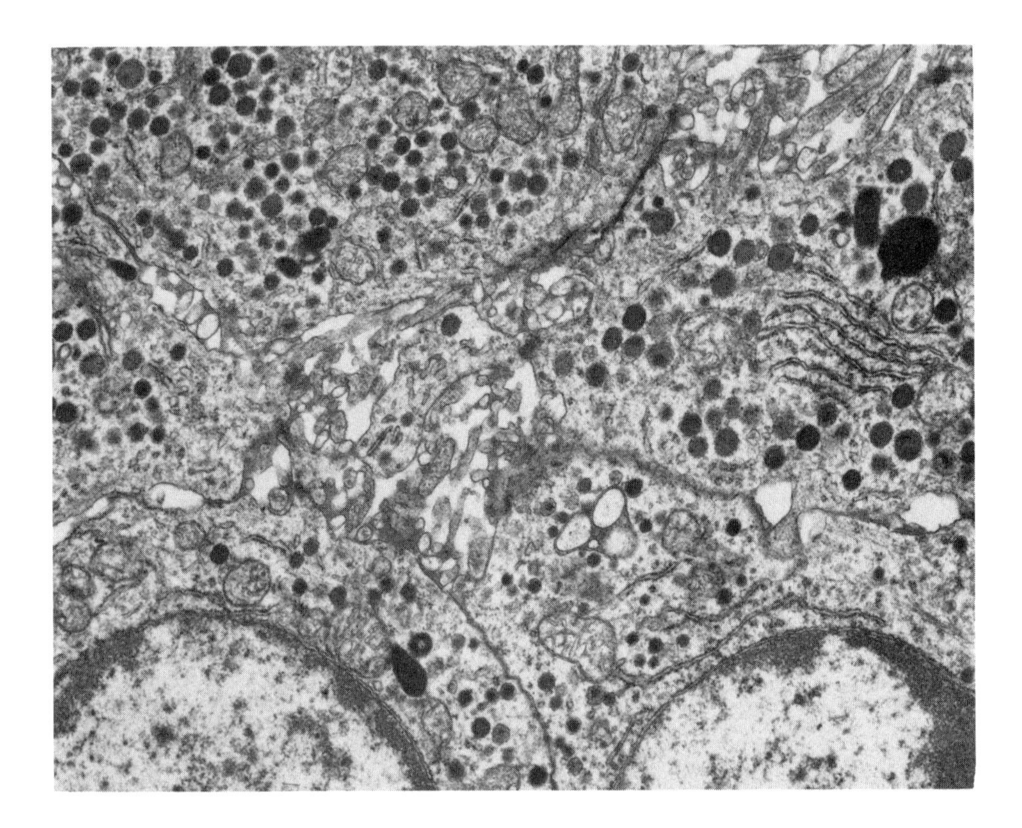

FIGURE 28. Carcinoid, electron microscopy, granules. Note the enormous number of mixed large and small neurosecretory granules. The larger granules are typically identified as carcinoid-type granules; the small as neurosecretory.

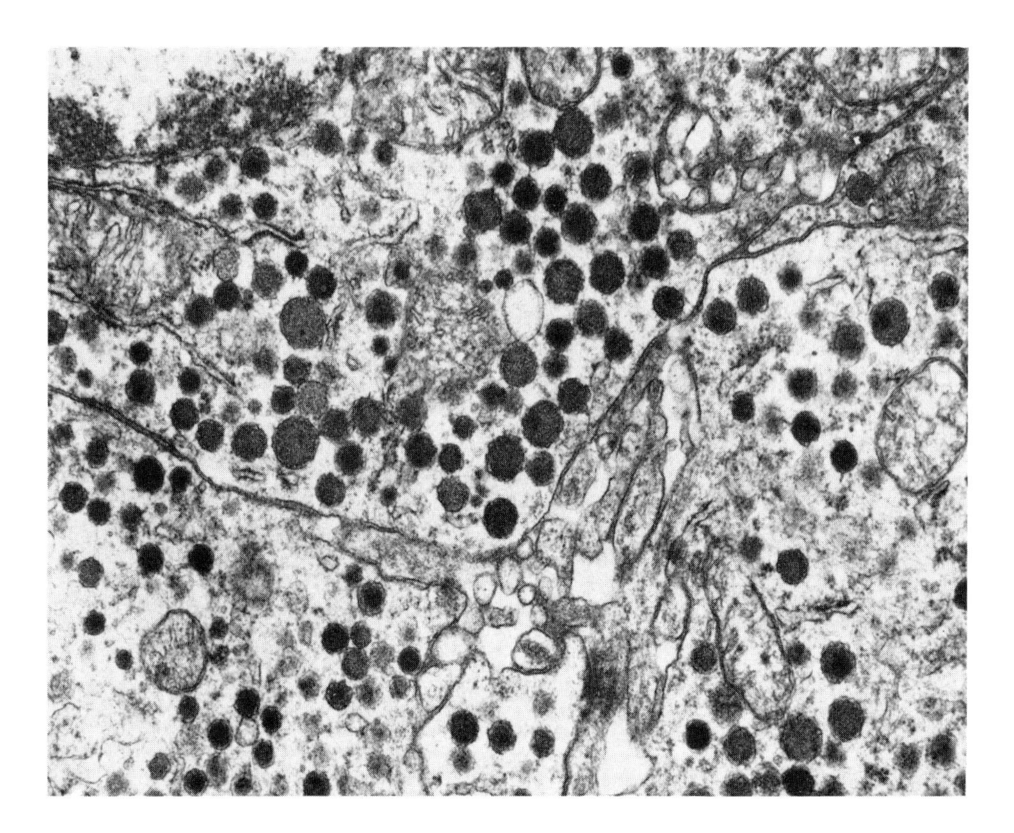

FIGURE 29. Bronchial carcinoid, electron microscopy. The very large uniformly electron dense granules illustrated here are readily identified by the Grimelius stain noted above. Many attain diameters of 200 to 400 nm.

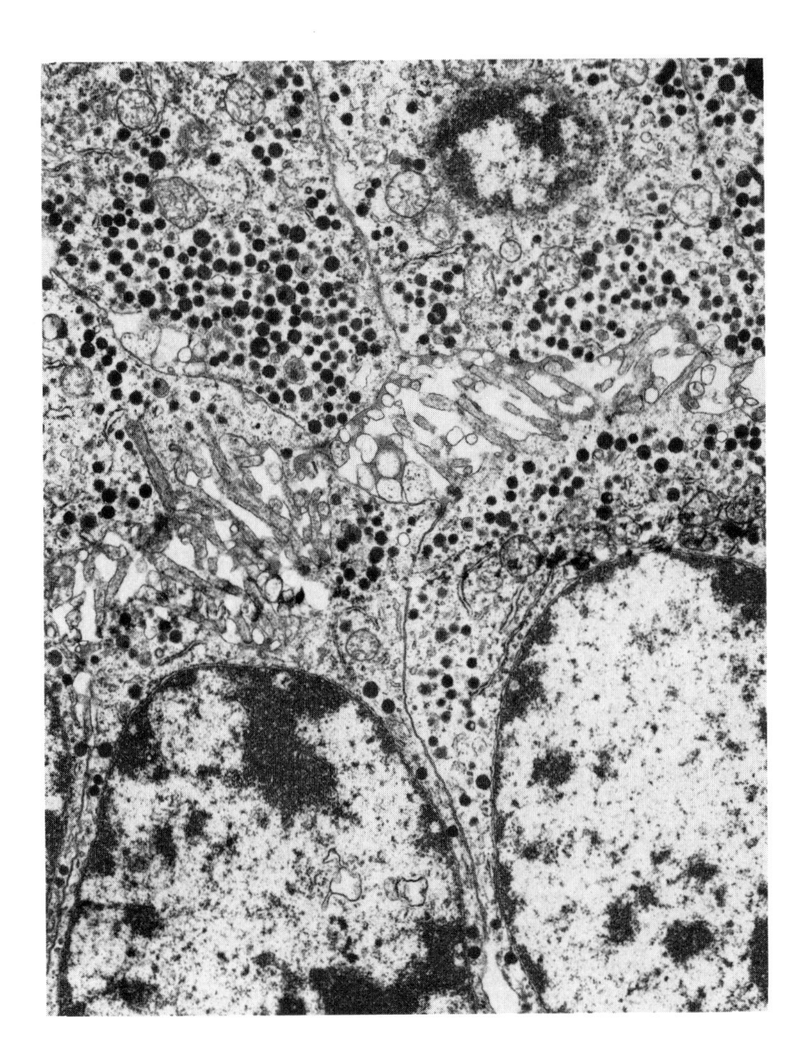

FIGURE 30. Bronchial carcinoid, glandular differentiation. Most bronchial carcinoids and many other carcinoids occurring in diverse anatomic sites show striking evidence of glandular differentiation as illustrated here. A lumen into which projects large numbers of microvilli traverses the center of the picture.

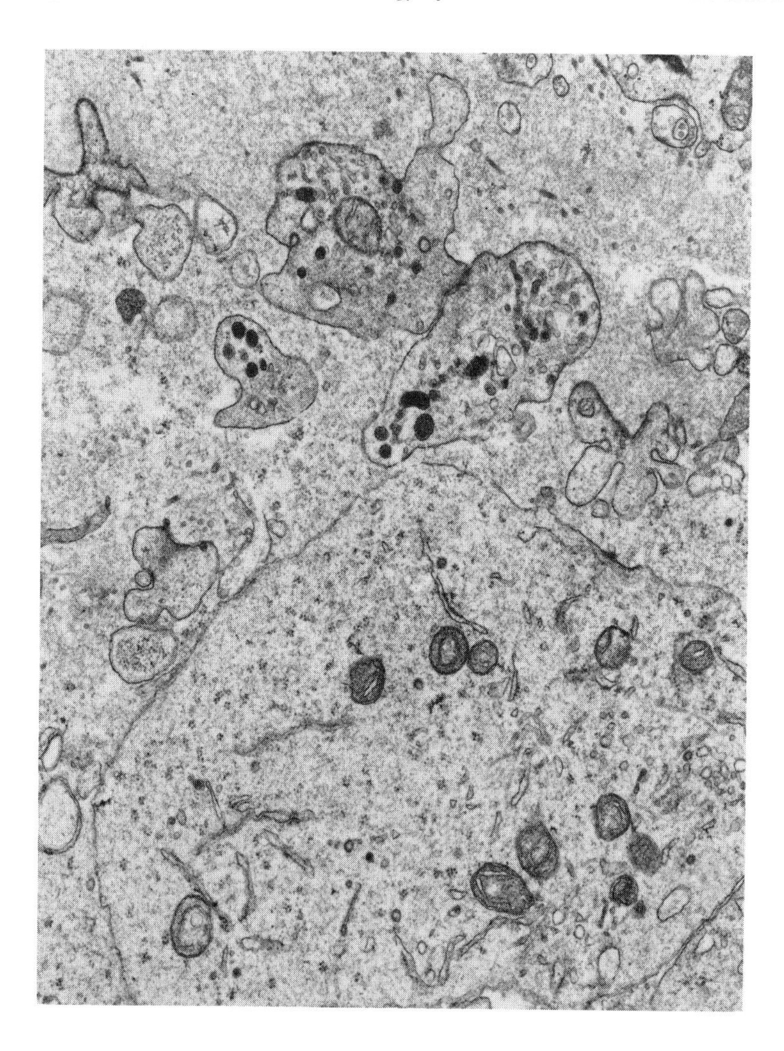

FIGURE 31. Oat cell (small cell) carcinoma, scant dense core granules. In contrast to the well-differentiated low-grade malignancy carcinoids noted above, the more common form of lung carcinoma of neuroendocrine type is small cell carcinoma, which generally has very few neurosecretory granules and these are located predominantly in neural processes as noted here. Glandular differentiation is generally not appreciated in this tumor.

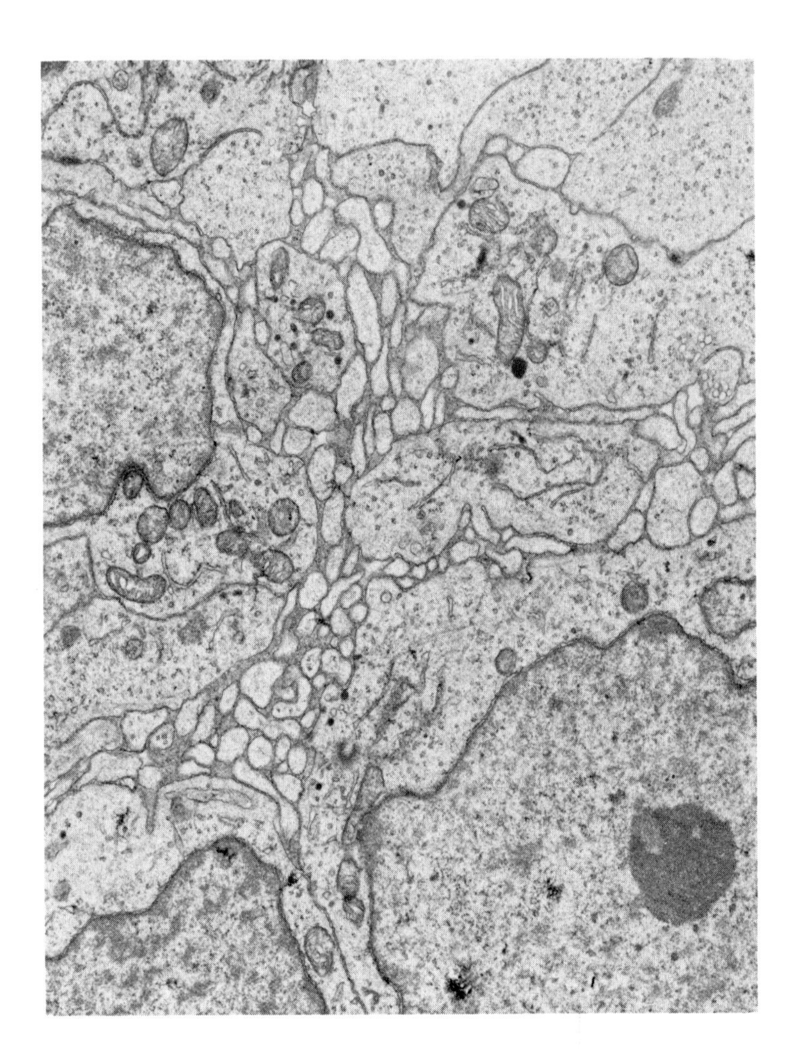

FIGURE 32. Oat cell carcinoma, neural processes. In many cases the numbers of neurosecretory granules in oat cell carcinoma are few and far between, and instead enormous numbers of slender neural processes not unlike those seen in neuroblastoma are noted instead, as seen here.

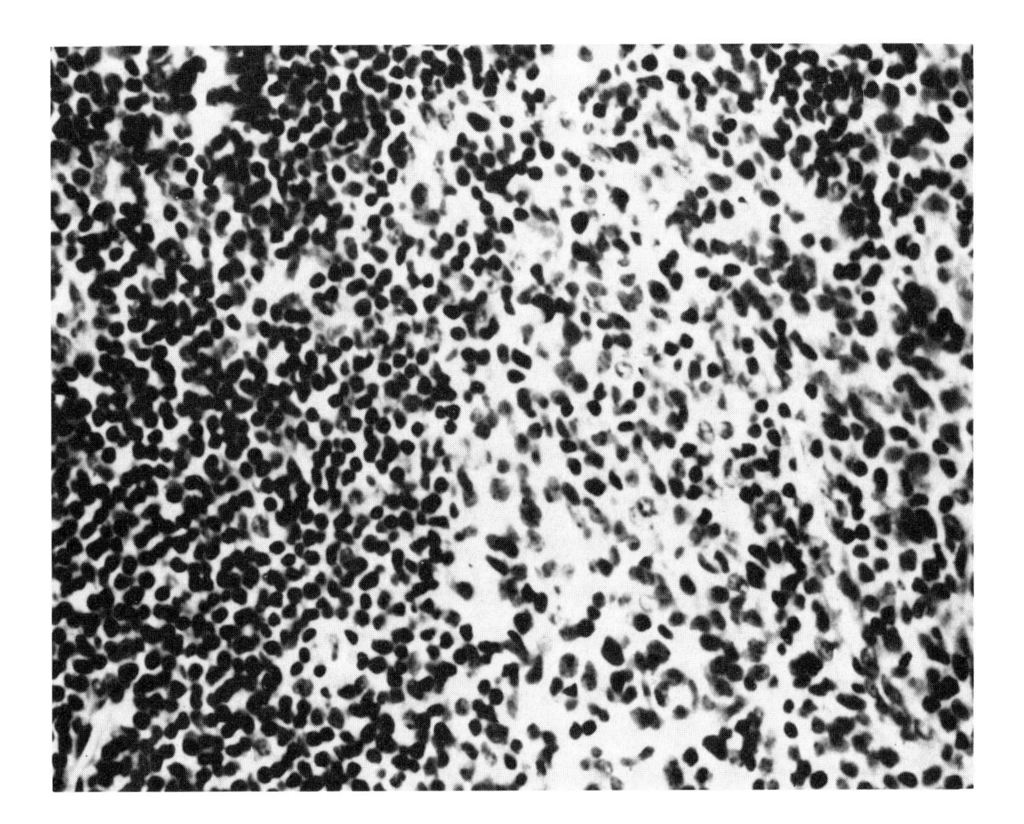

FIGURE 33. Aggressive carcinoid, mediastinal metastasis, light microscopy. In some cases intermediate forms of tumor are noted. The light microscopy of this tumor seen in a lymph node metastasis closely mimics that of oat cell carcinoma, where the tumor cells are similar in appearance to lymphocytes, seen to either side of the field illustrated here.

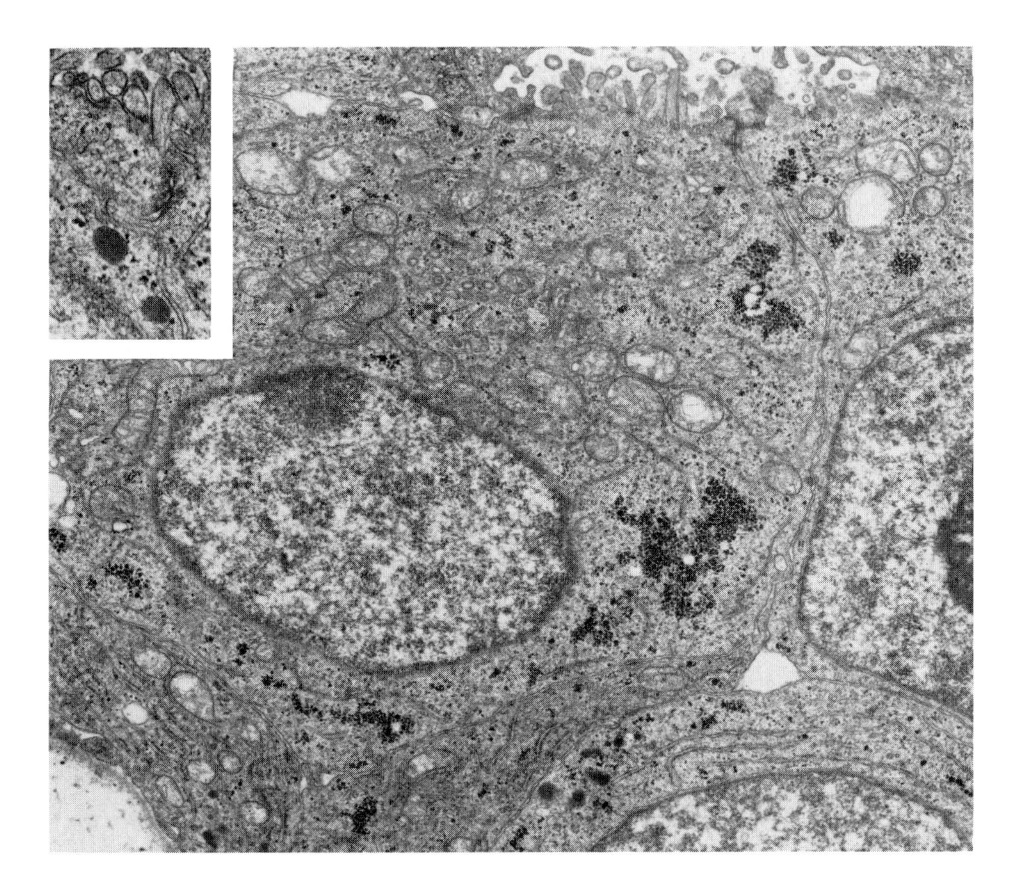

FIGURE 34. Aggressive carcinoid, mediastinal metastasis, glands, and microvilli. The same tumor illustrated in Figure 33 reveals upon electron microscopic examination clear cut evidence of glandular differentiation with microvilli protruding into the lumen at the top and a basal lamina at the bottom. This more closely approximates that seen in bronchial carcinoid. However, only scant granules are seen (noted in detail in the inset) much like that seen in oat cell carcinoma. This tumor shows intermediate ultrastructural characteristics and generally intermediate malignancy as well.

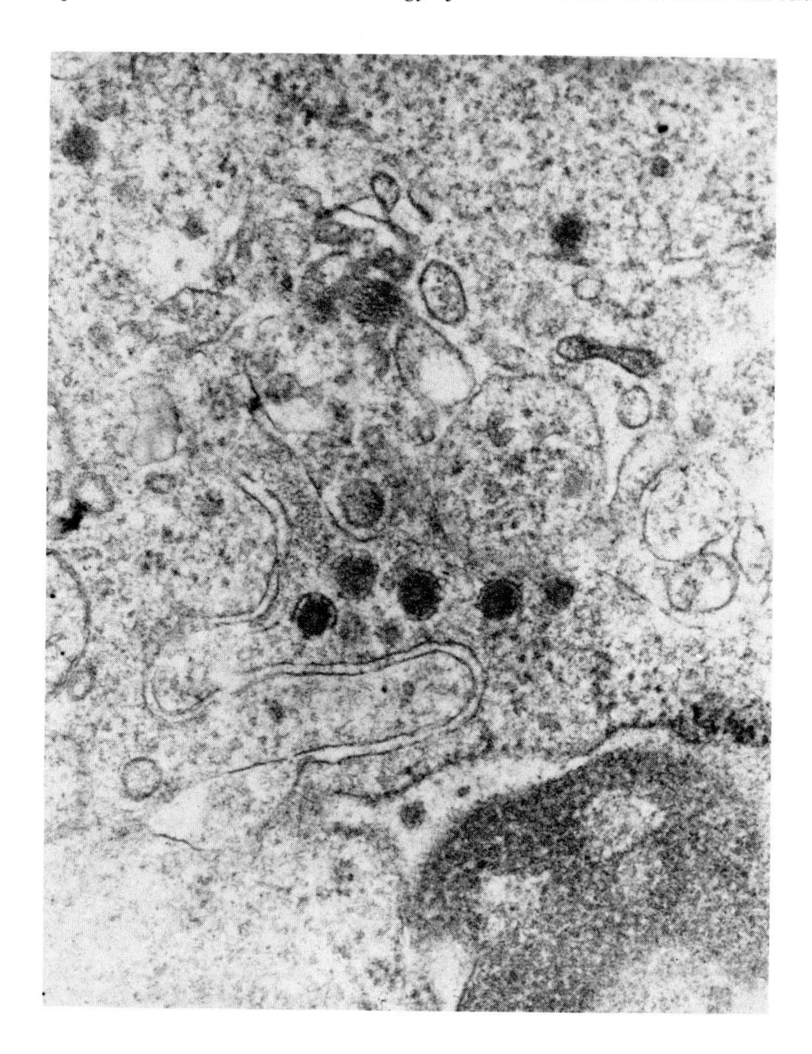

FIGURE 35. Mediastinal carcinoid, oat cell type granules. Carcinoids occurring in diverse locations may have a variety of granule types and still be termed carcinoid. This particular tumor arose with the thymus, is clearly of neuroendocrine origin, but shows only small neurosecretory type granules, and no large carcinoid type granules seen in bronchocarcinoid, as illustrated above.

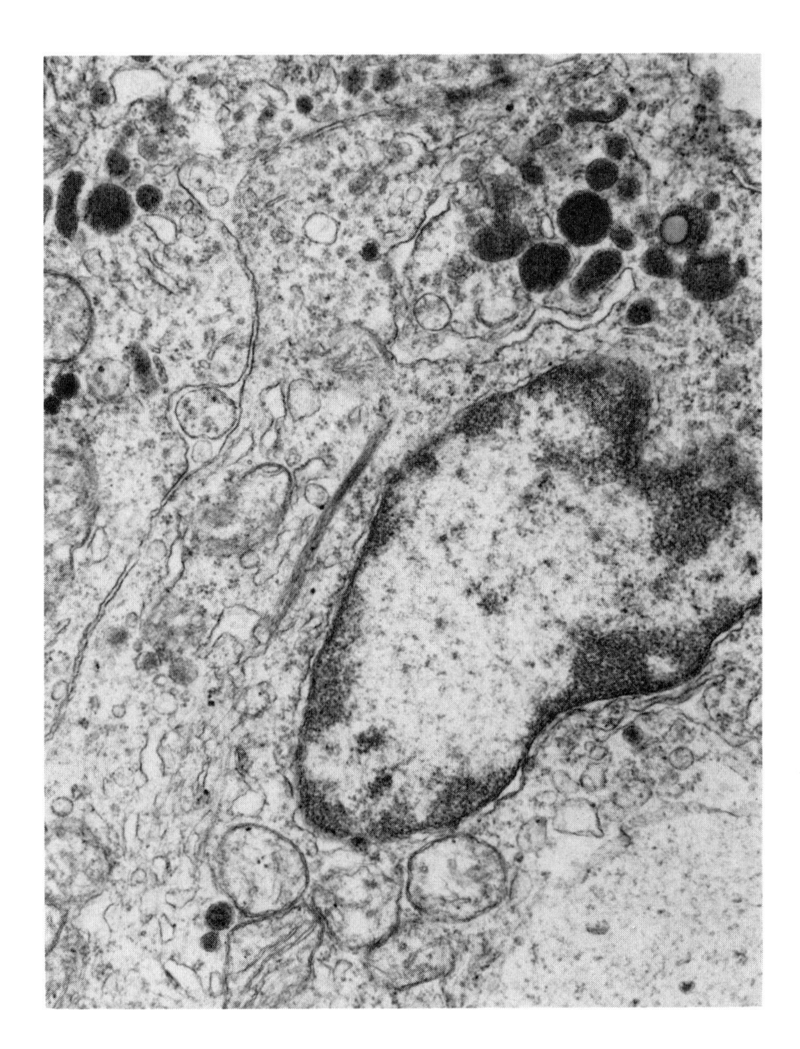

FIGURE 36. Mediastinal carcinoid, classic carcinoid granules. Other tumors such as the mediastinal carcinoid illustrated here in contrast may show very large granules which are generally less regular and less typical in appearance than the better differentiated low-grade malignancy bronchial carcinoids. Also note that scattered throughout the cell cytoplasm are individual small neurosecretory-type granules as well. A further finding is the presence of keratin-type filaments adjacent to the nucleus noted in the center of the field.

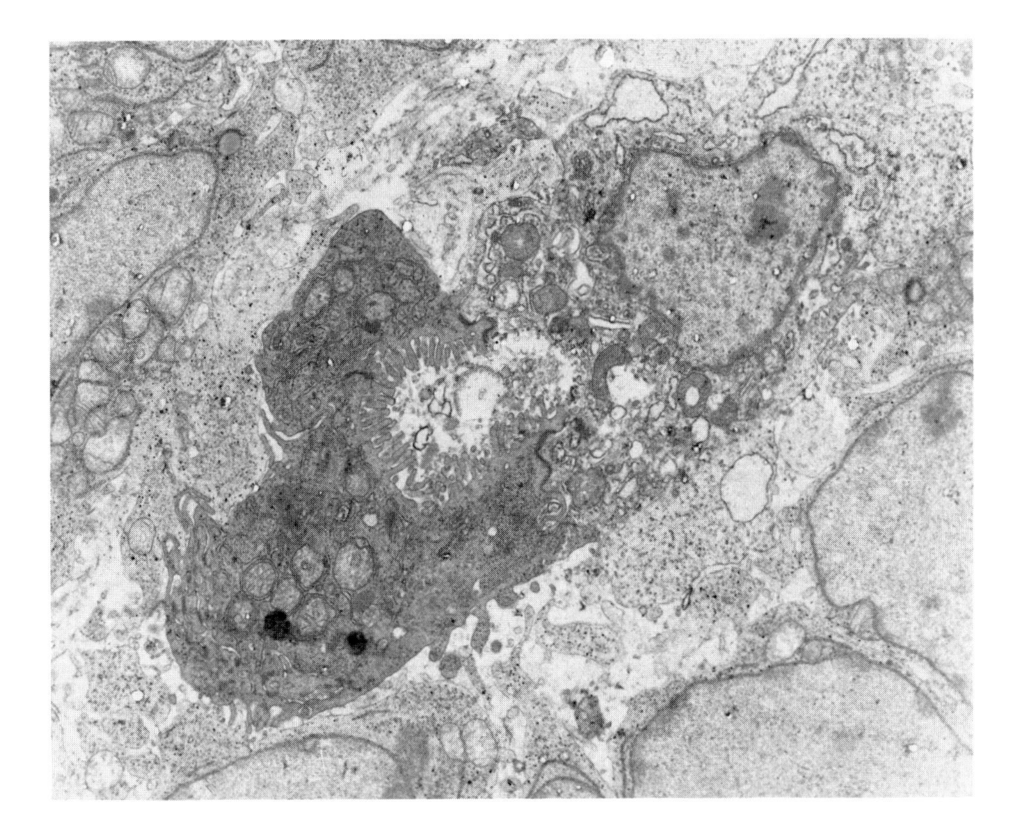

FIGURE 37. Hepatic carcinoid, ductal association. Rare carcinoids of the hepatobiliary tree have also been reported. The example illustrated here at low magnification appears to be arising from a small bile duct. A tumor cell noted in the corner is similar in appearance to the surrounding tumor cells, but is clearly part of the bile duct structure itself.

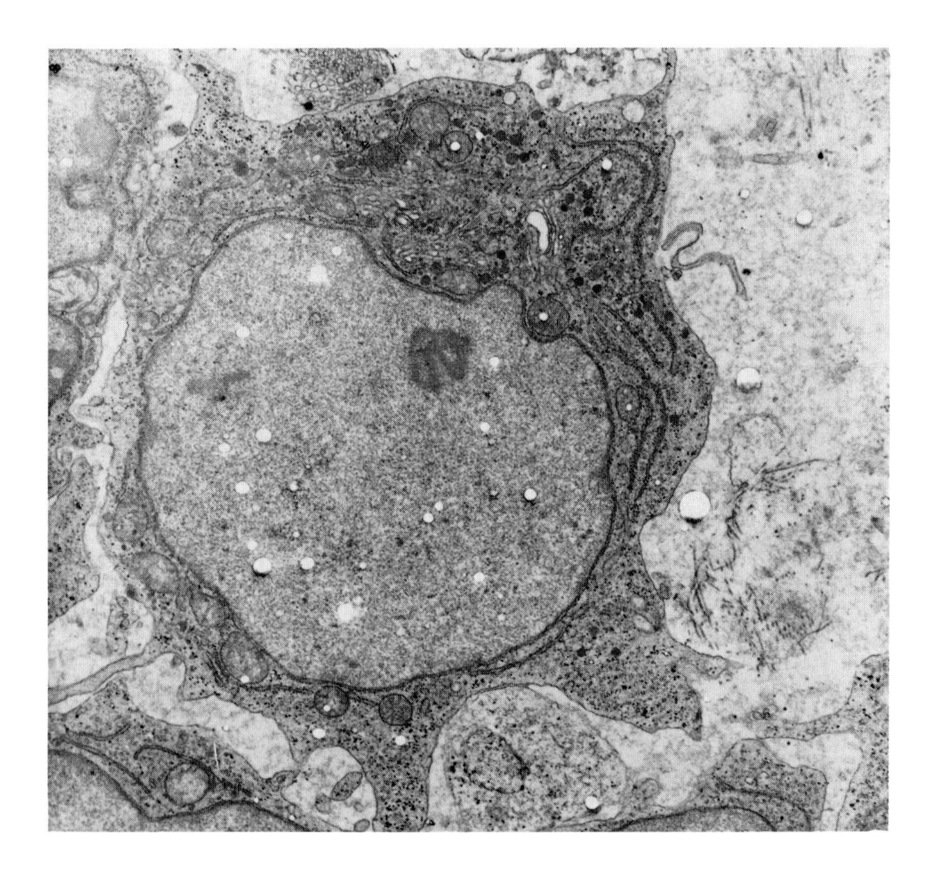

FIGURE 38. Hepatic carcinoid, detail. The same case illustrated in Figure 37 seen here at higher magnification shows a tumor cell containing a reasonably large number of neurosecretory-type granules of intermediate size.

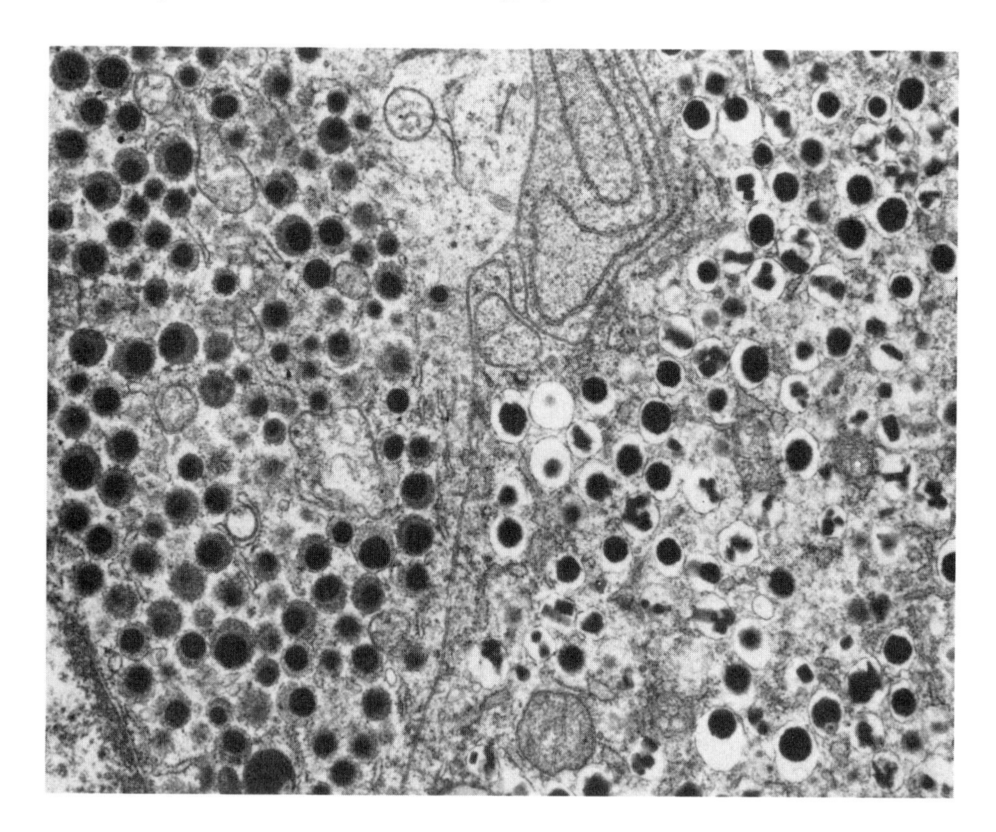

FIGURE 39. Normal islet of Langerhan's with alpha and beta granules. Note here that the alpha granules (left) and the beta granules (right) have characteristic ultrastructural appearances that clearly distinguish one from the other. Alpha granules contain an often eccentric electron dense nucleoid surrounded by less electron dense matrix which completely fills the membrane bound granule. In contrast, beta granules frequently contain crystalline material that appears to "float" within a lucent matrix bounded by a poorly fitting membrane. Other beta granules may appear more like the pheochromocytoma granules noted above with a noncrystalline, uniformly electron-dense core and loose fitting halo as well. Only the former, crystalline inclusions, are diagnostic of insulin-containing granules.

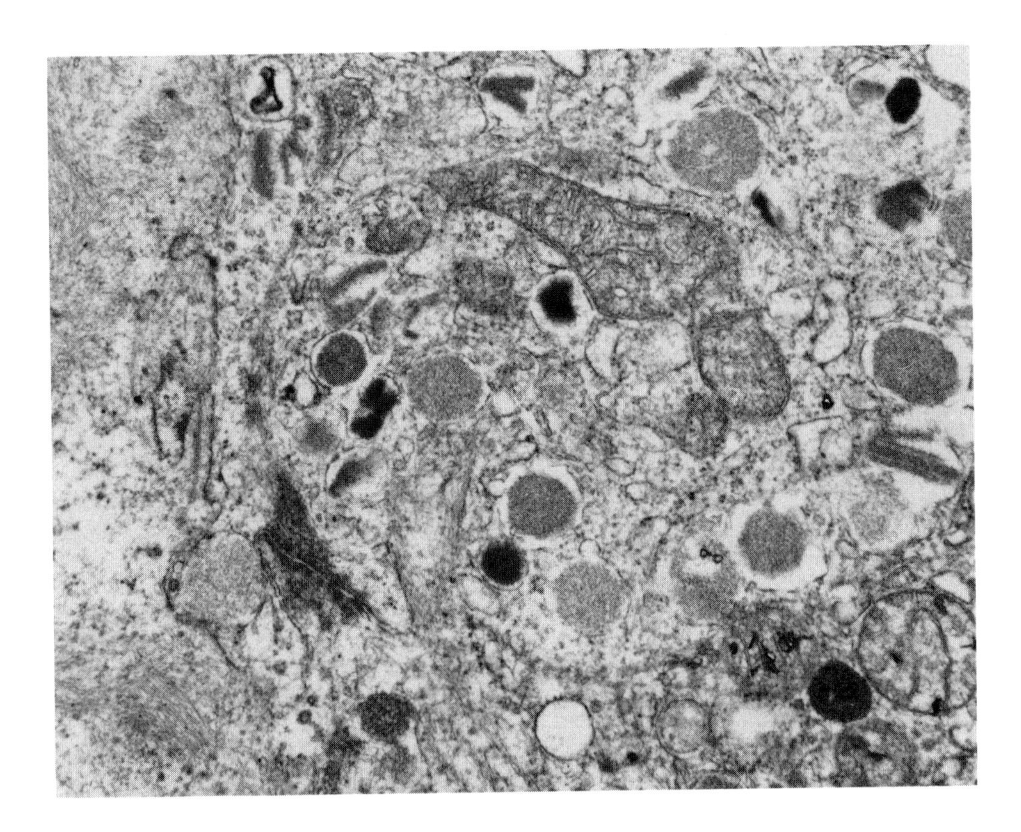

FIGURE 40. Insulinoma, beta granules, typical. The neoplastic counterpart of normal islet cells occurs among islet cell tumors. This insulin-secreting beta cell granule containing tumor shows insulin-containing granules virtually indistinguishable from those seen in the beta cells of normal islets illustrated in Figure 39.

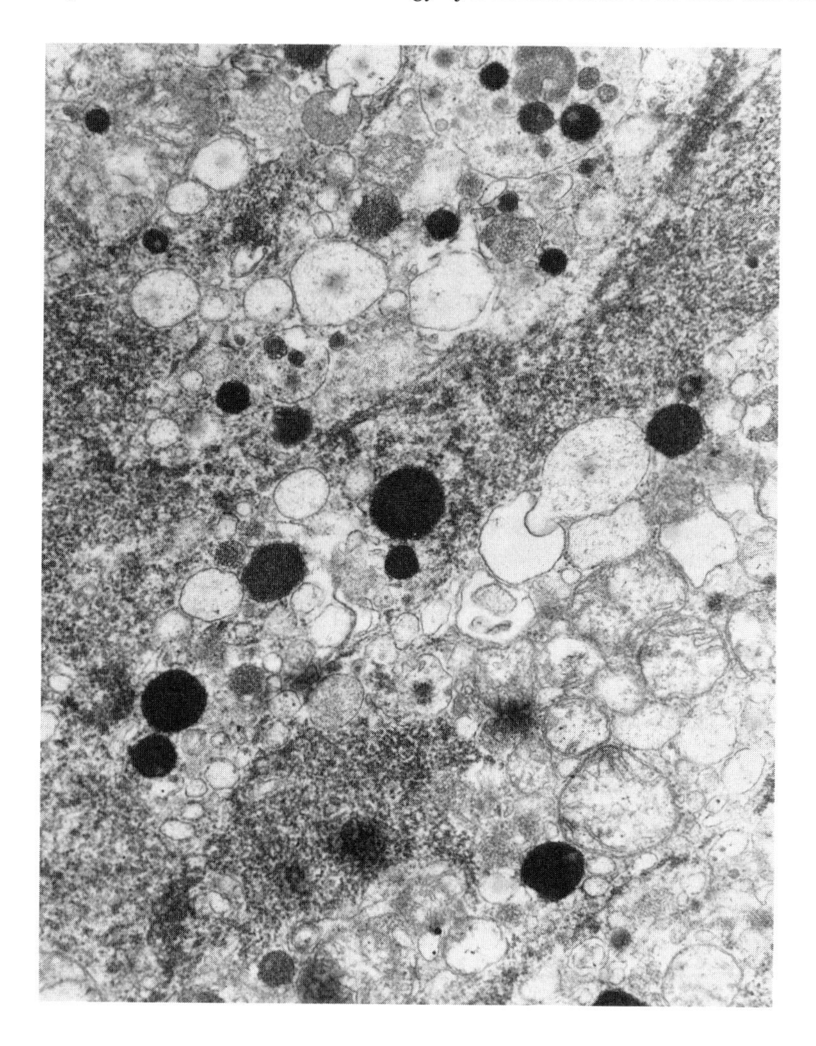

FIGURE 41. Insulinoma, atypical beta granules. Other cases of insulinoma may lack diagnostic ultrastructural characteristics of beta granules, as illustrated here. This documented biologically active insulin-secreting tumor lacked any evidence of typical beta granules as illustrated previously.

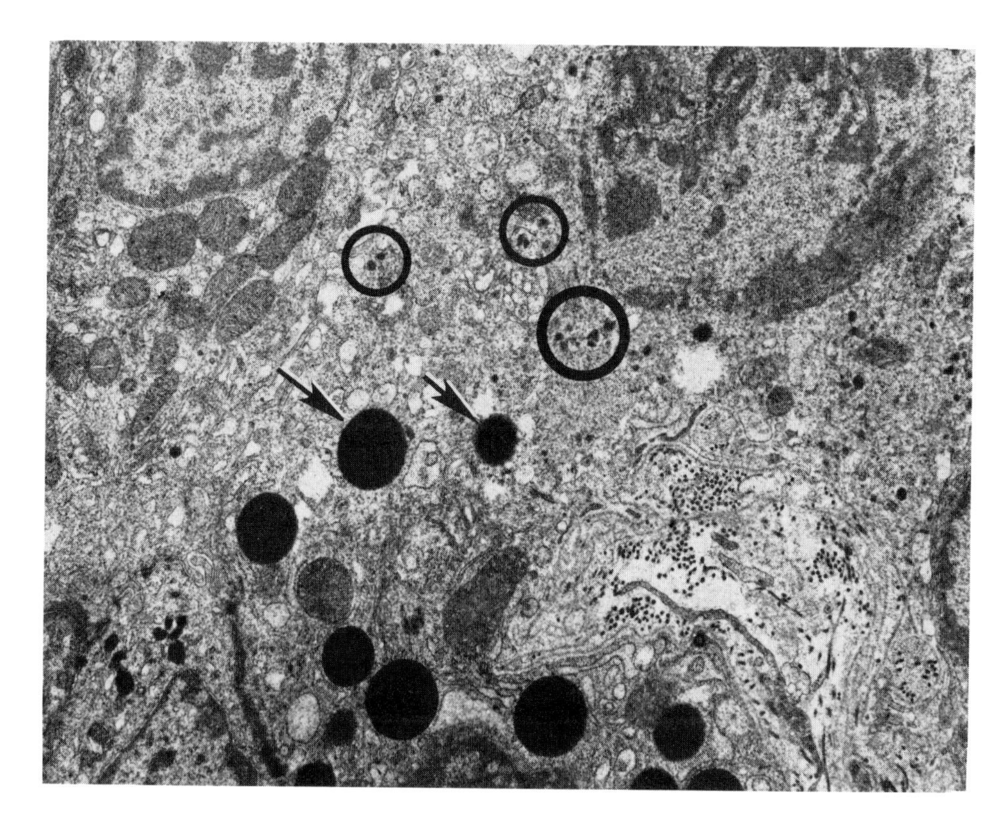

FIGURE 42. Insulinoma, beta, and zymogen granules. This low magnification electron micrograph illustrates the cytoplasm of a tumor cell containing both very small but electron-dense granules with irregular crystalline structure immediately adjacent to very large regular zymogen granules normally seen in the **exocrine** portion of the pancreas.

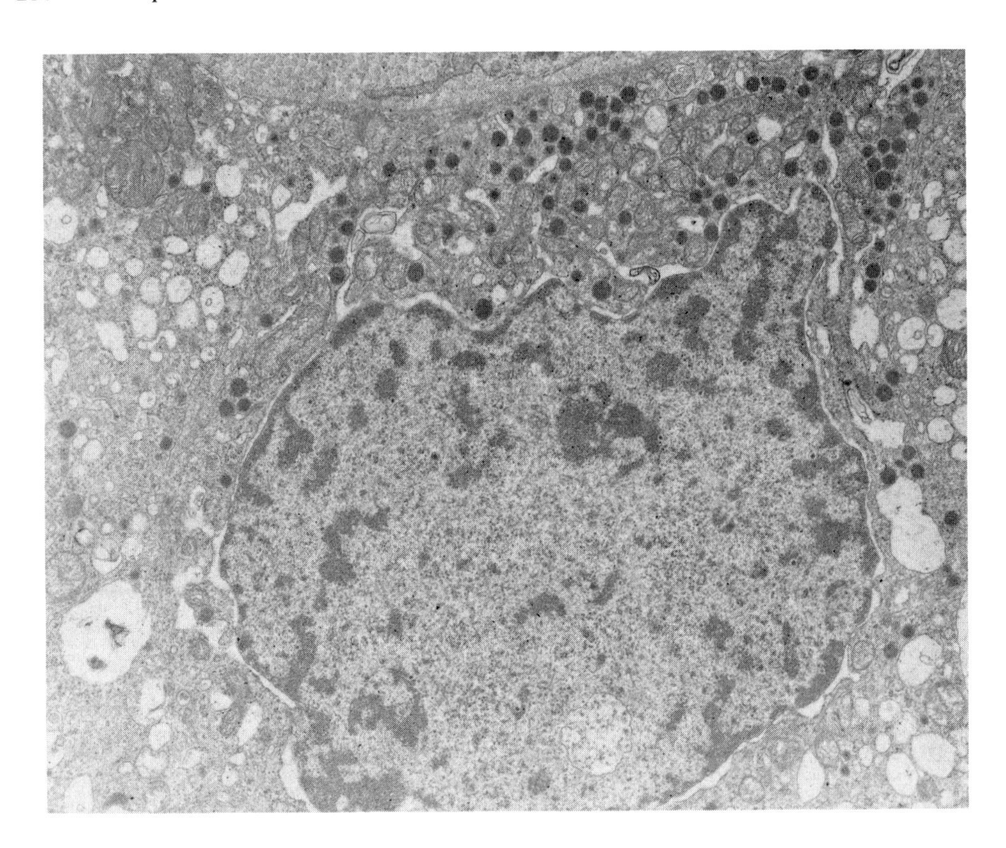

FIGURE 43. Glucagonoma, nonspecific alpha granules. In contrast to insulinomas, most other islet cell tumors contain granules which cannot be correlated precisely with content. This known biologically active glucagon-secreting tumor contains a large number of carcinoid-type granules illustrated here. However, no classic alpha granules such as those illustrated in the normal islet of Langerhan's (above) are seen in this tumor. This is typical of glucagonoma and other islet cell tumors.

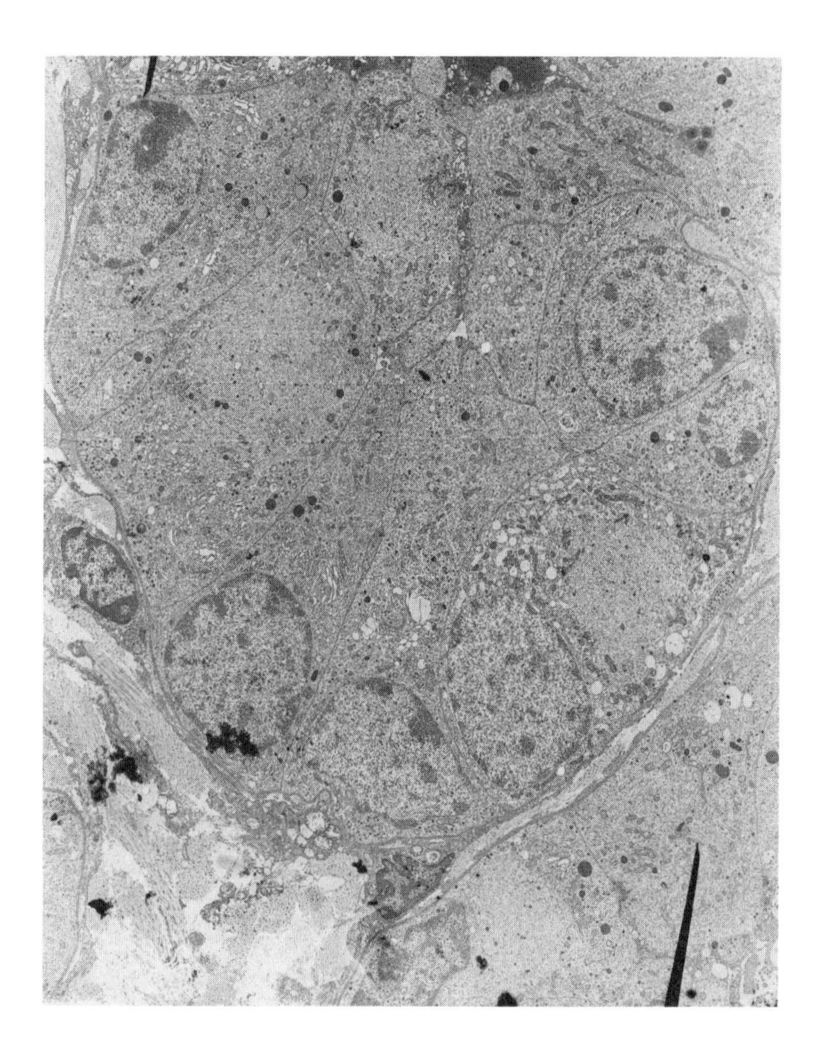

FIGURE 44. Gastrinoma, glandular differentiation. So-called Zollinger-Ellison syndrome patients have islet cell tumors of the pancreas or small bowel, often with glandular differentiation as illustrated here. These gastrin-secreting tumors likewise contain carcinoid-type granules which cannot be distinguished from the glucagon-secreting tumor, noted above or any other non-beta cell tumor for that matter.

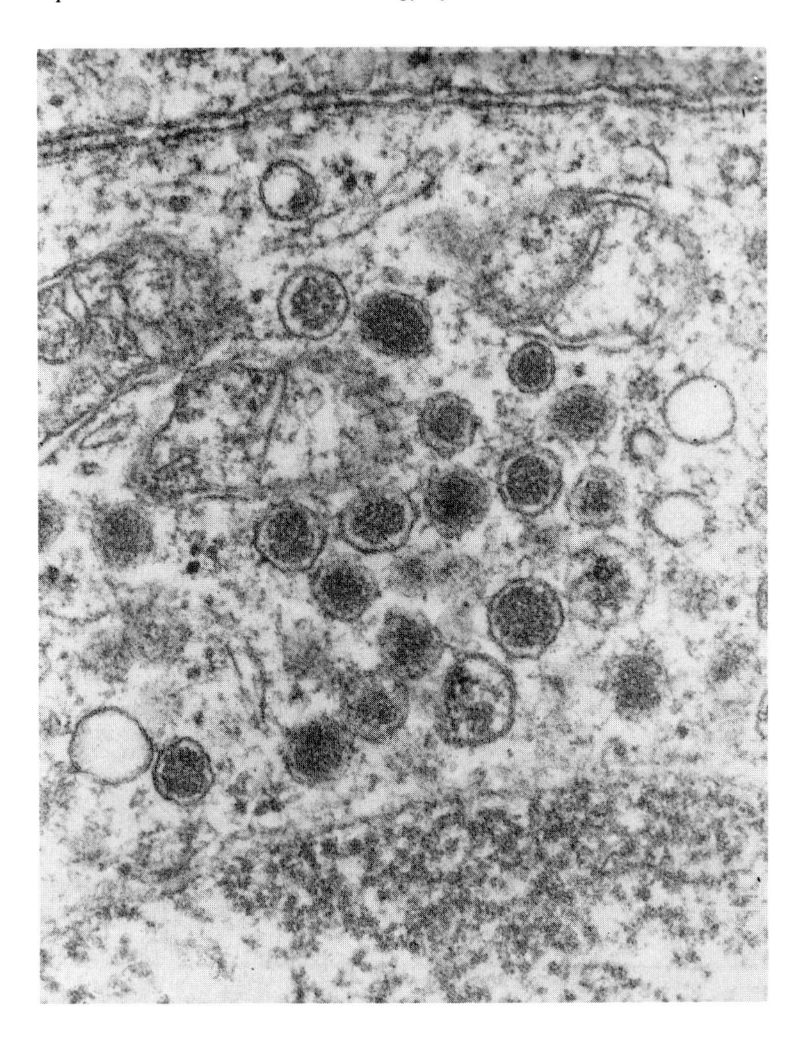

FIGURE 45. Gastrinoma, carcinoid-type granules, detail. At higher magnification, the granules of the gastrinoma noted above closely resemble the neurosecretory granules of neuroblastoma with a regular electron-dense core and scant lucent halo surrounding the core as opposed to the carcinoid granules illustrated in Figures 28 to 30.

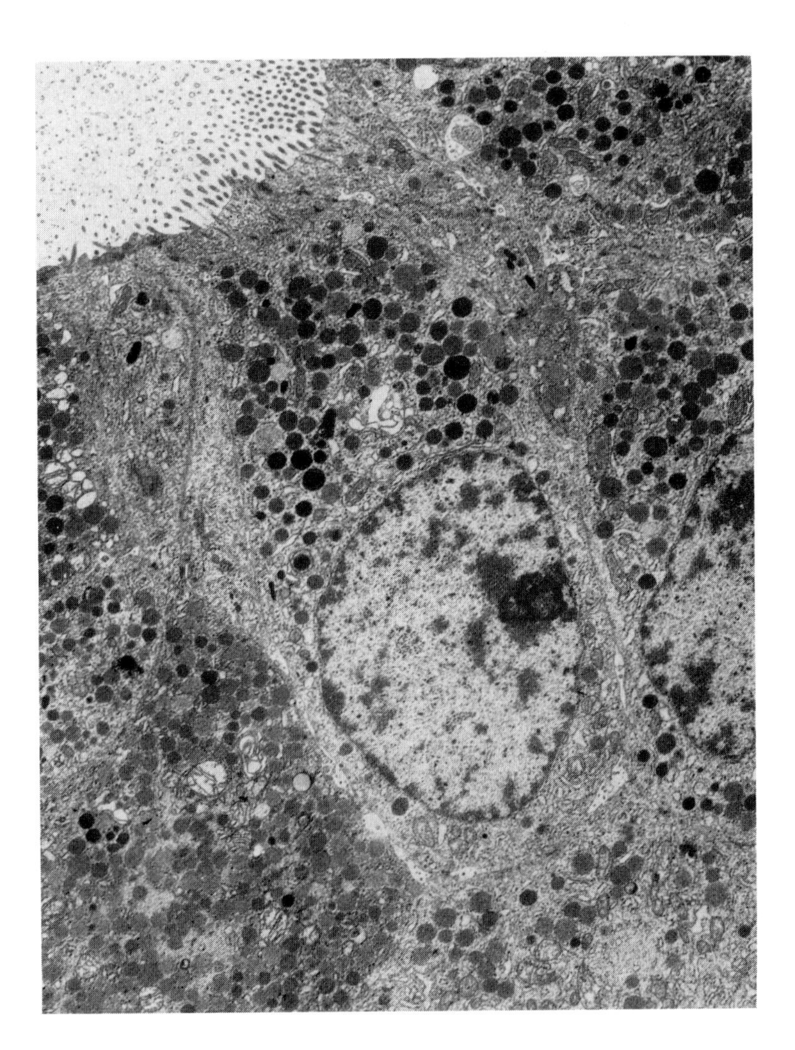

FIGURE 46. Somatostatinoma, glandular differentiation. A variety of islet cell tumors secreting other neuroendocrine substances, including somatostatin, have been described. As noted with glucagonoma and gastrinomas above, the secretory product cannot be discerned from the morphology of the granule. Here the granules more closely resemble the bronchial carcinoids noted in Figures 28 to 30, and, like that tumor, show striking glandular differentiation with microvilli protruding into a lumen above.

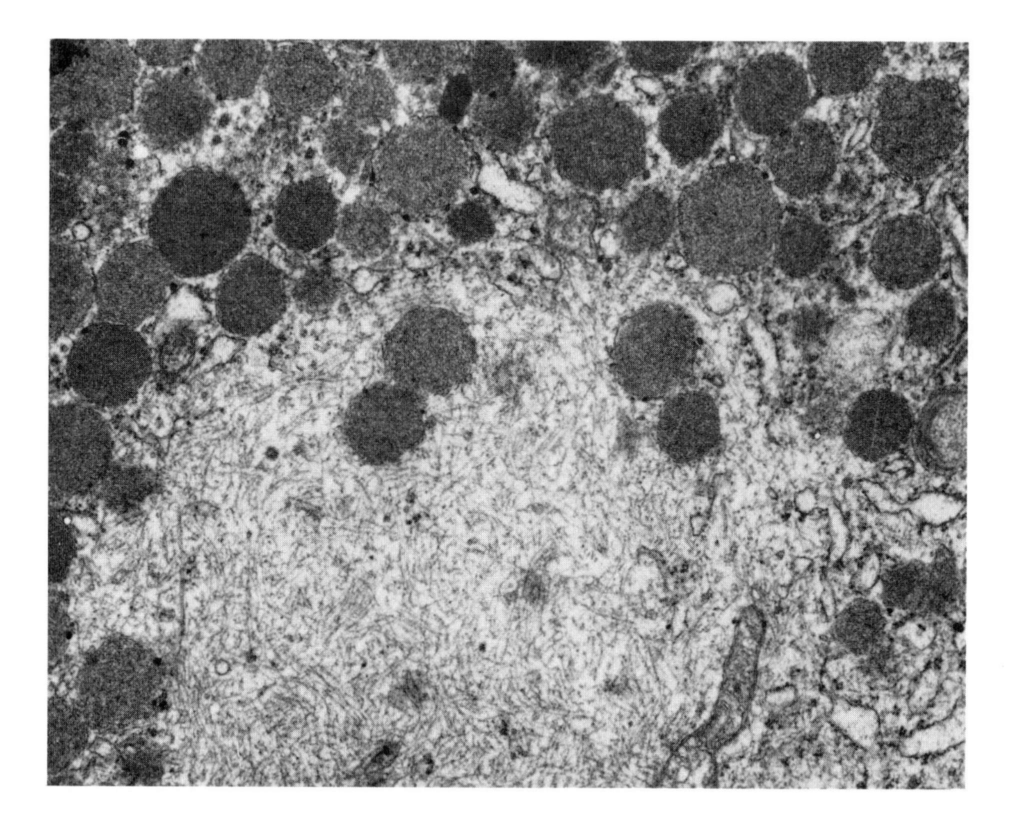

FIGURE 47. Somatostatinoma, detail. At high magnification the carcinoid-type granules are clearly seen as well as cytoplasmic fibrils which have been variously interpretated as cytoplasmic amyloid fibrils representing fibrillar secretory product or masses of cytoskeletal filaments. In either case this material is distinct from the extracellular amyloid illustrated in medullary carcinoma, Figures 25 and 26.

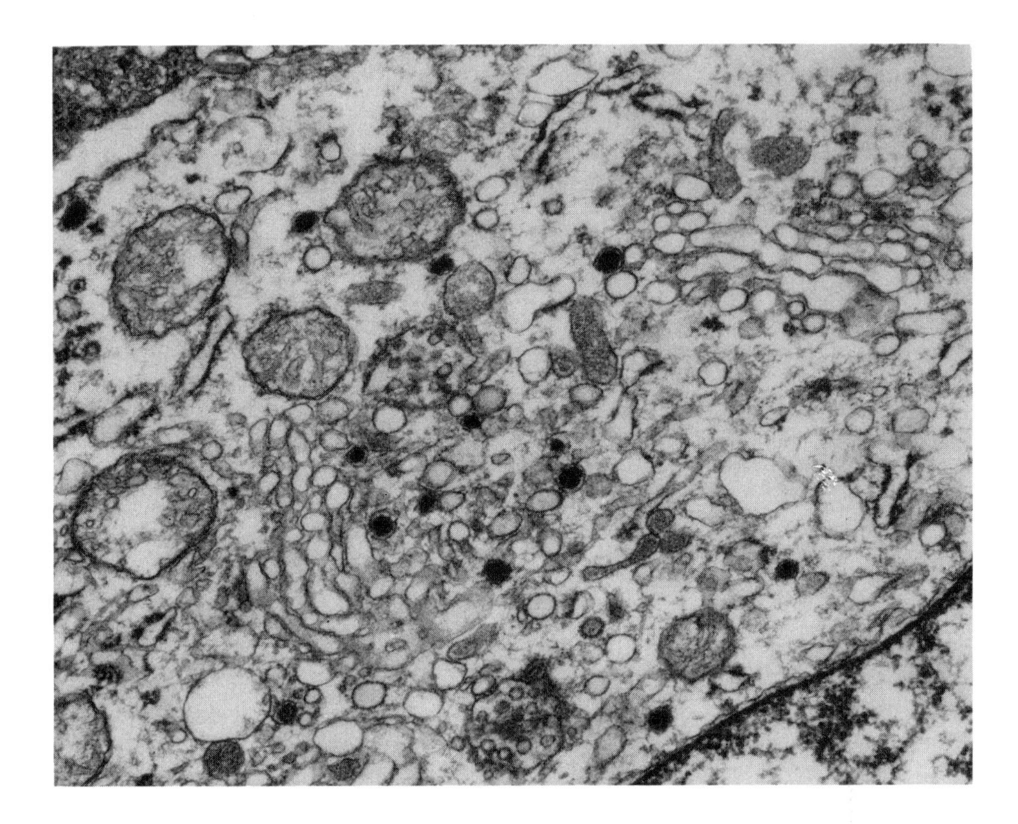

FIGURE 48. Merkel cell carcinoma, neurosecretory granules. Other than melanoma, the most common primary neuroendocrine neoplasm of skin is the so-called Merkel cell tumor or trabecular carcinoma of skin. The tumor cells typically contain classic neurosecretory, not carcinoid-type, granules as illustrated here in the Golgi region of a tumor cell.

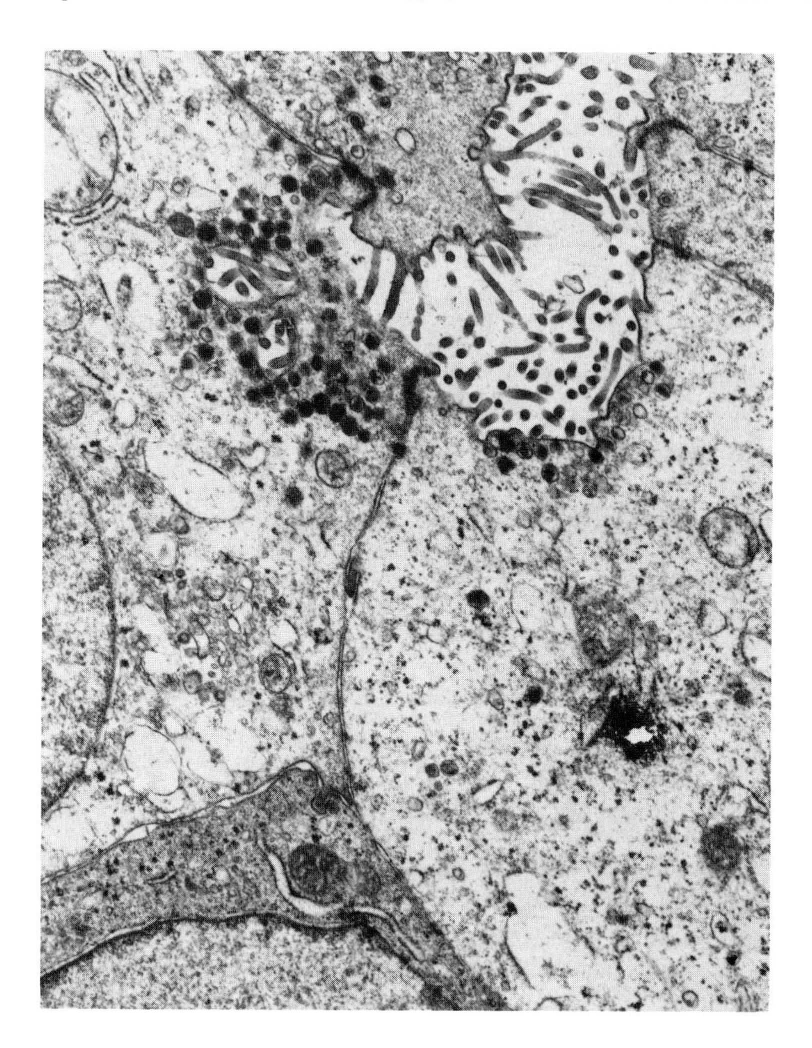

FIGURE 49. Neuroendocrine carcinoma, pleomorphic granules. This neuroendocrine carcinoma of skin shows glandular differentiation and apically arrayed small uniformly electron dense neuroendocrine granules. Microvilli are evident in this field. In other areas basally located neuritic-type processes were also noted.

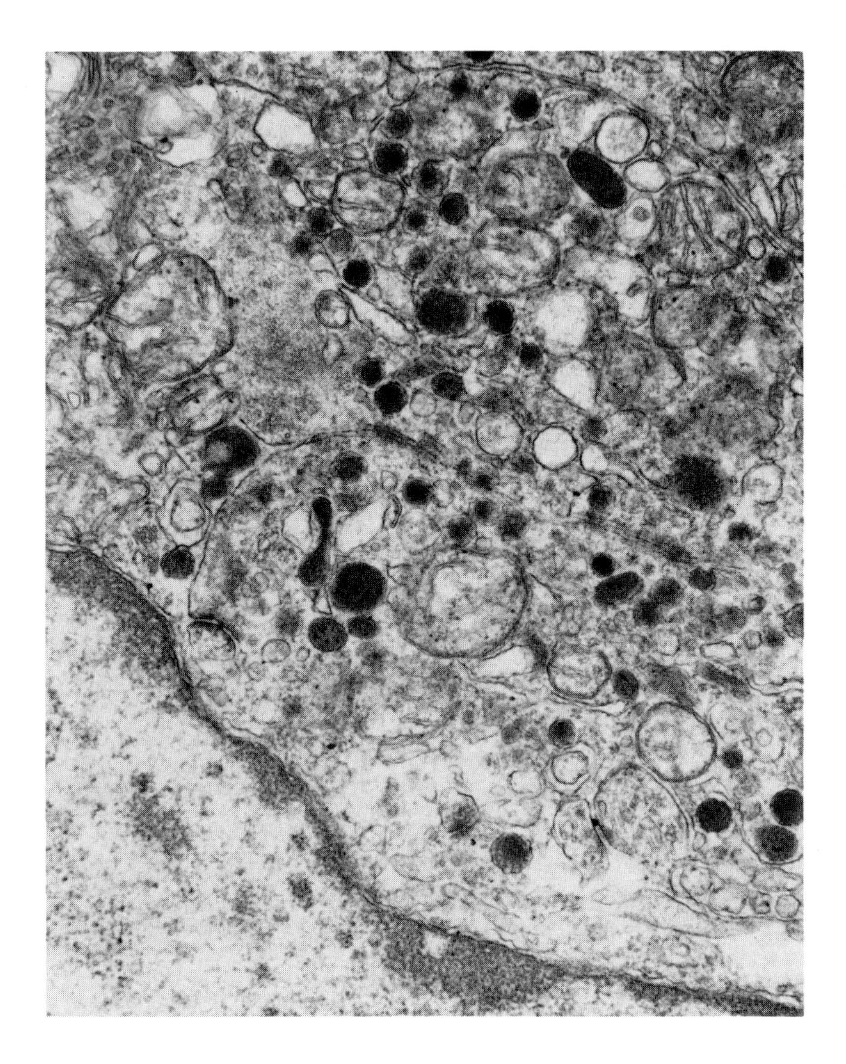

FIGURE 50. Neuroendocrine carcinoma, granules. This detail micrograph shows rather heterogeneous neuroendocrine-type granules in variable caliber cell processes. Immediately adjacent to the nucleus is seen a transected process attached to adjacent cell cytoplasm by a diminutive desmosomal-type attachment. In other areas keratin filaments are frequently observed as well in this group of tumors.

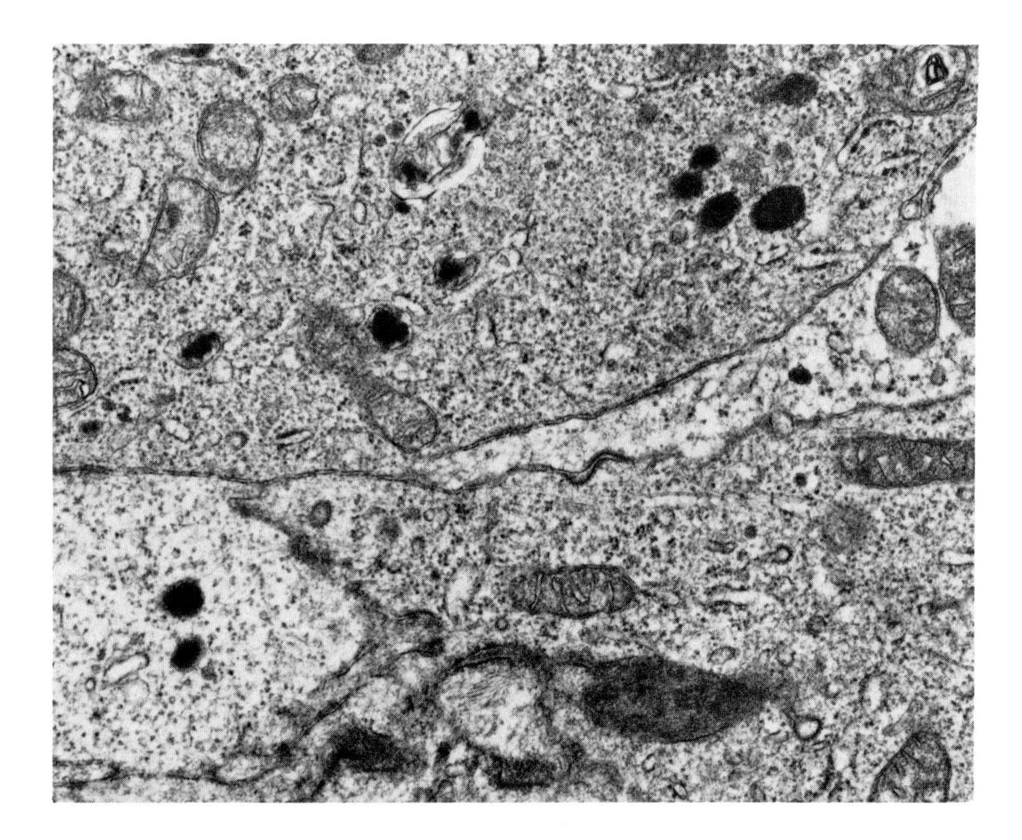

FIGURE 51. Primitive neuroectodermal tumor, dense core granules. The cytoplasm contains scattered, pleomorphic unit membrane bound granules resembling neuroendocrine granules. In addition, an irregular caliber cell process traverses the center of the micrograph. Within this process is noted a single smaller more typical appearing neuroendocrine granule.

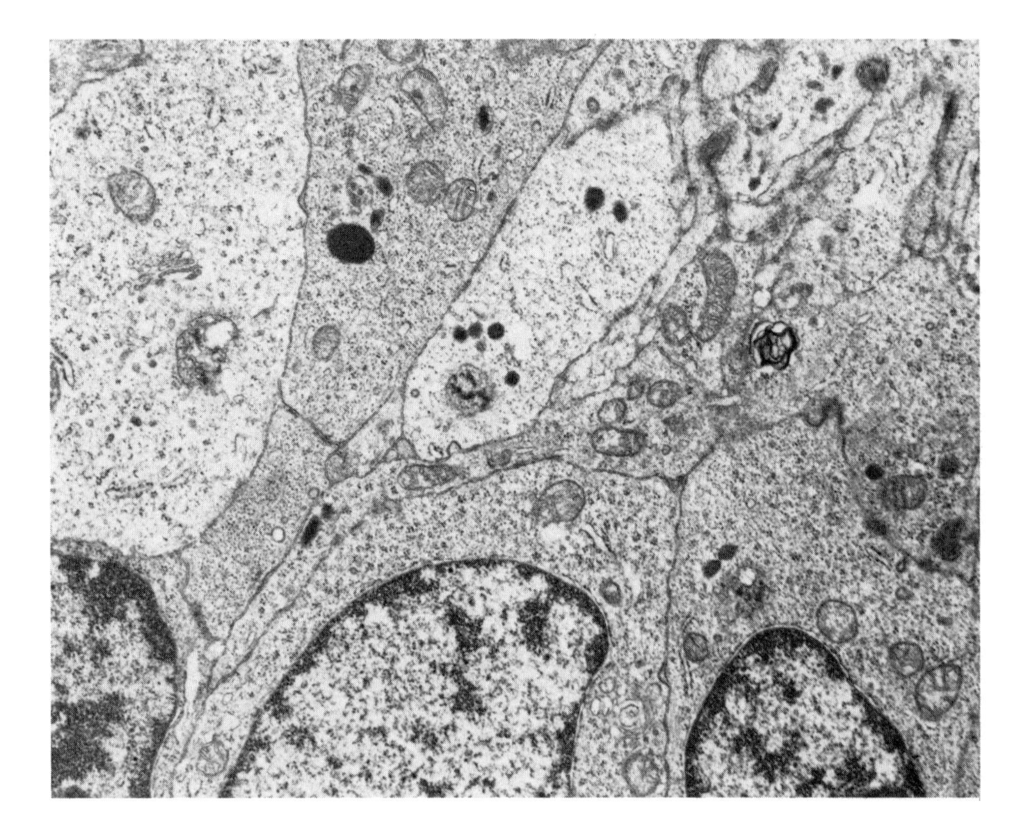

FIGURE 52. Primitive neuroectodermal tumor, neuritic processes. In better differentiated tumors, multiple irregular cell processes are found in multiple planes of section, as illustrated here. In this case more numerous and more convincing neuroendocrine granules are likewise identified.

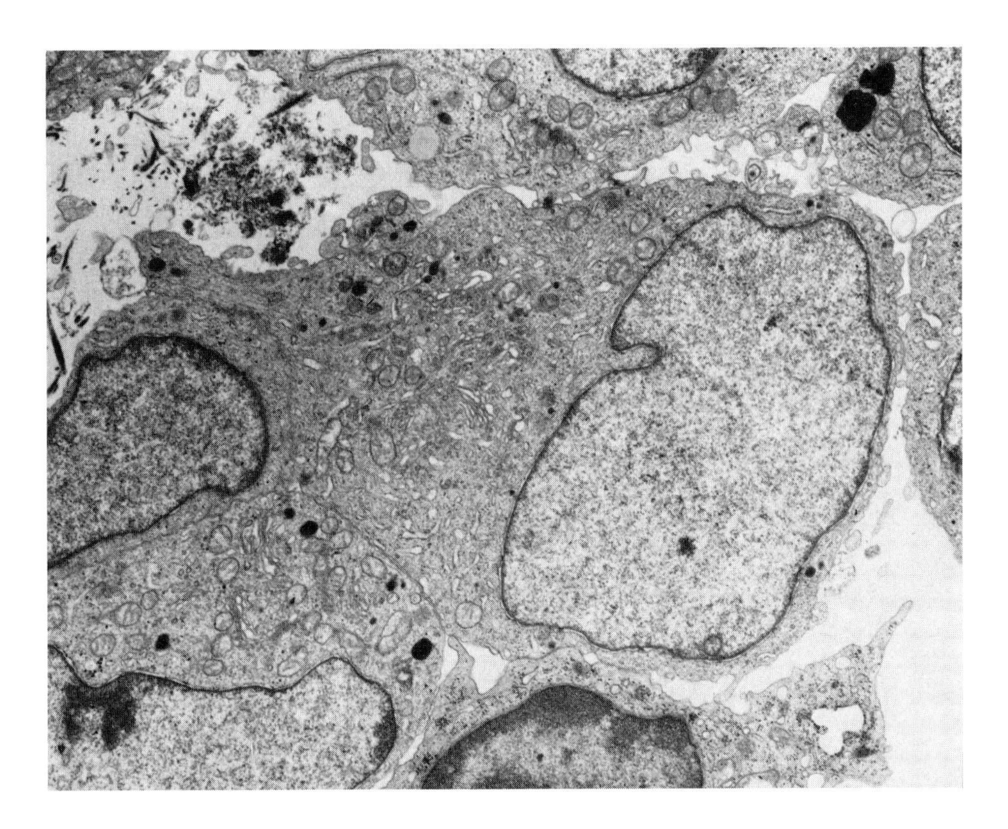

FIGURE 53. Primitive neuroectodermal tumor, hypertrophied Golgi and rough endoplasmic reticulum. In this particular micrograph one can appreciate an abundant Golgi apparatus with scattered cytoplasmic granules in the vicinity of the Golgi apparatus as well as in a process extending therefrom.

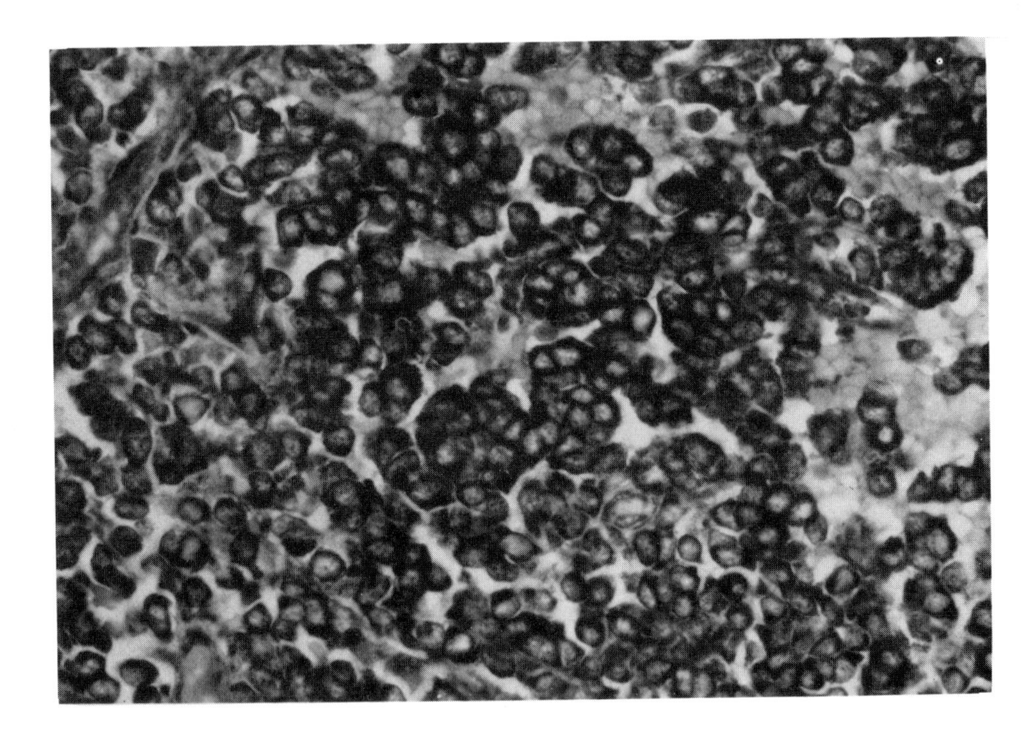

FIGURE 54. Primitive neuroectodermal tumor, neuron specific enolase (NSE) immunocytochemistry. The more primitive tumors such as that illustrated in Figure 51 are readily identified as neuroendocrine in origin by immunocytochemistry. The most commonly employed antibody, that against neuron specific enolase, routinely identifies that enzyme in the cytoplasm, not nucleus, as seen here in all classes of neuroendocrine tumors including the PNETs.

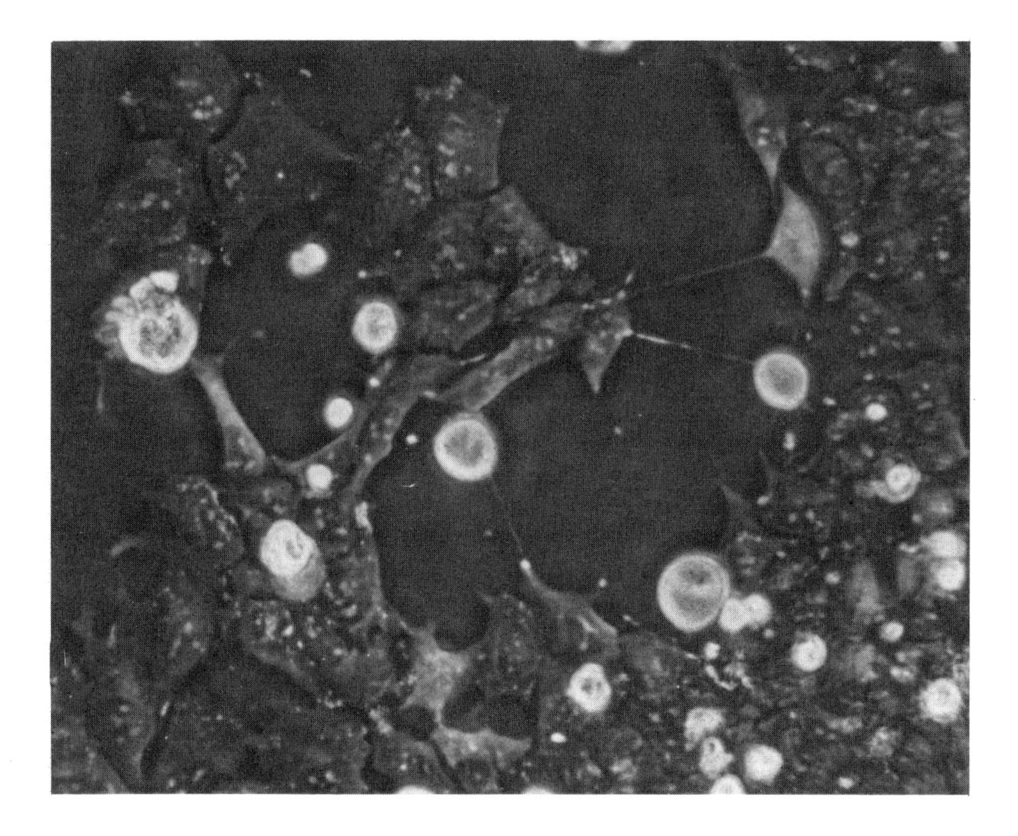

FIGURE 55. Primitive neuroectodermal tumor, *in vitro* neural differentiation. An additional diagnostic procedure demonstrating the neural character of these tumors is short term *in vitro* cultivation of tumor cells. Here erstwhile undifferentiated tumor cells after 3 d in culture have extended classic neural processes (or neurites) including bulbous swellings or varicosities along these processes. There are absolutely typical features of neural differentiation.

REFERENCES

1. **Gould, V. E.,** Neuroendocrine tumors in ''miscellaneous'' primary sites; clinical, pathologic, and histogenic implications, in *Progress in Surgical Pathology 1981,* Year Book Medical, Chicago, 1981.
2. **Pearse, A. G. E. and van Noorden, S.,** The functional cytology of the human adenohypophysis, *Can. Med. Assoc.,* 88, 462, 1963.
3. **Pearse, A. G. E.,** The cytochemistry and ultrastructure of polypeptide hormone-producing cells of the APUD series and the embryologic, physiologic, and pathologic implications of the concept, *J. Histochem. Cytochem.,* 17, 303, 1969.
4. **Fontaine, J. and Le Douarin, N. M.,** Analysis of endoderm formation in the avian blastoderm by the use of quali-chick chimaeras. The problem of the neuroectodermal origin of the cells of the APUD series, *J. Embryol. Exp. Morphol.,* 41, 209, 1977.
5. **Pictet, R. L., Rall, L. B., Phelps, P., Rutter, W. J.,** The neural crest and the origin of the insulin-producing and other gastrointestinal hormone-producing cells, *Science,* 191, 191, 1976.
6. **Sidhu, G. S.,** The endodermal origin of digestive and respiratory tract APUD cells. Histopathologic evidence and a review of the literature, *Am. J. Pathol.,* 96, 5, 1979.
7. **Gould, V. E. and De Lillis, R. A.,** The neuroendocrine cell system: its tumors, hyperplasias, and dysplasias, in *Principles and Practice of Surgical Pathology,* Silverberg, S. G., Ed., John Wiley & Sons, New York, 1983.
8. **Pearse, A. G. E.,** The diffuse neuroendocrine system and the APUD concept: related endocrine peptides in brain, intestine, pituitary, placenta, and anuran cutaneous glands, *Med. Biol.,* 55, 115, 1977.

9. **Pearse, A. G.,** The diffuse neuroendocrine system: peptides, amines, placodes, and the APUD theory, *Prog. Brain Res.*, 68, 25, 1986.
10. International histological classification of tumours of domestic animals. I, *Bull. W.H.O.*, 50, 1, 1974.
11. International histological classification of tumours of domestic animals, II, *Bull. W.H.O.*, 53, 137, 1976.
12. **Varndell, I. M., Tapia, F. J., De Mey, J., Rush, R. A., Bloom, S. R., and Polak, J. M.,** Electron immunocytochemical localization of enkephalin-like material in catecholamine-containing cells of the carotid body, the adrenal medulla, and in pheochromocytomas of man and other mammals, *J. Histochem. Cytochem.*, 30, 682, 1982.
13. **Triche, T. J. and Askin, F. B.,** Neuroblastoma and the different diagnosis of small-, round, blue cell tumors, *Hum. Pathol.*, 14, 569, 1983.
14. **Melicow, M. M.,** One hundred cases of pheochromocytoma (107 tumors) at the Columbia-Presbyterian Medical Center, 1926—1976; a clinicopathological analysis, *Cancer*, 40, 1987, 1977.
15. **Liu, T. H., Chen, G. S., Nan, C., and He, Z. G.,** Clinico-pathologic and ultrastructural characteristics of pheochromocytoma. An analysis of 55 cases, *Pathol. Res. Pract.*, 178, 355, 1984.
16. **Kakudo, K., Uematsu, K., Matsuno, Y., Mitsuobu, M., Toyosaka, A., Okamoto, E., and Fukuchi, M.,** Malignant pheochromocytoma with ACTH production, *Acta Pathol. Jpn.*, 34, 1403, 1984.
17. **Lamove, J., Memoli, V. A., Terzakis, J. A., Sommers, S. C., and Goulde, V. E.,** Pheochromocytoma producing immunoreactive ACTH with Cushing's syndrome, *Ultrastruct. Pathol.*, 6, 41, 1984.
18. **Grimley, P. M. and Glenner, G. G.,** Histology and ultrastructure of carotid body paragangliomas. Comparison with the normal gland, *Cancer*, 20, 1473, 1967.
19. **Aleshire, S. L., Glick, A. D., Cruz, V. E., Bradley, C. A., and Parl, F. F.,** Neuroblastoma in adults. Pathologic findings and clinical outcome, *Arch. Pathol. Lab. Med.*, 109, 352, 1985.
20. **Beckwith, J. B. and Perrin, E. V.,** *In situ* neuroblastomas: a contribution to the natural history of neural crest tumors, *Am. J. Pathol.*, 43, 1963, 1989.
21. **Triche, T. J., Askin, F. B., and Kissane, J. M.,** Neuroblastoma, Ewing's sarcoma, and the differential diagnosis of small-, round-, blue-cell tumors, *Major Probl. Pathol.*, 18, 145, 1986.
22. **Jaffe, N.,** Neuroblastoma: review of the literature and an examination of factors contributing to its enigmatic character, *Cancer Treat Rev.*, 3, 61, 1976.
23. **Clark, W. H., Jr., Mastrangelo, M. J., Ainsworth, A. M., Berd, D., Bellet, R. E., and Bernardino, E. A.,** Current concepts of the biology of human cutaneous malignant melanoma, *Adv. Cancer Res.*, 24, 267, 1977.
24. **Carstens, P. H. B. and Kuhns, J. G.,** Ultrastructural confirmation of malignant melanoma, *Ultrastruct. Pathol.*, 2, 147, 1981.
25. **Falck, B., Hillarp, N. A., Thieme, G., et al.,** Fluorescence of catecholamines and related compounds condensed with formaldehyde, *J. Histochem. Cytochem.*, 10, 348, 1962.
26. **Lindvall, O. and Bjorklund, A.,** The glyoxylic acid fluorescence histochemical method: a detailed account of the methodology for the visualization of central catecholamine neurons, *Histochemistry*, 39, 97, 1974.
27. **Inoshita, T. and Youngberg, G. A.,** Fluorescence of melanoma cells. A useful diagnostic tool, *Am. J. Clin. Pathol.*, 78, 311, 1982.
28. **Dalal, B. I. and Slinger, R. P.,** Formaldehyde-induced fluorescence in melanomas and other lesions, *Arch. Pathol. Lab. Med.*, 109, 551, 1985.
29. **Cochran, A. J., Wen, D. R., Hershman, H. R., and Gaynor, R. B.,** Detection of S-100 protein as an aid to the identification of melanocytic tumors, *Int. J. Cancer*, 30, 295, 1982.
30. **Springall, D. R., Gu, J., Cocchia, D., Michetti, F., Levene, A., Levene, M. M., Marangos, P. J., Bloom, S. R., and Polak, J. M.,** The value of S-100 immunostaining as a diagnostic tool in human malignant melanomas, *Virchows Arch. A*, 400, 331, 1983.
31. **Van Duinen, S. G., Ruiter, D. J., Hageman, P., Vennegoor, C., Dickersin, G. R., Scheffer, E., and Rumke, P.,** Immunohistochemical and histochemical tools in the diagnosis of amelanotic melanoma, *Cancer*, 53, 1566, 1984.
32. **Enzinger, F. M.,** Clear-cell sarcoma of tendons and aponeuroses. An analysis of 21 cases, *Cancer*, 18, 1163, 1965.
33. **Kindblom, L., Lodding, P., and Angervall, L.,** Clear-cell sarcoma of tendons and aponeuroses. An immunohistochemical and electron microscopic analysis indicating neural crest origin, *Virchows Arch. A*, 401, 109, 1983.
34. **Mukai, M., Torikata, C., Iri, H., Mikata, A., Kawai, T., Hanaoka, H., Yakumaru, K., and Kageyama, K.,** Histogenesis of clear cell sarcoma of tendons and aponeuroses, *Am. J. Pathol.*, 114, 264, 1984.
35. **Chung, E. B., Enzinger, F. M.,** Malignant melanoma of soft parts. A reassessment of clear cell sarcoma, *Am. J. Surg. Pathol.*, 7, 405, 1983.
36. **Dabbs, D. J. and Bole, J. W.,** Superficial spreading malignant melanoma with neurosarcomatous metastasis, *Am. J. Clin. Pathol.*, 82, 109, 1984.

37. **Mazur, M. T. and Katzenstein, A. A.,** Metastatic melanoma: the spectrum of ultrastructural morphology, *Ultrastruct. Pathol.,* 1, 337, 1980.
38. **Erlandson, R. A.,** Peripheral nerve sheath tumors, *Ultrastruct. Pathol.,* 9, 113, 1985.
39. **Mennemeyer, R. P., Hammar, S. P., Tytus, J. S., Hallman, K. O., Raisis, J. E., and Bockus, D.,** Melanotic schwannoma. Clinical and ultrastructural studies of three cases with evidence of intracellular melanin synthesis, *Am. J. Surg. Pathol.,* 3, 3, 1979.
40. **Font, R. L. and Truong, L. D.,** Melanotic schwannoma of soft tissues. Electron microscopic observations and review of literature, *Am. J. Surg. Pathol.,* 8, 129, 1984.
41. **Pearse, A. G. E.,** Islet cell precursors are neurones, *Nature,* 295, 96, 1982.
42. **Pearse, A. G. E.,** The APUD cell concept and its implications in pathology, *Pathol. Annu.,* 9, 27, 1974.
43. **Williams, E. D.,** Histogenesis of medullary carcinoma of the thyroid, *J. Clin. Pathol.,* 19, 114, 1966.
44. **Hazard, J. B.,** The C cells (parafollicular cells) of the thyroid gland and medullary thyroid carcinoma, *Am. J. Pathol.,* 88, 214, 1977.
45. **Pearse, A. G. E. and Polak, J. M.,** Cytochemical evidence for the neural crest origin of mammalian ultimobranchial C-cells, *Histochemie,* 27, 96, 1971.
46. **DeLellis, R. A., Nunnemacher, G., and Wolfe, H. J.,** C-cell hyperplasia. An ultrastructural analysis, *Lab Invest.,* 36, 237, 1977.
47. **Hazard, J. B., Hawk, W. A., and Crile, G., Jr.,** Medullary (solid) carcinoma of the thyroid: a clinicopathologic entity, *J. Clin. Endocrinol. Metab.,* 19, 152, 1959.
48. **LiVolsi, V. A.,** Editorial: Mixed thyroid carcinoma: a real entity?, *Lab. Invest.,* 57, 237, 1987.
49. **Holm, R., Sobrinho-Simoes, M., Nesland, J. M., Sambade, C., and Johannessen, J. V.,** Medullary thyroid carcinoma with thyroglobulin immunoreactivity. A special entity?, *Lab. Invest.,* 57, 258, 1987.
50. **Warren, W. H., Memoli, V. A., and Gould, V. E.,** Immunohistochemical and ultrastructural analysis of bronchopulmonary neuroendocrine neoplasms. I. Carcinoids, *Ultrastruct. Pathol.,* 6, 15, 1984.
51. **Godwin, J. D., II,** Carcinoid tumors. An analysis of 2837 cases, *Cancer,* 36, 560, 1975.
52. **Williams, E. D. and Sandler, M.,** The classification of carcinoid tumours, *Lancet,* 238, 1963.
53. **Salyer, W. R., Salyer, D. C., and Eggleston, J. C.,** Carcinoid tumors of the thymus, *Cancer,* 37, 958, 1976.
54. **Marchevsky, A. M. and Dikman, S. H.,** Mediastinal carcinoid with an incomplete sipple's syndrome, *Cancer,* 43, 2497, 1979.
55. **Patchefsky, A. S., Solit, R., Phillips, L. D., Craddock, M., Harrer, W. V., Cohn, H. E., and Kowlessar, O. D.,** Hydroxindole-producing tumors of the pancreas. Carcinoid-islet cell tumor and oat cell carcinoma, *Ann. Int. Med.,* 77, 53, 1972.
56. **Ali, M., Fayemi, A. O., and Braun, E. V.,** Malignant apudoma of the liver with symptomatic intractable hypoglycemia, *Cancer,* 42, 686, 1978.
57. **Warner, T. F. C. S., Seo, I. S., Madura, J. A., Polak, J. M., and Pearse, A. G. E.,** Pancreatic-polypeptide-producing apudoma of the liver, *Cancer,* 46, 1146, 1980.
58. **Mills, S. E., Cooper, P. H., Walker, A. N., and Kron, I. L.,** Atypical carcinoid tumor of the lung. A clinicopathologic study of 17 cases, *Am. J. Surg. Pathol.,* 6, 643, 1982.
59. **Gould, V. E., Linnoila, R. I., Memoli, V. A., and Warren, W. H.,** Neuroendocrine cells and neuroendocrine neoplasms of the lung, *Pathol. Annu.,* 18, 287, 1983.
60. **Nomori, H., Shimosato, Y., Kodama, T., Morinaga, S., Nakajima, T., and Watanabe, S.,** Subtype of small cell carcinoma of the lung: morphometric, ultrastructural, and immunohistochemical analyses, *Hum. Pathol.,* 17, 604, 1986.
61. **Brereton, H. D., Mathews, M. M., Costa, J., Kent, C. H., and Johnson, R. E.,** Mixed anaplastic small-cell and squamous-cell carcinoma of the lung, *Ann. Intern. Med.,* 88, 805, 1978.
62. **Hashimoto, T., Fukuoka, M., Nagasawa, S., Tamai, S., Kusunoki, Y., Kawahara, M., Furuse, K., Sawamura, K., and Fujimoto, T.,** Small cell carcinoma of the lung and its histological origin. Report of a case, *Am. J. Surg. Pathol.,* 3, 343, 1979.
63. **Churg, A., Johnston, W. H., and Stulbarg, M.,** Small cell squamous and mixed small cell squamous-small cell anaplastic carcinomas of the lung, *Am. J. Surg. Pathol.,* 4, 255, 1980.
64. **Saba, S. R., Azar, H. A., Richman, A. V., Solomon, D. A., Spurlock, R. G., Mardelli, I. G., and Kasnic, G.,** Dual differentiation in small cell carcinoma (oat cell carcinoma) of the lung, *Ultrastruct. Pathol.,* 2, 131, 1981.
65. **Jass, J. R.,** Small cell anaplastic carcinoma of the lung with glandular and squamous differentiation, *Am. J. Surg. Pathol.,* 5, 307, 1981.
66. **McDowell, E. M. and Trump, B. F.,** Pulmonary small cell carcinoma showing tripartite differentiation in individual cells, *Hum. Pathol.,* 12, 286, 1981.
67. **Geller, S. A. and Gordon, R. E.,** Peripheral spindle-cell carcinoid tumor of the lung with type II pneumocyte features. An ultrastructural study with comments on possible histogenesis, *Am. J. Surg. Pathol.,* 8, 145, 1984.
68. **Sobin, L. H.,** The histologic classification of lung tumors: the need for a double standard, *Hum. Pathol.,* 14, 1020, 1983.

69. **Rosai, J., Levine, G., Weber, W. R., and Higa, E.,** Carcinoid tumors and oat cell carcinomas of the thymus, *Pathol. Annu.,* 11, 201, 1976.
70. **Barsky, S. H., Linnoila, I., Triche, T. J., and Costa, J.,** Hepatocellular carcinoma with carcinoid features, *Hum. Pathol.,* 9, 892, 1984.
71. **Payne, C. M., Nagle, R. B., Paplanus, S. H., and Graham, A. R.,** Fibrolamellar carcinoma of liver: a primary malignant oncocytic carcinoid?, *Ultrastruct. Pathol.,* 10, 539, 1986.
72. **Judge, D. M., Dickman, P. S., and Trapukdi, S.,** Nonfunctioning argyrophilic tumor (APUDoma) of the hepatic duct: simplified methods of detecting biogenic amines in tissue, *Am. J. Clin. Pathol.,* 66, 40, 1976.
73. **Schein, P. S., DeLellis, R. A., Kah, C. R., Gorden, P., and Kraft, A. R.,** Islet cell tumors: current concepts and management, *Ann. Intern. Med.,* 79, 239, 1973.
74. **Bordi, C. and Tardini, A.,** Electron microscopy of islet cell tumors, *Prog. Surg. Pathol.,* 1, 135, 1980.
75. **Dahms, B. B., Landing, B. H., Blaskovics, M., and Roe, T. F.,** Nesidioblastosis and other islet cell abnormalities in hyperinsulinemic hypoglycemia of childhood, *Hum. Pathol.,* 11, 641, 1980.
76. **Jaffe, R., Hashida, Y., and Yunis, E. J.,** Pancreatic pathology in hyperinsulinemic hypoglycemia of infancy, *Lab. Invest.,* 42, 356, 1980.
77. **Weidenheim, K. M., Hinchey, W. W., and Campbell, W. G., Jr.,** Hyperinsulinemic hypoglycemia in adults with islet-cell hyperplasia and degranulation of exocrine cells of the pancreas, *Am. J. Clin. Pathol.,* 79, 14, 1983.
78. **Bani, D., Sacchi, T. B., and Biliotti, G.,** Nesidioblastosis and intermediate cells in the pancreas of patients with hyperinsulinemic hypoglycemia, *Virchows Arch. B,* 48, 18, 1985.
79. **McGavran, M. Y., Unger, R. H., Recang, L., Polk, H. C., Kilo, C., and Levin, M. E.,** A glucagon-secreting alpha-cell carcinoma of the pancreas, *N. Eng. J. Med.,* 274, 1408, 1966.
80. **Creutzfeldt, W., Arnold, R., Creutzfeldt, C., and Track, N. S.,** Pathomorphologic, biochemical, and diagnostic aspects of gastrinomas (Zollinger-Ellison syndrome), *Hum. Pathol.,* 6, 47, 1975.
81. **Ganda, O. P., Weir, G. C., Soeldner, J. S., Legg, M. E., Chick, W. L., Patel, Y. C., Ebeid, A. M., Gabbay, K. H., and Reichlin, S.,** "Somatostatinoma": a somatostatin-containing tumor of the endocrine pancreas, *N. Eng. J. Med.,* 296, 963, 1977.
82. **Toker, C.,** Trabecular carcinoma of the skin, *Arch. Dermatol.,* 105, 107, 1972.
83. **Tang, C. and Toker, C.,** Trabecular carcinoma of the skin: further clinicopathologic and ultrastructural study, *Mt. Sinai J. Med.,* 46, 516, 1979.
84. **Sibley, R. K., Rosai, J., Foucar, E., Dehner, L. P., and Bosl, G.,** Neuroendocrine (Merkel cell) carcinoma of the skin. A histologic and ultrastructural study of two cases, *Am. J. Surg. Pathol.,* 4, 211, 1980.
85. **Gomez, L. G., Silva, E. G., Di Maio, S., and Mackay, B.,** Association between neuroendocrine (Merkel cell) carcinoma and squamous cell carcinoma of the skin, *Am. J. Surg. Pathol.,* 7, 171, 1983.
86. **Gould, V. E., Moll, I., Lee, I., and Franke, W. W.,** Biology of disease. Neuroendocrine (Merkel) cells of the skin: hyperplasias, dysplasias, and neoplasms, *Lab. Invest.,* 4, 334, 1985.
87. **Sibley, R. K., Dehner, L. P., and Rosai, J.,** Primary neuroendocrine (Merkel cell?) carcinoma of the skin. I. A clinicopathologic and ultrastructural study of 43 cases, *Am. J. Surg. Pathol.,* 9, 95, 1985.
88. **Sibley, R. K. and Dahl, D.,** Primary neuroendocrine (Merkel cell?) carcinoma of the skin. II. An immunocytochemical study of 21 cases, *Am. J. Surg. Pathol.,* 9, 109, 1985.
89. **Yoshida, Y., Takei, T., Hattori, A., Kaku, T., Yokokawa, K., and Mori, M.,** Merkel cell tumor of the skin. Ultrastructural and immunohistochemical studies, *Acta Pathol. Jpn.,* 34, 1433, 1984.
90. **Gould, V. E., Memoli, V. A., Dardi, L. E., Sobe, H. J., Somers, S. C., and Johannessen, J. V.,** Neuroendocrine carcinomas with multiple immunoreactive peptides and melanin production, *Ultrastruct. Pathol.,* 2, 199, 1981.
91. **Dehner, L. P.,** Peripheral and central primitive neuroectodermal tumors. A nosologic concept seeking a consensus, *Arch. Pathol. Lab. Med.,* 110, 997, 1986.
92. **Shuangshoti, S.,** Primitive neuroectodermal (neuroepithelial) tumour of soft tissue of the neck in a child: demonstration of neuronal and neuroglial differentiation, *Histopathology,* 10, 651, 1986.
93. **Nesbitt, K. A. and Vidone, R. A.,** Primitive neuroectodermal tumor (neuroblastoma) arising in sciatic nerve of a child, *Cancer,* 37, 1562, 1976.
94. **Hashimoto, H., Enjoji, M., Nakajima, T., Kiryu, H., and Daimaru, Y.,** Malignant neuroepithelioma (peripheral neuroblastoma). A clinicopathologic study of 15 cases, *Am. J. Surg. Pathol.,* 7, 309, 1983.
95. **Seemayer, T. A., Thelmo, W. L., Bolande, R. P., and Wiglesworth, F. W.,** Peripheral neuroectodermal tumors, in *Perspectives in Pediatric Pathology,* Vol. 5, Rosenberg, H. and Bolander, R. P., Eds., Masson, New York, 1979.
96. **Schmidt, D., Harms, D., and Burdach, S.,** Malignant peripheral neuroectodermal tumours of childhood and adolescence, *Virchows Arch. A,* 406, 351, 1985.
97. **Bolen, J. W. and Thorning, D.,** Peripheral neuroepithelioma: a light and electron microscopic study, *Cancer,* 46, 2456, 1980.

98. **Voss, B. L., Pysher, T. J., and Humphrey, G. B.,** Peripheral neuroepithelioma in childhood, *Cancer,* 54, 3059, 1984.
99. **Askin, F. B., Rosai, J., Sibley, R. K., Dehner, L. P., and McAlister, W. H.,** Malignant small cell tumor of the thoracopulmonary region in childhood. A distinctive clinicopathologic entity of uncertain histogenesis, *Cancer,* 43, 2438, 1979.
100. **Lipinski, M., Braham, K., Philip, I., Wiels, J., Philip, T., Goridis, C., Lenoir, G. M., and Tursz, T.,** Neuroectoderm-associated antigens on Ewing's sarcoma cell lines, *Cancer Res.,* 47, 183, 1987.
101. **Lipinski, M., Hirsch, M., Deagostini-Bazin, H., Yamada, O., Tursz, T., and Goridis, C.,** Characterization of neural cell adhesion molecules (NCAM) expressed by Ewing and neuroblastoma cell lines, *Int. J. Cancer,* 40, 81, 1987.
102. **Cavazzana, A. O., Miser, J. S., Jefferson, J., and Triche, T. J.,** Experimental evidence for a neural origin of Ewing's sarcoma of bone, *Am. J. Pathol.,* 127, 507, 1987.
103. **Thiele, C. J., KcKeon, C., Triche, T. J., Ross, R. A., Reynolds, C. P., and Israel, M. A.,** Differential proto-oncogene expression characterizes histopathologically indistinguishable tumors of the peripheral nervous system, *J. Clin. Invest.,* 80, 804, 1987.
104. **Triche, T. J.,** Diagnosis of small round cell tumors of childhood, *Bull. Cancer,* 75, 1988.
105. **Woodruff, J. M.,** Arthur Purdy Stout and the evolution of modern concepts regarding peripheral nerve sheath tumors, *Am. J. Surg. Pathol.,* 10, 63, 1986.
106. **DiCarlo, E. F., Woodruff, J. M., Bansal, M., and Erlandson, R. A.,** The purely epithelioid malignant peripheral nerve sheath tumor, *Am. J. Surg. Pathol.,* 10, 478, 1986.

CONCLUSION

The information provided in this book should aid cancer researchers with a background in a variety of different disciplines to more efficiently utilize differences in the biology and pathology of tumors to arrive at a better understanding of the mechanisms of carcinogenesis and the behavior of well-defined tumor types.

Cancer research, in general, has failed to succeed in elucidating the mechanisms of carcinogenesis, providing sufficient tools for cancer prevention/inhibition, or in providing a cure for cancer. The major problem of all such research is an unfortunately widespread attitude among cancer researchers to try and simplify their test systems and/or to extrapolate data generated in one test system to cancer in general.

We have to understand that cancer is not one disease but rather the umbrella of many subclasses of cancerous diseases. Although these subclasses may share one or the other feature, their mechanisms of induction, development and reaction to treatment may be vastly different. The only way to obtain information on tumor type specific differences in the mechanisms of induction and biological behavior of tumors is to thoroughly define the system under study. One simple way of doing this is to classify the experimentally induced tumors in animals, the tumors in human patients and the *in vitro* tumor systems by the most advanced morphological techniques including electron microscopy, immunochemistry, and histochemistry. Admittedly, this approach will not generate a cure for cancer overnight. However, the unsuccessful research efforts of the past amply demonstrate that large-scale fast screenings, whether they be in the area of carcinogenesis or cancer therapy, are unlikely to yield meaningful data.

INDEX

A

B

C

D

E

F

G

L

M

N

O

P

Q

R

S

T

U

V

W

Z